MAGNETIC SYSTEMS

WITH

COMPETING INTERACTIONS

(Frustrated Spin Systems)

MAGNETIC SYSTEMS
WITH
COMPETING INTERACTIONS

(Frustrated Spin Systems)

edited by

H T Diep

Groupe de Physique Statistique
Université de Cergy-Pontoise
Cergy-Pontoise, Cedex
France

World Scientific
Singapore • New Jersey • London • Hong Kong

Published by

World Scientific Publishing Co. Pte. Ltd.

P O Box 128, Farrer Road, Singapore 9128

USA office: Suite 1B, 1060 Main Street, River Edge, NJ 07661

UK office: 73 Lynton Mead, Totteridge, London N20 8DH

MAGNETIC SYSTEMS WITH COMPETING INTERACTIONS
(FRUSTRATED SPIN SYSTEMS)

ISBN: 981-02-1715-3

Printed in Singapore by Uto-Print

PREFACE

Magnetic systems with competing interactions have been first investigated four decades ago. Well-known examples include the Ising model on the antiferromagnetic triangular lattice studied by G. H. Wannier in 1950 and the Heisenberg helical structure discovered independently by A. Yoshimori and J. Villain in 1959. However, extensive investigations on magnetic systems with competing interactions have really started with the concept of frustration introduced almost at the same time by G. Toulouse and J. Villain in 1977 in the context of spin glasses. Therefore, the title of the book should have been **Frustrated Spin Systems**. The frustration was initially defined for Ising spin systems. It was later generalized to vector spin systems. The frustration is generated by the competition of different kinds of interaction and/or by the lattice geometry. As a result, in the ground state a number of spins in a frustrated Ising system behave as free spins. In the case of frustrated vector spin systems, the ground-state configuration is usually non-collinear. These ground-state properties (degeneracy and symmetry) give rise to spectacular and often unexpected behaviors at finite temperatures.

Many properties of frustrated systems are still not well understood at present. Recent studies shown in this book reveal that many established theories have never before encountered so many difficulties as they do now in dealing with frustrated systems. In some sense, frustrated systems are excellent candidates to test approximations and to improve theories.

The contents of this book cover recent developments in the problem of frustration effects in spin systems. The chapters of this book are written by researchers who have actively contributed to the field. Many results are from recent works of the authors. Some new, yet unpublished results and a number of unsettled questions are also included.

The book is intended for post-graduade students as well as researchers in statistical physics, magnetism, materials science and various domains where real systems can be described with the spin language. Explicit demonstrations of formulae and full arguments leading to important results are given where it is possible to do so. Pedagogical effort has been made to make each chapter to be self-contained, comprehensible for researchers who are not really involved in the field. Basic methods are given in detail.

The systems shown in this book include two- and three-dimensional systems of Ising, XY or Heisenberg spins. For practical purposes, each chapter treats a different aspect of the problem. The book is organised as follows.

The first three chapters treat frustrated vector spin systems. The two following chapters study frustrated Ising spin systems. In these five chapters, only systems periodically defined on a lattice are considered. Frustration effects are thus studied without interaction disorder. The remaining two chapters are devoted to random magnets and experimental studies of frustrated spin systems, respectively.

I summarize in the following the contents of each chapter.

Chapter I is devoted to critical properties of frustrated XY and Heisenberg spin systems. A review on various models studied so far is presented. The nature of the phase transition is discussed by mean-field theory and Monte Carlo simulations. These methods are described in detail. The utility of mean-field theory is demonstrated, in particular the description of ordered states in terms of the spin density $s(\mathbf{r})$ written as a Fourier expansion, and analysis of Landau-type free energies $F[s(\mathbf{r})]$. Such studies make clear the symmetries broken at phase transitions and can lead to very good agreement with experimental results on magnetic phase diagrams. Most of the results discussed in this chapter concern frustration caused by lattice geometry (mostly triangular), or the effects of further-neighbor interactions. The authors emphasize phase-transition phenomena in three-dimensional (3D) systems, induced by varying temperature or an applied magnetic field. In particular, magnetic phase diagrams which exhibit unusual multicritical-point structures are examined. Comparison with experimental results is emphasized, for the most part from work on the quasi-1D hexagonal insulators ABX_3 (such as $CsNiCl_3$).

Chapter II deals with the renormalization of the effective theories relevant for classical and quantum frustrated Heisenberg models. Both Landau-Ginzburg-Wilson and Non-Linear-Sigma models are studied by the $\epsilon = 4 - D$, $\epsilon = D - 2$ and large N expansions. Since no agreement is found betweeen these perturbative approaches, the authors discuss the possible scenarios for the physics of the classical systems in $D = 3$. In particular, a detailed discussion is given on the controversial issue of the nature of the phase transition in canted spin systems. In the quantum case, a study is made of the phase transition that occurs in two dimensions at zero temperature between a semi-classical Néel ordered phase and a quantum disordered phase as the spin magnitude is varied. It is shown that one of the most effective signature at finite size of the transition is the existence and the scaling of a tower of states that collapse onto the ground state in the thermodynamical limit. At finite temperature, it is shown that the behavior of the correlation length as a function of the temperature gives also a test of the nature of the order at zero temperature.

Chapter III treats some quantum Heisenberg antiferromagnets in low dimension at zero temperature. The so-called $J_1 - J2$ model in one dimension and on the square lattice are studied by finite-size scaling analysis. The model on a square lattice has attracted most attention as a rather crude model of the effects of doping on copper oxide planes. The square sites correspond to the copper sites in the two-dimensional plane of copper and oxygen that is the common structural feature of the family of superconducting copper oxides. A shorter review of work concerning more specifically with triangular and Kagomé lattices is also given. The interest of the last is that there is a good experimental realization. Special attention is paid to the case of spin 1/2 for which quantum fluctuations should be most noticeable. The aim of this chapter is to discuss calculations of the last few years which attempt to give correct answers to the question of what happens to a quantum Heisenberg antiferromagnet when one introduces interactions or geometry such that in a classical picture not all antiferromagnetic bonds may be satisfied.

Chapter IV shows the frustration effects in exactly solved two-dimensional Ising models. The systems considered in this chapter are periodically defined (without bond disorder). The frustration due to competing interactions will itself induce disorder in the spin orientations. The results obtained can be applied to physical systems that can be described by a spin language. After a detailed presentation of 16- and 32-vertex models, applications are made to some selected systems which possess most of the spectacular features due to the frustration such as high ground-state degeneracy, reentrance, successive phase transitions and disorder solutions. In some simple models, up to five transitions separated by two reentrant paramagnetic phases are found. A conjecture is made on the origin of the paramagnetic reentrant phase. The nature of ordering as well as the relation between the considered systems and the random-field Ising model are discussed. The relevance of disorder solutions for the reentrance phenomena is also pointed out.

Chapter V deals mainly with the Ising model on the antiferromagnetic triangular and stacked triangular lattices. Ground-state properties and the nature of the phase transition are studied by various methods, as functions of the spin magnitude S and nearest- and next-nearest-neighbor interactions. It is shown in this chapter that the symmetry of spin ordering is strongly dependent S. Furthermore, due to the frustration, there exist "free" spins or "free" linear-chains, on which internal fields are canceled out, in some frustrated Ising spin systems. These free spins and free linear-chains play an important role as for spin orderings. Existence of these free spins and free linear-chains is explicitly shown in this chapter and the role of them is discussed. Another characteristic feature of frustrated Ising spin systems is that various metastable states exist in these systems. Existence of metastable states is closely related to the degeneracy of ground state and also to the excited states. These metastable states may give rise to a first order phase transition as found in some models introduced in this chapter. The effects of the far-neighbor interactions in the Ising model on the antiferromagnetic triangular and stacked triangular lattices are clarified.

Chapter VI is devoted to the problem of reentrant spin glass. The previous chapters of this book have largely focused on the novel and interesting competing effects induced in magnetic systems via the presence of frustration. Frustration is also ubiquitous in random systems. For example, a change in concentration of some species in a magnetic ferromagnet alloy can introduce randomly located antiferromagnetic bonds and, consequently, generate "random frustration". In the limit of large random frustration a magnetic system cannot accomodate a percolating ordered magnetic cluster at zero temperature labelled by a magnetic Bragg peak at a well defined wavevector $\vec{q}$. However, in that case, a randomly frustrated magnet often displays a cooperative freezing transition into a spin glass state. In the case of weak disorder and small random frustration, there is competition between conventional magnetic order and the spin glass ordering. In that regime, one observes experimentally the reentrant spin glass (RSG) behavior. In the RSG regime, a system develops partial collinear ferromagnetic (or antiferromagnetic) order at some temperature T_c. For a range of temperature $T_g < T < T_c$, the system behaves largely like a conventional unsaturated magnet. However, below T_g, the system shows magnetic hysteresis and displays properties analogous to those found in conventional spin glasses, hence the name "reentrant spin glasses". The physics at the origin of this two-transition behavior has remained for twenty years a subject of intense controversy. Chapter VI reviews the progress made in the past five years.

Chapter VII describes experimental results on geometrically-frustrated magnetic systems. In particular, results on stacked triangular lattice antiferromagnets and helimagnets are shown and compared to theoretical predictions presented in Chapters I and II. Materials containing antiferromagnetically-coupled magnetic moments which reside on geometrical units can inhibit a frustrated state at low temperatures. The materials of interest may undergo phase transitions with novel properties to an unusual ordered state, or they may not undergo a conventional phase transition at all. This chapter reviews recent experimental progress in understanding the phases and phase transitions displayed by several such magnetic materials. Much of it will focus on neutron-scattering studies, as such studies have played a central role in characterizing the properties of these antiferromagnets and their phase transitions. The review presented here is not intended to be exhaustive in nature, but rather to give an overview of several materials with which the author is familiar, and to discuss some of the experimental challenges which must be dealt with in such studies.

As a number of issues treated in this book are still debated, I alert the reader that the authors of each chapter have taken the liberty to express their viewpoint on each unsettled issue. This concerns specially the nature of the phase transition in canted spin systems discussed in Chapters I, II and VII. Since the domains treated in this book are currently investigated, it is clear that in a few years from now, progress will be made and some of the unsettled questions discussed in this book will be understood.

H. T. Diep
University of Cergy-Pontoise, France.
Spring 1994.

CONTENTS

Chapter II. The Renormalization Group Approach to Frustrated Heisenberg Spin Systems

Patrick Azaria and Bertrand Delamotte

Chapter V. Properties and Phase transitions in Frustrated Ising Spin Systems

O. Nagai, T. Horiguchi and S. Miyashita

Chapter VII. Experimental Studies of Geometrically-Frustrated Magnetic Systems

Bruce D. Gaulin

Magnetic System with Competing Interaction
Ed. H. T. Diep
©1994 World Scientific Publishing Co.

CHAPTER I

CRITICAL PROPERTIES OF FRUSTRATED VECTOR SPIN SYSTEMS

M.L. Plumer, A. Caillé, and A. Mailhot*
*Centre de Recherche en Physique du Solide et Département de Physique
Université de Sherbrooke, Sherbrooke, Québec, Canada J1K 2R1*

and

H.T. Diep
*Groupe de Physique Statistique, Université de Cergy-Pontoise
49, Avenue des Genottes, B.P. 8428, 95806 Cergy-Pontoise, France*

1. Introduction

Most physical systems in condensed matter are *frustrated* in the sense that there usually exists several competing interactions, each favoring a different type of ordered state.[1] Such competition can often be revealed by changing a parameter of the system (such as temperature, pressure, magnetic field, etc.), which serves to enhance the effect of a particular interaction and drive the system into a different ordered state. With such a broad definition, it is useful to begin with an overview and review of earlier work concerning those aspects which are relevant to the contents of this chapter. For this purpose, we refer the reader to a number of earlier review articles[2-9]. Most of the results discussed in this chapter are our own and concern frustration caused by lattice geometry (mostly triangular), or the effects of further-neighbor interactions, in classical vector (not Ising) spin systems. We emphasize phase-transition phenomena in three-dimensional (3D) systems, induced by varying temperature or an applied magnetic field. In particular, magnetic phase diagrams which exhibit unusual multicritical-point structures are examined by mean-field theory and Monte Carlo simulations. Comparison with experimental results is emphasized, for the most part from work on the quasi-1D hexagonal insulators ABX_3 (such as $CsNiCl_3$).

The triangular lattice with antiferromagnetic (AF) near-neighbor (NN) interactions appears to be the simplest example of lattice-geometry frustration. (Such lattices are called *tripartite*, in contrast with unfrustrated *bipartite* geometries, such as the square lattice.) It is not possible for all three spins at the corners of a triangle to satisfy the optimum antiparallel configuration which would minimize the energy of individual pair interactions. The resulting compromise in the case of vector spins is the so-called 120° structure, as shown in Fig. 1. In frustrated systems, local minimization of the energy is not compatible with the global energy minimum (or minima). A formal definition proposed by Toulouse[1] (and also Villain[10]) in the study of spin glasses states that a geometry is frustrated if the sign the product of exchange interactions J_i around a plaquette C

$$\Phi_C = sign[\prod_{i \in C} -J_i] \tag{1}$$

is negative (where $J_i > 0$ implies AF interactions). An AF triangular plaquette is thus frustrated as it involves a product of three $J_i > 0$. The triangular lattice is *fully frustrated* since all plaquettes satisfy this rule. The principal effect of this frustration is that it gives rise to a *noncolinear* magnetic order, such as the 120° (period-3) spin structure, and often a high degree of *degeneracy*. (Conventional wisdom states that a system requires both frustration and some degree of randomness in order to exhibit spin-glass behavior, although this view has been challenged[10,11].) The simple hexagonal lattice, which concerns most of our work, is constructed by stacking the triangular layers and is not frustrated along the c axis. Results of our work on magnetic phase transitions involving bct, fcc and hcp lattices will also be discussed. These and other frustrated lattice types are reviewed in the next section.

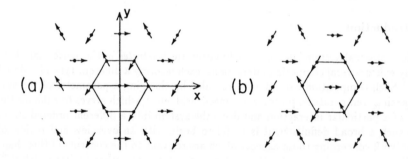

FIG. 1. Schematic of the two degenerate chiral states of the 120° spin structure where a) shows the $\mathbf{Q} = +4\pi/(3a)\hat{\mathbf{x}}$ state and b) shows the $\mathbf{Q} = -4\pi/(3a)\hat{\mathbf{x}}$ state.

There is a variety of effects resulting from the competition of NN with next-nearest neighbor (NNN) interactions. A simple example is given by three spins occupying consecutive sites along a single axis. If the interaction between the end spins (NNN) J_2 is AF then it is impossible for the system to be satisfied by either ferromagnetic or by AF near-neighbor coupling, so the spins are frustrated. 3D systems are typically composed of ferromagnetically coupled chains of this type. In the case of one-component spins, such models are referred to as the axial next-nearest-neighbor Ising (ANNNI) model and exhibit rich phase-transition phenomena.[12,13] For vector spins the competition between interactions can yield an incommensurate modulation and a *helical* polarization of the spin density. As will be seen later, the 120° spin structure can also be described as a helical spin density. The *symmetry of the order parameter*[14] is the same in both cases. According to the concept of universality of phase transitions, these two helical antiferromagnets should be in the same universality class. In the next section, a review is presented of the disparate proposals that have been made regarding the critical behavior of this class of models. The addition of AF NNN interactions in the triangular plane is also of interest (see Sec. 4). For relatively weak values of J_2, a period-2 structure is stabilized and the transition to the paramagnetic state is, rather surprisingly, first order. A continuous transition to an incommensurate state is observed at larger values of J_2.

A focus of the present review is the competition introduced by anisotropy, either due to single-ion effects, an applied magnetic field, distortions of the triangular array, or distortions of the simple hexagonal stacking of triangular layers. These effects can compete with each other in addition to interactions which favor the helical spin structure. Entropy terms generally favor the helical configuration so that competitions between temperature and anisotropy effects can give rise to complicated phase diagrams involving a number of different ordered states and which often exhibit novel multicritical-point structures. Although the phase transition at zero field is independent of the sign of the exchange coupling between triangular planes, it is very important if a field is present. In real physical systems anisotropy always exists to some degree but may not play an important role. An example is systems with relatively strong planar anisotropy so that in many cases a good approximation is the *planar* model where spins are assumed to lie strictly in a plane and all out-of-plane fluctuations are neglected.

In the large class of hexagonal ABX_3 compounds, the exchange interaction along the c axis is typically 100 times greater than the coupling in the triangular plane. Short-range magnetic order along the c-axis chains is thus well established at temperatures much higher than the onset of long-range order. Such anisotropy is not usually considered relevant to the concept of universality but, as discussed in Sec. 4, may play an important role in systems which exhibit unstable critical behavior. ABX_3 compounds also exhibit a wide variety of spin anisotropies, ranging from Ising to axial Heisenberg to planar, and are relatively easy to study experimentally.[5] The large collection of published data presents a challenge to theorists interested in frustrated magnetic systems.

Although mean-field models neglect the effects of critical fluctuations, which are especially strong in quasi-1D systems, they appear to yield qualitatively correct results for the magnetic structures and phase diagrams in most cases. Our approach has been to develop Landau-type free energies, based on symmetry arguments, expressed as an expansion in powers of the spin density. Quite flexible models result if each term in the expansion which is invariant with respect to the symmetry operations of the crystallographic space group is assigned an independent (and unknown) coefficient. Starting from a particular Hamiltonian, an equivalent free energy may be *derived* through molecular-field theory where only the Hamiltonian parameters are involved. The advantage of these formalisms is that analytic results can be obtained in many cases and the competition between interactions made clear. Such free energies are also similar in structure to Landau-Ginzburg-Wilson Hamiltonians (LGW) and can be useful in the study of critical-fluctuation effects by renormalization-group methods.

Monte-Carlo (MC) simulations have proven especially useful in the study of critical behavior in frustrated systems due to the delicate nature of applying standard renormalization-group techniques (see next section). Much progress has been made in recent years to improve MC algorithms and remarkably accurate results have been obtained for the critical exponents in unfrustrated systems.[15] Relatively few MC studies have attempted to estimate critical exponents in frustrated systems with the same degree of accuracy. A substantial investment in computer time is required due to larger fluctuation effects which arise as a consequence of the competing interactions. In some cases, however, the order of the phase transition can be clearly established. A much less formidable task is to determine the general structure of phase diagrams (e.g., magnetic-field *vs* temperature) by MC simulations. Comparison with mean-field results is emphasized in Sec. 4.

The remainder of this chapter is organized as follows. A summary of various geometrically frustrated models and predictions concerning their critical behavior is made in Sec. 2. The development of a Landau-type free energy based on symmetry arguments, as well as a discussion of MC simulations on frustrated spin systems, is presented in Sec. 3. A variety of magnetic phase diagrams from mean-field theory and MC simulations on the simple hexagonal lattice is presented in Sec. 4 for the planar as well as (axial) Heisenberg models with an applied magnetic field and also with NNN interactions. Results are compared with experimental data for a number of ABX_3 compounds. Other problems of interest are considered in Sec. 5 and our concluding remarks are made in Sec. 6.

2. Various models and predictions

A review is given here of a variety of geometrically frustrated models, defined by the Hamiltonian and lattice structure, and the predictions that have been put forth regarding their associated critical behavior in 3D systems. (The critical behavior of many 2D models is reviewed in Refs.[2- 8].) In only a few cases is there general agreement between theoretical predictions, numerical results and experimental data. It is not intended to be a complete survey but rather a highlight of those models which have been studied extensively in recent years and which have some relevance to our own work reviewed in later sections. The discussion is restricted to vector spin systems with either two ($n=2$) or three ($n=3$) components. The relevant Hamiltonian in cases of layered systems (derived for hexagonal symmetry in the Sec. 3) is of the general form

$$\mathcal{H} = \frac{1}{2} \sum_{ij} J_{ij} \mathbf{s}_i \cdot \mathbf{s}_j + D \sum_i (s_{iz})^2 - \mathbf{H} \cdot \sum_i \mathbf{s}_i \qquad (2)$$

where J_{ij} is the exchange interaction, D is the single ion anisotropy and $\mathbf{H}$ is an applied magnetic field. Some authors[2] prefer to work with exchange anisotropy of the type $J_{ij}^z s_{iz} s_{jz}$ in place of the D-term. It appears that the ordered states and critical behavior are independent of which form is chosen, provided that the anisotropies are small. (Exchange anisotropy must be used for spin 1/2 systems where the single-ion contribution is just a constant). The Heisenberg model is defined by (2) with $D = 0$. For a positive anisotropy coefficient $D > 0$, the energy is minimized by a planar configuration $s\perp\hat{z}$. If D is large in this case, it is a good approximation to assume the planar (or XY) model. For $D < 0$, this term is minimized by an axial configuration $s\|\hat{z}$ but may be in competition with the exchange term. If D is not too large, we call this the axial Heisenberg model. The very large D limit in this case is the Ising model. We restrict our discussion to antiferromagnets so that dipole effects are relatively weak due to cancellations. The case of ferromagnetic chains in a triangular array with weak AF coupling has been considered.[16] Anisotropy effects are considered in Secs. 4 and 5.

For later purposes, we define here the Fourier transform of the exchange integral

$$J_Q = \sum_R J(\mathbf{R})e^{i\mathbf{Q}\cdot\mathbf{R}}. \tag{3}$$

The wave vector $\mathbf{Q}$ which minimizes this function determines the modulation of the magnetically ordered state which first appears as the temperature is lowered. In the absence of anisotropy, this state persists to temperatures $T = 0^+$.

In cases where a single Fourier component is sufficient, it is convenient to describe magnetic order in terms of a spin density written as

$$s(\mathbf{r}) = \mathbf{S}e^{i\mathbf{Q}\cdot\mathbf{r}} + \mathbf{S}^*e^{-i\mathbf{Q}\cdot\mathbf{r}} \tag{4}$$

where $\mathbf{S} = \mathbf{S}_a + i\mathbf{S}_b$ is the complex polarization vector. This expression can be re-written as

$$s(\mathbf{r}) = 2\mathbf{S}_a\cos(\mathbf{Q}\cdot\mathbf{r}) - 2\mathbf{S}_b\sin(\mathbf{Q}\cdot\mathbf{r}). \tag{5}$$

2.1. *Triangular lattices and helical spin structures*

The exchange energy for the NN spins occupying the corners of a triangle is given by

$$E_J = 2J_\perp[\mathbf{s_1}\cdot\mathbf{s_2} + \mathbf{s_2}\cdot\mathbf{s_3} + \mathbf{s_1}\cdot\mathbf{s_3}]. \tag{6}$$

where $J_\perp$ is the NN exchange interaction. For $J_\perp > 0$, minimization of this expression yields $\mathbf{s_1} + \mathbf{s_2} + \mathbf{s_3} = 0$, e.g., the 120° spin structure as shown in Fig. 1. (In general, a cluster with all $p \geq 2$ spins interconnected by antiferromagnetic exchange interactions will minimize the energy with a zero magnetization,[17] $\sum_{i=1}^{p} \mathbf{s}_i = 0$.) It is also useful to consider the spin density representation. Evaluation of the expression (3) with $\mathbf{R}$ taken as the six nearest-neighbors to a given site in the triangular plane yields[16] (with axes defined as in Fig. 1)

$$J_Q = 2J_\perp[\cos(aQ_x) + 2\cos(\tfrac{1}{2}aQ_x)\cos(bQ_y)] \tag{7}$$

where a is the lattice constant, and $b = \sqrt{3}a/2$. For $J_\perp > 0$, minimization of this expression with respect to $\mathbf{Q}$ yields the period-three modulation $\mathbf{Q} = \pm 4\pi/(3a)\hat{\mathbf{a}}$, which can also be written as $\mathbf{Q} = \frac{1}{3}\mathbf{G}$, where $\mathbf{G}$ is a reciprocal lattice vector. A helical polarization is characterized by $S_a = S_b$ and $\mathbf{S}_a \perp \mathbf{S}_b$. In particular, one can take $\mathbf{S}_a = S\hat{\mathbf{x}}/\sqrt{2}$ and $\mathbf{S}_b = S\hat{\mathbf{y}}/\sqrt{2}$. These specifications of $\mathbf{S}$ and $\mathbf{Q}$ yield the 120° spin structure shown in Fig. 1, as found in many ABX_3 compounds. An important aspect of these results is that there are two energetically equivalent structures distinguished by their *chirality*, or sign of the wave vector $\mathbf{Q}$, as depicted in Fig. 1.

Helical spin structures also occur in rare-earth metals such as Ho, Dy, and Tb which have hcp structures[18] as well as certain bct transition metals.[19] In these materials, however, the in-plane interaction $J_\perp$ is ferromagnetic and frustration arises due to NNN c axis antiferromagnetic interactions. The modulation is thus in the $\hat{\mathbf{c}}$ direction and is determined by

$$J_Q = 2J_1 \cos(cQ) + 2J_2 \cos(2cQ) \tag{8}$$

where c is the spacing between consecutive layers. Minimization yields Q=0 for $|J_1/4J_2| > 0$ and $\cos(Q) = -J_1/4J_2$ otherwise. The latter case gives the observed incommensurate structures which are stabilized by relatively large values of J_2 arising from the long-range RKKY origin of the exchange coupling. Again, two degenerate chiralities ($\pm\mathbf{Q}$) are a result of this analysis. A strong anisotropy $D > 0$ in the rare-earths force the spins to lie in the basal plane so that $\mathbf{Q}\perp\mathbf{S}$, defining a proper helix, or *helimagnet*. This is in contrast with the XY 120° spin structure where $\mathbf{Q}, \mathbf{S}_a$, and $\mathbf{S}_b$ all lie in the basal plane.

The helical spin structure arises as a consequence of frustration. This is in contrast with a geometrically unfrustrated bipartite, such as the square, lattice. In this latter case one has $\mathbf{Q} = \frac{1}{2}\mathbf{G} = (\pi/a)\hat{\mathbf{x}} + (\pi/a)\hat{\mathbf{y}}$ so that $\sin(\mathbf{Q}\cdot\mathbf{r}) = 0$. From (5) it is clear that for unfrustrated spin systems only a single vector $\mathbf{S}_a$ is required to characterize the magnetic order whereas two vectors $\mathbf{S}_a$ and $\mathbf{S}_b$ are necessary in the frustrated case. (This difference gives rise to the two distinct chiral states.) In the case of XY spins, it is not possible to transform from one of these states to another by a global spin rotation. These considerations form the basis for arguments regarding the nature of the symmetry of the order parameter space V. In the unfrustrated case of the XY model, the single spin vector $\mathbf{S}_a$ is free to rotate in the plane so that $V = SO(2) = S_1$, the two-dimensional rotation group. In 3D, such transitions belong to the standard XY universality class. In the frustrated case, however, there is an additional two-fold discrete degeneracy[9,20] Z_2 associated with the two chiral states so that $V = Z_2 \times SO(2)$. In the unfrustrated Heisenberg case, the rotational degeneracy of the single vector $\mathbf{S}_a$ in the three-dimensional spin space maps-out the surface of a sphere in Euclidean three-space and is isomorphic to the two-dimensional sphere S_2. The two orthogonal vectors required to describe frustrated helical spin structures, however, map-out a three-dimensional solid sphere, isomorphic to the three-dimensional rotation group $SO(3)$. Note that there is no discrete degeneracy in the frustrated Heisenberg case since it *is* possible to relate the two chiral states by a global spin rotation.

There have been numerous renormalization-group studies in $4 - \epsilon$ dimensions on equivalent LGW Hamiltonians appropriate for the examination of the critical behavior of phase transitions to helical spin ordering. These are of the form (see Sec. 3)

$$\mathcal{H} = r\mathbf{S} \cdot \mathbf{S}^* + \nabla\mathbf{S} \cdot \nabla\mathbf{S}^* + U_1(\mathbf{S} \cdot \mathbf{S}^*)^2 + U_2 \,|\, \mathbf{S} \cdot \mathbf{S} \,|^2 \,. \qquad (9)$$

Note that frustration introduces *two* fourth-order terms. A positive coefficient $U_2 > 0$ is seen to stabilize a helical polarization of the spin density, $\mathbf{S} = S(\hat{\mathbf{x}} + i\hat{\mathbf{y}})/\sqrt{2}$, since in that case the energy is minimized by $\mathbf{S} \cdot \mathbf{S} = 0$. A linear polarization is realized for $U_2 < 0$. One reason for the plethora of such studies is that in addition to describing helical spin systems[21-27], such Hamiltonians are also relevant to the dipole phase of superfluid ^3He,[28] Josephson-junction arrays in a transverse field,[29] as well as the fully frustrated bipartite lattice[30] (Villain model[10]). Although some results suggested a continuous transition within standard universality classes,[21,22] others found a first-order transition.[23,24,26,28] The most recent study makes no definite conclusions regarding the nature of the phase transition in 3D for the XY or Heisenberg models.[27] It appears that low-order expansions in ϵ may be unreliable in the case of frustrated systems.

Universality classes are in principle determined by the space dimensionality and symmetry of the order parameter. The identification of new $V = Z_2 \times SO(2)$, and $V = SO(3)$ symmetries in the frustrated XY and Heisenberg models, respectively,[9,31,32] supports the possibility that these systems belong to new universality classes,[9,31,32] provided that the transitions are not first order. In addition to symmetry arguments and renormalization-group studies, Kawamura has found evidence for the existence of new *chiral* universality classes from MC-simulation results for critical exponents in 3D[9] (which are, in principle, unique to a given universality class). Values are given in the table below, together with those of the standard universality classes $O(n)$,[33,34] with α determined by the scaling relation $\alpha = 2 - 2\beta - \gamma$.

universality class	α	β	γ	ν
n=2	-0.012(8)	0.3485(35)	1.315(7)	0.671(5)
n=3	-0.126(11)	0.368(4)	1.390(10)	0.710(7)
n=4	-0.22	0.39	1.47	0.74
n=2 chiral	0.34(6)	0.253(10)	1.13(5)	0.54(2)
n=3 chiral	0.24(8)	0.30(2)	1.17(7)	0.59(2)
mean-field	0	$\frac{1}{2}$	1	$\frac{1}{2}$
mean-field tricritical	$\frac{1}{2}$	$\frac{1}{4}$	1	$\frac{1}{2}$

MC finite-size scaling results for the chiral $n = 3$ system which corroborate these values have recently been reported.[35-37] Experimental support for the existence of the new universality classes through measurements of critical exponents for stacked triangular antiferromagnets, as well as rare-earth helimagnets, is discussed in Ref.[9]. Cases of larger n, as well as for increased number of spanning vectors m (where $m = 2$ for helical spin systems), have been examined.[38] There are also a set of *chirality* exponents associated with the chirality order parameter given in Ref.[9]. For a given plaquette of a triangular plane, this may be defined as

$$\kappa = 2/(3\sqrt{3})[s_1 \times s_2 + s_2 \times s_3 + s_3 \times s_1]. \tag{10}$$

For the 120° spin structure, $\kappa = \pm 1$. The possibility for experimental determination of these exponents is discussed later.

   In contrast, Azaria, Delamotte and Jolicoeur[34] maintain that there is no new universality class for these frustrated spin systems based on a study of the nonlinear sigma model in $2 + \epsilon$ dimensions (with $\epsilon \to 1$) for the Heisenberg case (also see Refs.[27] and [39]). Determination of the true critical behavior at the small values of n of interest is seen by these authors to be inaccessible by perturbation expansions in $\epsilon = 4 - D$. A scenario for the n-D phase diagram emerges in which there is a tricritical line dividing regions where first-order and continuous transitions are expected. For large n the transition is known to be continuous at D=4. At n=3, the line lies between the continuous region at D=2 and the first-order region at D=4, but its precise location is not known. If the transition is continuous, then the prediction is $O(4)$ universality (also see Refs.[21] and [22]), with exponents as given in the table above (n=4). There appears to be no particular reason that the tricritical line crosses $D = 3$. If this is the case, however, the mean-field tricritical exponents given in the table are expected. These values are not so different from those given by Kawamura for the $n = 2$ chiral universality class. Conclusions regarding the critical behavior in the case $n = 2$ were not made by Azaria *et al* but it is tempting to speculate either first-order or tricritical behavior. Kawamura, as well as Azaria *et al*, have each found experimental results in support of their claims. The frustrated XY model is somewhat unusual since discrete chiral degeneracy is involved at the transition in addition to the continuous rotational symmetry. In generalized Villain models,[8] as well as hybrid XY-Ising models,[32] it is possible to tune parameters so that there are two successive transitions as the temperature is lowered, associated with the breaking of S_1 and Z_2 symmetries, respectively. As will be discussed later, such behavior is also observed in the stacked triangular XY AF in the presence of an applied magnetic field. It is not clear to what extent the excitations associated with these two symmetries are coupled and influence the nature of the phase transition at zero field.

   One of the assumptions of the nonlinear sigma model is equivalent to forcing the two spin vectors S_a and S_b to be perpendicular. This can be generalized to objects composed of m orthogonal n-dimensional vectors of the Stiefel manifold[40] $V_{n,m} = O(n)/O(n-m)$. The frustrated models discussed above correspond to $m = 2$ and unfrustrated systems have $m = 1$. MC results on this model appear to support the nonlinear sigma model analysis in the $2 + \epsilon$ expansion[40] where finite-size scaling behavior suggests possible mean-field tricritical exponents in the case of $n = 3$ and some evidence for a first-order transition for $n = 2$.

An already confusing picture is further complicated by examination of MC results on the NNN helimagnet[19] where a number of values for $\mid J_1/4J_2 \mid$ were chosen so that $90° < Q < 180°$ (with J_1 antiferromagnetic) was commensurate. (Equivalent values in the range $0° < Q < 90°$ are realized for J_1 ferromagnetic.) In the XY case, systems with smaller values of Q (105° and 120°) yielded a single first-order transition at which both the chirality and spin order parameter exhibited discontinuities. The strength of these discontinuities weakened as Q increased. At larger values of Q (150° and 165°), however, it appears that spin ordering occurs continuously at a higher temperature than the onset of chirality, which remains a first-order transition. The critical value of Q distinguishing these two behaviors occurs at *approximately* $Q_c = 3\pi/4$. A possible explanation is proposed in Ref.[19] which is related to observation that the second term in (8) vanishes at $3\pi/4$. The rare-earth helimagnets appear to be characterized by J_1 ferromagnetic within this simple NNN model[41] so that a first-order transition is expected for $45° < Q < 90°$. Although strongly temperature dependent, neutron diffraction data indicate the following values for Q at the Néel point: 50° for Ho,[42] 43° for Dy,[43] and 20° for Tb.[44] Weak first-order transitions have been observed in Ho[45] and in Dy[46] but the transition in Tb appears to be continuous.[47] Except possibly for Dy, where the value of Q is very close to the critical point $Q_c \simeq \pi/4$, these results are consistent with the MC simulations. With the long-range nature of the RKKY interactions in the rare-earths, it is probable that third and higher-neighbor interactions can modify the value of Q_c. These results are of particular interest in view of commensurability effects examined by Garel and Pfeuty[22] in $4 - \epsilon$. They note that in systems with a periodicity of four ($4\mathbf{Q} = \mathbf{G}$), there is an additional term in the LGH Hamiltonian of the form (also see Refs.[16, 41])

$$\mathcal{H}_V = V[(\mathbf{S} \cdot \mathbf{S})^2 + c.c] \tag{11}$$

which drives the transition from being continuous (O(4) criticality) to first order. For MC simulations of the NNN helimagnet with Heisenberg spins,[19] a single continuous transition (independent of Q) was found, with estimates for the critical exponents (at $Q = 120°$) $\alpha = 0.32(3)$ and $\nu = 0.57(2)$, consistent with those obtained by Kawamura. Differences between these MC results for the XY case and those of Kawamura are puzzling (particularly in view of recently obtained MC data of higher accuracy[9]) since the NNN helimagnet and stacked triangular antiferromagnet should have the same critical behavior. Further comments are made in the Sec. 3 regarding the reliability of MC simulations on frustrated systems. Possible avenues for clarifying the nature of the critical behavior in these helical spin systems are explored in Sec. 4.

2.2. Fcc, hcp and rhombohedral antiferromagnets: order by disorder

Fcc and hcp antiferromangets, formed by different stacking sequences of the basic tetrahedral unit, exhibit a high degree of ground-state degeneracy. This is illustrated in Fig. 2: spins 1 and 3, as well as spins 2 and 4, are antiparallel ($s_1 + s_2 + s_3 + s_4 = 0$). For XY spins, the angle α between these two sublattices can take on any value. For

Heisenberg spins there is an additional degree of freedom, the angle β between the plane formed by spins 1,3 and the plane formed by spins 2,4. Thus in addition to the degeneracy associated with a global spin rotation (as in the case of the triangular lattice), there are additional infinite degeneracies associated with α and β degrees of freedom. Henley[48] has applied the concept of breaking ground-state degeneracies by thermal fluctuations, "order by disorder"[49], to the fcc (hcp) tetrahedron. In the case of vector spins, spin-wave analysis demonstrates that a colinear configuration, $\alpha = \beta = 0$ is selected. There are three ways to chose the antiparallel spin pair on a tetrahedron, so that in addition to global rotational symmetry there is discrete 3-fold degeneracy. The symmetry of the order parameter is thus $V = Z_3 \times S_1$ and $V = Z_3 \times S_2$ in the XY and Heisenberg cases, respectively. Henley then speculated that since the 3-state Potts model, with $V = Z_3$, exhibits a first-order transition (for a recent review see, e.g. Ref.[50]), then such behavior should also be observable in the fcc (and hcp) antiferromagnet. Confirmation of the colinear spin structure, as well as the (weak) first-order character of the phase transition, has been found in MC-simulation results on both lattice types and for XY and Heisenberg spins.[51]

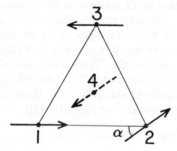

FIG. 2. Schematic of a tetrahedron with vector spins 1 and 3, as well as 2 and 4, being antiparallel.

The rhombohedral lattice can be constructed from a repetition of three equally spaced (along the c axis) triangular layers,[52] two of which are displaced relative to the reference plane at $u_0 = 0$ by $u_\pm = \pm(\frac{2}{3}b\hat{y} + \frac{1}{3}c\hat{z})$. (Yet another stacking sequence of anisotropic tetrahedra). The ordering wavevector in this case is determined by[3,53] (compare (7))

$$\begin{aligned} J_Q &= 2J[\cos(aQ_x) + 2\cos(\tfrac{1}{2}aQ_x)\cos(bQ_y)] \\ &+ 2J'\{\cos(\tfrac{1}{3}cQ_z)[\cos(\tfrac{2}{3}aQ_y) + 2\cos(\tfrac{1}{2}aQ_x)\cos(\tfrac{1}{3}bQ_y)] \\ &+ \sin(\tfrac{1}{3}cQ_z)[2\sin(\tfrac{1}{3}aQ_y)\cos(\tfrac{1}{2}aQ_x) - \cos(\tfrac{2}{3}aQ_y)]\} \end{aligned} \tag{12}$$

where $J > 0$ is the in-plane coupling and $J' > 0$ is the interaction between planes. For $J' = 0$, the 120° structure is stabilized. For $J' \geq 3J$, a simple AF c axis modulation $Q = (3\pi/c)\hat{z}$ is stabilized. For values $J' < 3J$, however, a degenerate incommensurate modulation occurs, where the degeneracy exists along lines in Q-space. It is speculated in Ref.[53] that this degeneracy is at least partially removed by thermal fluctuations but that with such a high degree of degeneracy, long-range order may not occur. It is known that NNN coupling selects a particular ground-state. Physical realizations of this model include characterization of the crystal structure of the β-phase of solid O_2[53] (with a mapping onto XY spins) as well as the quasi-2D Heisenberg AF $CuCrO_2$[52]. The hcp lattice composed of anisotropic tetrahedra may also be described in terms of displaced triangular layers with intraplane J and interplane J' interactions. The ground-state degeneracy is identical to that of the rhombohedral lattice[54] and the critical phenomena are expected to be the same.

MC simulations on anisotropic Hamiltonians similar to (2) exhibit rather complex behavior for the hcp AF.[54,55] A first-order transition to an incommensurate helical phase occurs in the case of planar anisotropy ($D > 0$). For Ising anisotropy ($D < 0$), the transition becomes continuous, with an intermediate Ising-like (non-chiral) phase of short-range order. The Heisenberg point ($D = 0$) thus appears to be tricritical, in principle exhibiting the exponents given in the table of Sec. 2.1. These results appear to support a possible scenario put forth by Azaria *et al*, but this conclusion is contingent upon the claim that the observed Ising phase does not exhibit long-range order. If the Ising phase is long-range ordered, then $D = 0$ represents a multicritical, not a simple tricritical, point. Such multicritical behavior is discussed in detail in Sec. 4. It is not clear to what extent such MC simulations on finite size systems are influenced by the type of incommensurability inherent in these systems. The authors of Ref.[55] also note that chirality and spin order occur at the same temperature.

2.3. Kagomé and pyrochlore antiferromagnets: more disorder

The 2D kagomé antiferromagnet exhibits an unusually high degree of degeneracy. The lattice is composed of corner-sharing triangles, the centers of which form a honeycomb lattice, as shown in Fig. 3. In recent years there have been many studies of possible types of vector-spin ordering in this system.

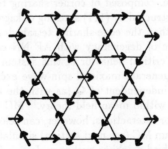

FIG. 3. Kagomé lattice showing the $\sqrt{3} \times \sqrt{3}$ spin structure.

The ground-state and low-temperature properties are nicely reviewed in a recent work by Reimers and Berlinsky.[56] Any configuration which satisfies the constraint that the three spins of each *individual* triangle have zero magnetization, $s_1 + s_2 + s_3 = 0$ (120°

structure), is one of the infinite number of degenerate ground states. Since there are three possible choices for such a configuration, *each* plaquette has a symmetry of the 3-state Potts model. Within a harmonic treatment of excitations, there is an entire branch of zero-energy modes in the first zone so that the existence of long-range order at $T = 0$ has been a question of some interest. A high-temperature series expansion in J/T selects the modulation $Q = (\frac{2}{3}, \frac{2}{3})$, the $\sqrt{3} \times \sqrt{3}$ (period-3) structure shown in Fig. 3, only at 8th order.[57] This structure also shows an alternating (antiferromagnetic) chiral order of neighboring triangles (as in Fig. 1) and thus discrete chiral degeneracy. In contrast, the nearly degenerate $Q = 0$ phase exhibits a period-2 structure (and uniform chirality) and can be stabilized by ferromagnetic third NN coupling.[58] (Note that the kagomé is a non-Bravais lattice so that Q itself does not characterize the type of magnetic order.[59] For example, there is also a colinear state with $Q = 0$.) Low-temperature expansions in the case of Heisenberg spins appear to favor *coplanar nematic* order by disorder.[60] Evidence for the saturation of such order at $T \to 0$ was found in MC simulations[60] which calculate the chiral nematic correlation function

$$g(\mathbf{r}_{\alpha\beta}) = \frac{3}{2} < (\boldsymbol{\kappa}_\alpha \cdot \boldsymbol{\kappa}_\beta)^2 > -\frac{1}{2}, \tag{13}$$

where $\boldsymbol{\kappa}$ is the chirality (10) of the triangular plaquette α. Some evidence for the low-T $\sqrt{3} \times \sqrt{3}$ spin ordering was also found. These results have been confirmed, as well as evidence for 3-state Potts and chiral order, in more recent and extensive MC simulations.[56] A number of experimental systems are available, such as $SrCr_{8-x}Ga_{4+x}O_{19}$, with alternating kagomé and triangular layers, which are quasi-2D materials and contain defects[61] (the effects of which have recently been considered theoretically[62]). FeF_3 forms a simple stacked kagomé lattice which exhibits $Q = 0$ spin order, presumably due to further-neighbor in-plane coupling. MC simulations of this 3D lattice with only NN interactions appear to show no long-range spin order.[63]

An extension of this structure which brings frustration to the third spatial dimension is the pyrochlore lattice, composed of corner sharing tetrahedra and a 16-site unit cell. (It can also be constructed by (111) stacking of kagomé layers.) This lattice is more sparsely connected than the edge-sharing tetrahedra of the fcc counterpart and exhibits a higher degree of degeneracy with AF NN coupling. Reimers *et al* have recently examined the critical behavior of this system both theoretically and experimentally. The two degenerate maximal spin-wave excitations for the Heisenberg model in this case are independent of Q (as with the kagomé AF) so that no long-range spin order occurs within mean-field theory.[17] MC simulations corroborate this result.[64] Further neighbor interactions, however, can select $Q = 0$ or incommensurate modulations. Histogram MC (see next section) simulations were performed on the Heisenberg model with third-neighbor exchange $J_3 = -J_1/10$ so that the $Q = 0$ phase is stabilized.[65] An effort was made to determine the order of the transition with results suggesting a continuous one. Finite-size scaling estimates for critical exponents are

$$pyrochlore: \quad \alpha = 0.6(1), \quad \beta = 0.18(2), \quad \gamma = 1.1(1), \quad \nu = 0.38(3). \tag{14}$$

These results led the authors to speculate the existence of a new universality class, encouraged by the suggestion of the new chiral universality class in the case of the stacked triangular AF and NNN helimagnet. The MC results also indicate that among the degenerate ground states, a colinear phase is selected by thermal fluctuations. For the same reasons as in the case of the fcc Heisenberg AF discussed above, the symmetry of the order parameter is expected to be $V = Z_3 \times S_2$. Neutron diffraction data on the pyrochlore form of FeF_3 suggests a continuous phase transition with an estimated value for the order-parameter exponent $\beta = 0.18(2)$, in agreement with the MC results. It is not clear, however, to what extent anisotropy is present in this system and the relevance of the Heisenberg model. We also note that these results are not in agreement with the first-order transition found in the MC data for the fcc lattice, in principle a system with the same critical behavior since V is the same. A subsequent re-analysis of the MC data of Ref.[65] suggests the possibility that the transition in the pyrochlore AF can be seen to be first order through asymptotic volume dependence in several thermodynamic quantities.[66]

This work on the pyrochlore AF has illustrated two important aspects of our own studies of frustrated spin systems. The first is that an understanding of results from mean-field models is usually helpful in the later examination of critical phenomena. Secondly, MC results can be misleading. Both of these observations arise from the competition for long-range order from the nearly degenerate states. In an earlier MC study of the pyrochlore lattice, only a small NNN AF interaction was included in addition to the NN AF coupling.[63] A very unusual set of critical exponents was found, including an estimate for $\gamma < 1$. After a mean-field analysis including third NN interactions, through minimization of the exchange function $J(\mathbf{Q})$, it was observed in the $J_2 - J_3$ phase diagram that the chosen value of J_2 put the system in a region where three types of ordered states are very nearly degenerate (see Fig. 3 of Ref.[65]). Fluctuations must have been enormous, making the extraction of reliable critical exponents nearly impossible. A highly unusual set of critical exponents has also been estimated from early MC simulations on the fully frustrated cubic lattice[67] for $n = 1, 2$, and 3, probably a system with as much frustration as the pyrochlore AF. In the case $n = 3$ an estimate $\gamma = 0.80$ was found. It appears to be an empirical rule, however, that γ must be greater than unity.[68] In addition, the other exponents do not obey hyperscaling and also violate scaling relations such as $\gamma = \nu(2 - \eta)$. Such inconsistencies suggest that the transition may be first order in this system.[69] As in the pyrochlore case, a mean-field analysis may be useful in understanding why traditional MC methods would yield unusual results.

3. Mean-field Landau models and Monte Carlo simulations

For exchange models, mean-field theory[70] is an approximation in which the effective field acting on an individual spin, $\sum_j J_{ij}s_j$, is replaced by its thermal *average* $\sum_j J_{ij}\langle s_j \rangle$. Differences between the states of these neighboring spins, the fluctuations away from the mean, are ignored. Critical exponents are then independent of n and D and differ significantly from values of the standard universality classes (see table in Sect. 2). Mean-field estimates of T_N are universally overestimated; critical fluctuations inhibit the onset of long-range order. Such effects become more important at lower dimensionality: D=1 systems exhibit no long-range order in contrast with mean-field results. In spite of these deficiencies mean-field theories are usually reliable in predicting the correct symmetry of long-range order when it occurs. Landau-type theories usually begin with an expansion of the free energy in powers of this mean spin density $\langle s(r) \rangle$ and as such are mean-field models. We distinguish here between microscopically derived Landau-type free energy functionals $F[s(r)]$, which begin with a Hamiltonian such as (2), and phenomenological models where only symmetry arguments are used. The latter approach, although less elegant, is much more general and is capable of mimicing some effects due to critical fluctuation.

Apart from renormalization-group calculations, numerical simulations are an increasing popular method for the study of critical phenomena and are limited only by available (ever increasing) computer power.[70] The quality of Monte Carlo results is determined by the size of the system and by the number of MC steps used to perform thermal averages. Accurate determination of critical exponents is impeded by critical slowing down at the temperatures of interest near T_N. This effect can be quite serious at large lattice sizes L when standard MC simulations are performed and the relaxation time grows with the correlation length $\xi \simeq L$ as $\tau \sim \xi^z$, where $z \simeq 2$ is typical for many systems. Inventive new MC techniques devised for overcoming this problem appear not to be effective for frustrated spin systems. Less demanding is the determination of general features of magnetic phase diagrams. For this purpose, as well as for finite-size scaling analysis to estimate critical exponents, the *histogram* MC method discussed below is particularly useful.

3.1. Phenomenological Landau-type free energy

The usual Landau free energy represents a type of Taylor expansion in powers of the order parameter associated with a phase transition.[70,71] Each term in the expansion must be invariant with respect to the relevant symmetry operations[71,72] of the system. As an example, it is instructive to give symmetry arguments which can be used to formulate the Hamiltonian (2) for crystals with hexagonal symmetry. The idea is to start with the most general type of (symmetric) quadratic interaction (omitting site labels for simplicity)

$$\mathcal{H} = \sum_{ab} J_{ab} s_a s_b, \tag{15}$$

where $a, b = x, y, z$, and to retain only terms which are invariant with respect to the *generators* of the symmetry group of the crystal.[73] Note that terms with odd powers of s are excluded by the requirement of time-reversal symmetry. Spin vectors are *axial* vectors, written as a cross product of two *polar* vectors $s \sim r \times j$, where j is the current density. Rotations have the same effect on both polar and axial vectors, however, reflection planes do not.[72] Spin vectors are thus invariant with respect to inversion symmetry. For this example, we consider only a generic hexagonal point-group symmetry with generators C_6^+, C_{2y} (where $\hat{y}$ is perpendicular to a basal-plane axis as in Fig. 1) and follow the arguments as given by Kittel[74] in his examination of the elastic energy. Equation (15) contains six independent terms, which can be expressed as

$$\mathcal{H} = J_{xx}s_x^2 + J_{yy}s_y^2 + J_{zz}s_z^2 + J_{xy}s_xs_y + J_{yz}s_ys_z + J_{xz}s_xs_z. \tag{16}$$

Consider first the requirement that $\mathcal{H}$ be invariant with respect to the two-fold rotation $C_{2y}(s_x, s_y, s_z) \to (-s_x, s_y, -s_z)$, giving

$$\mathcal{H} = J_{xx}s_x^2 + J_{yy}s_y^2 + J_{zz}s_z^2 - J_{xy}s_xs_y - J_{yz}s_ys_z + J_{xz}s_xs_z. \tag{17}$$

Equality of equations (16) and (17) thus demands that $J_{xy} = J_{yz} = 0$. The requirement of six-fold rotational symmetry about the c axis then gives

$$\mathcal{H} = \tfrac{1}{4}J_{xx}(s_x^2 + 3s_y^2 + 2\sqrt{3}s_xs_y) + \tfrac{1}{4}J_{yy}(3s_x^2 + s_y^2 - 2\sqrt{3}s_xs_y)$$
$$+ J_{zz}s_z^2 + \tfrac{1}{2}J_{xz}(s_x + \sqrt{3}s_y)s_z. \tag{18}$$

so that equality of equations (17) and (18) yields $J_{xx} = J_{yy}$ and $J_{xz} = 0$. There are thus *two* independent quadratic invariants for hexagonal crystals, namely $s_x^2 + s_y^2$ and s_z^2. These may be added together so that one of the invariants can be chosen as $\mathbf{s} \cdot \mathbf{s}$. Such symmetry arguments lead to Eq.(2). Since J_{ij} represents the exchange interaction, one has $i \neq j$. The Zeeman term is rotationally and time-reversal invariant.

Precisely the same types of arguments can be used to generate an expansion of the free energy functional to fourth order in the spin density, relevant for the study magnetic phase transitions in hexagonal (and tetragonal) crystals:[16,75]

$$F[\mathbf{s}(\mathbf{r})] = \tfrac{1}{2}\int d\mathbf{r}d\mathbf{r}'A(\tau)\,\mathbf{s}(\mathbf{r}) \cdot \mathbf{s}(\mathbf{r}') + \tfrac{1}{2}D\int d\mathbf{r}\,[s_z(\mathbf{r})]^2$$
$$+ \tfrac{1}{4}\int d\mathbf{r}_1 d\mathbf{r}_2 d\mathbf{r}_3 d\mathbf{r}_4\, B(\mathbf{r}_1,\mathbf{r}_2;\mathbf{r}_3,\mathbf{r}_4)\,\mathbf{s}(\mathbf{r}_1) \cdot \mathbf{s}(\mathbf{r}_2)\,\mathbf{s}(\mathbf{r}_3) \cdot \mathbf{s}(\mathbf{r}_4)$$
$$+ \tfrac{1}{4}E\int d\mathbf{r}\,[s_z(\mathbf{r})]^2\,\mathbf{s}(\mathbf{r}) \cdot \mathbf{s}(\mathbf{r}) + \tfrac{1}{4}G\int d\mathbf{r}\,[s_z(\mathbf{r})]^4 - \mathbf{H} \cdot \int d\mathbf{r}\,\mathbf{s}(\mathbf{r}) \tag{19}$$

where $\tau = \mathbf{r} - \mathbf{r}'$ and anisotropy of the single-ion type has been assumed. To fourth order, only anisotropy between the basal plane and the c axis appears. Anisotropy *within* the basal plane occurs at sixth order in the spin density.[76] The dominant origin of anisotropy is through spin-orbit effects which are usually very small. Note that the fourth-order isotropic term is *nonlocal*. Using the expression (4) for $\mathbf{s}(\mathbf{r})$ in (19) gives (in the absence of a magnetic field)

$$F = A_Q\mathbf{S} \cdot \mathbf{S}^* + D\,|\,S_z\,|^2 + B_1(\mathbf{S} \cdot \mathbf{S}^*)^2 + \tfrac{1}{2}B_2\,|\,\mathbf{S} \cdot \mathbf{S}\,|^2$$
$$+ \tfrac{1}{2}[A_Q(\mathbf{S} \cdot \mathbf{S} + \mathbf{S}^* \cdot \mathbf{S}^*) + D(S_z^2 + S_z^{*2})]\Delta_{2Q,G}$$
$$+ \tilde{B}_1(\mathbf{S} \cdot \mathbf{S}^*)[\mathbf{S} \cdot \mathbf{S} + \mathbf{S}^* \cdot \mathbf{S}^*]\Delta_{2Q,G} + \tfrac{1}{4}\tilde{B}_2[(\mathbf{S} \cdot \mathbf{S})^2 + (\mathbf{S}^* \cdot \mathbf{S}^*)^2]\Delta_{4Q,G} \tag{20}$$

where fourth-order anisotropy terms have been omitted for simplicity and[77]

$$B_1 = B_{Q,-Q;Q,-Q}, \quad B_2 = B_{Q,Q;-Q,-Q}, \quad \tilde{B}_1 = B_{-Q,Q;Q,Q}, \quad \tilde{B}_2 = B_{Q,Q;Q,Q} \quad (21)$$

with

$$B_{q_1,q_2;q_3,q_4} = \Delta_{q_1+q_2+q_3+q_4,G} \sum_{\mathbf{R}_1,\mathbf{R}_2,\mathbf{R}_3} B(\mathbf{R}_1,\mathbf{R}_2,\mathbf{R}_3)e^{i(q_1\cdot R_1+q_2\cdot R_2+q_3\cdot R_3)}. \quad (22)$$

Inversion symmetry has been used in deriving this result as well as the assumption that $B\{r_i\}$ depends only on differences between pairs of coordinates, e.g.,

$$B(\mathbf{r}_1,\mathbf{r}_2;\mathbf{r}_3,\mathbf{r}_4) = B(\mathbf{r}_1 - \mathbf{r}_4, \mathbf{r}_2 - \mathbf{r}_4, \mathbf{r}_3 - \mathbf{r}_4). \quad (23)$$

Note that the terms B_1 and B_2 in (20) are the same as the fourth-order terms that appeared in the LGW Hamiltonian (9), and that the Umklapp term $\tilde{B}_2$ is the same as in (11).

The number of independent fourth-order terms within this model decreases in the absence of frustration. Consider the case $2\mathbf{Q} = \mathbf{G}$. Since $-\mathbf{Q} = \mathbf{Q} + \mathbf{G}$ it follows from the above relations that $B_1 = B_2 = \tilde{B}_1 = \tilde{B}_2 \equiv B_0$. For frustrated systems, there is no relationship between these coefficients that is based on symmetry arguments. Note, however, that approximate relationships of this type may exist in the case of quasi-1D materials. These systems are not frustrated along the chains so that $2\mathbf{Q}_{\parallel} = \mathbf{G}_{\parallel}$. With $2\mathbf{Q} = \mathbf{G}$, the polarization vector $\mathbf{S}$ may be taken as real and the free energy can be written as

$$F = A_Q S^2 + DS_z^2 + B_0 S^4 \quad (24)$$

where $S^2 = \mathbf{S} \cdot \mathbf{S}^*$ and the renormalization $S^2 \rightarrow \frac{1}{2}S^2$ has been used. We adopt the usual Landau-theory convention by introducing temperature dependence in A_Q only through

$$A_Q = a(T - T_Q) \quad (25)$$

and $B_0 > 0$ is assumed for stability. In the case of planar anisotropy, $D > 0$, a configuration $\mathbf{S} \perp \hat{z}$ minimizes F and the transition occurs at $T_N = T_Q$. In the case of axial anisotropy, $D < 0$, a configuration $\mathbf{S} \| \hat{z}$ is preferred and the transition occurs at $T_N = T_Q - D/a$. There are no other transitions as the temperature is decreased further in this unfrustrated system.

For frustrated systems, such as those with $3\mathbf{Q} = \mathbf{G}$, the relevant free energy is similar to (24) but the difference is significant:

$$F = A_Q S^2 + D|S_z|^2 + B_1 S^4 + \frac{1}{2}B_2 |\mathbf{S} \cdot \mathbf{S}|^2. \quad (26)$$

As discussed in Sec. 2.1, a positive coefficient B_2 favors a helical polarization of the spin density, $\mathbf{S} \cdot \mathbf{S} = 0$. In the case of planar anisotropy, this state is realized at all temperatures below $T_N = T_Q$. Such a state is not, however, favored by axial anisotropy. Just below the Néel temperature, second-order terms are dominant and the configuration $\mathbf{S} \| \hat{\mathbf{z}}$ minimizes F, as in the unfrustrated system at $T_{N1} = T_Q - D/a$. At lower temperatures, the effect of the B_2-term becomes relevant and a second (continuous) transition occurs to a state where both S_z and $S_\perp$, with $\mathbf{S}_\perp \perp \hat{\mathbf{z}}$, are nonzero (an elliptical polarization)[76,78] at $T_{N2} = T_{N1} - |D| b/a$ where $b = B/(2B_2)$ and $B = B_1 + 2B_2$. The existence of two ordered states is a consequence of D *and* the competition from the B_2-term, which is due to frustration.

3.2. *Mean-field derivation of a Landau free energy*

A convenient derivation of the Landau-type free energy based on the mean-field approximation begins with the Bogoliubov inequalitiy[70,79,80]

$$F \leq F_0 + \langle \mathcal{H} - \mathcal{H}_0 \rangle_0, \tag{27}$$

where $\mathcal{H}_0$ is a trial Hamiltonian which is taken to be of the form

$$\mathcal{H}_0 = -\sum_i \mathbf{h}_i \cdot \mathbf{s}_i, \tag{28}$$

with $\mathbf{h}_i$ being the *mean* field. Using a slightly generalized version of the Hamiltonian (2)

$$\mathcal{H} = \tfrac{1}{2} \sum_{ij} \sum_{ab} J_{ij}^{ab} s_{ia} s_{jb} - \sum_i \mathbf{H} \cdot \mathbf{s}_i \tag{29}$$

gives

$$F \leq F_0 + \tfrac{1}{2} \sum_{ij} \sum_{ab} J_{ij}^{ab} \langle s_{ia} \rangle \langle s_{jb} \rangle - \sum_i \mathbf{H} \cdot \langle \mathbf{s}_i \rangle + \sum_i \mathbf{h}_i \cdot \langle \mathbf{s}_i \rangle. \tag{30}$$

The *equilibrium* condition $\partial F / \partial \langle s_{ia} \rangle = 0$ may then be used to obtain an expression for the mean field

$$h_{ia} = -\sum_j \sum_b J_{ij}^{ab} \langle s_{jb} \rangle + H_a. \tag{31}$$

The thermodynamic average of the spin variable is given by

$$\langle s_{ia} \rangle = \frac{tr \; s_{ia} e^{\beta \mathbf{h}_i \cdot \mathbf{s}_i}}{tr \; e^{\beta \mathbf{h}_i \cdot \mathbf{s}_i}}, \tag{32}$$

where $\beta = 1/(k_B T)$. At equilibrium, $\langle \mathbf{s}_i \rangle \| \mathbf{h}_i$, so that for a system with angular momentum j,

$$\langle s_{ia} \rangle = (h_{ia}/h_i) \frac{\sum_{m=-j}^{j} m e^{\beta h_i m}}{\sum_{m=-j}^{j} e^{\beta h_i m}} \tag{33}$$

giving

$$\langle s_{ia} \rangle = (h_{ia}/h_i)\{u \coth(u\beta h_i) - v \coth(v\beta h_i)\} \tag{34}$$

where $u = (2j + 1)/2j$ and $v = 1/2j$. At temperatures close to T_N where $h \sim \langle s \rangle$ is small, this expression can be expanded in powers of βh_i. Inversion of the resulting equation yields

$$h_{ia} = k_B T\{a\langle s_{ia} \rangle + b\langle s_{ia} \rangle \langle s_i \rangle \cdot \langle s_i \rangle + \cdots\}, \tag{35}$$

where $a = 3j/(j + 1)$ and $b = (a^4/45)[(2j + 1)^4 - 1]/(2j)^4$. (Results at sixth order are given in Ref.[41].) The free energy is then obtained by integration[79] of

$$\partial F/\partial \langle s_{ia} \rangle = \sum_j \sum_b J_{ij}^{ab} \langle s_{jb} \rangle - H_a$$
$$+ k_B T\{a\langle s_{ia} \rangle + b\langle s_{ia} \rangle \langle s_i \rangle \cdot \langle s_i \rangle + \cdots\} \tag{36}$$

to give, in a continuous representation,

$$F[s(\mathbf{r})] = \frac{1}{2}\sum_{ab} \int d\mathbf{r}d\mathbf{r}' J^{ab}(\tau) s_a(\mathbf{r})s_b(\mathbf{r}') - \mathbf{H} \cdot \int d\mathbf{r}\, s(\mathbf{r})$$
$$+ k_B T\{\frac{1}{2}a \int d\mathbf{r}\, s(\mathbf{r}) \cdot s(\mathbf{r}) + \frac{1}{4}b \int d\mathbf{r}\, [s(\mathbf{r}) \cdot s(\mathbf{r})]^2 + \cdots\}, \tag{37}$$

where the thermal-average notation $\langle \cdots \rangle$ has been omitted for convenience.

The connection with the Hamiltonian (2) is made by the identification

$$J_{ij}^{ab} = \{J_{ij}(1 - \delta_{ij}) + 2D\delta_{ij}\delta_{az}\}\delta_{ab}. \tag{38}$$

For simplicity, we have chosen to treat the single-ion anisotropy term also within a mean-field approximation $s_{iz}s_{iz} \rightarrow s_{iz}\langle s_{iz} \rangle$ as done by White.[81] Although this application is not exactly consistent with the notion that the mean field acting on a site is an average of interactions with neighboring spins, it yields qualitatively correct results. An exact treatment of the D-term may be found in Ref.[80].

The above expression for the mean-field free energy is similar in structure to the phenomenological Landau result (19). Second-order terms are seen to be identical, with

$$A(\mathbf{r}) = ak_b T\delta(\mathbf{r}) + J(\mathbf{r}). \tag{39}$$

This justifies further the temperature dependence in (25). The fourth-order term is notably *local* in form, in contrast with the more general Landau formulation. Thus, all biquadratic terms have a common coefficient $B = bk_B T$, where the approximation $B \simeq bk_B T_N$ is usually made since the expansion is valid only at small $\langle s(\mathbf{r}) \rangle$. Mean-field theory is thus seen to be more restricitve than the phenomenological approach in the treatment of frustrated spin systems. In some cases, qualitatively different results are found. It is this mean-field formalism, however, that is used in the following sections to compare with MC results in determining the effects of fluctuations. For classical spin systems, where $j \to \infty$, we use $a = 3$ and $b = \frac{9}{5}$.

3.3. Monte Carlo simulations

In an exact formulation, thermodynamic averages are performed as a weighted sum over all phase space $\{s\}$

$$\langle A \rangle = \sum_{\{s\}} A[s] \, P[s, T] \tag{40}$$

where the probability for the occurence of a configuration in thermal equilibrium is given by the Boltzman factor

$$P[s, T] = \frac{e^{-\beta E[s]}}{\sum_{\{s\}} e^{-\beta E[s]}}. \tag{41}$$

Enumeration of all possible configurations is impractical for all but the smallest of systems since the number grows exponentially with its size.[70] Fortunately, one can make use of the fact that thermodynamic systems are nearly always in a state close to equilibrium and implement *importance sampling*. This involves generating a set of phase-space configurations from a Markov chain defined by the transition probability $W[s, s']$ between states $\{s\}$ and $\{s'\}$, where the new configuration depends only on the preceding one. Unfortunately, because such states are highly correlated, many configurations must be generated to ensure fluctuations are properly accounted. In the Metropolis algorithm, the probability $\{s\} \to \{s'\}$ is given by

$$\begin{aligned} W[s, s'] &= e^{-\beta(E[s']-E[s])}, &&for \ E[s'] > E[s] \\ &= 1 &&, \ for \ E[s'] < E[s], \end{aligned} \tag{42}$$

which obeys ergodicity as well as detailed balance. New configurations are typically generated randomly, spin by spin, so that all states are accessible (and ergodicity is achieved). The algorithm is very popular and exceedingly simple to implement. Thermodynamic averages are made simply by

$$\bar{A} = (1/N) \sum_{n=1}^{N} A[n] \tag{43}$$

where N is the number of times a new configuration n is generated (Monte Carlo Steps: MCS) and $A[n]$ is the value of the quantity at that step. The great advantage to this method is that the partition function $Z = \sum_{\{s\}} e^{-\beta E[s]}$ need not be calculated.

Errors associated with the averaged quantities are difficult to assess. Since each new configuration is correlated with the previous state, the usual standard-deviation estimate of *statistical* error

$$(\delta \bar{A})^2 = \frac{\chi}{N}, \quad \chi = \langle A^2 \rangle - \langle A \rangle^2 \tag{44}$$

is inappropriate. A more suitable quantity is the variance

$$\langle \delta^2 A \rangle = (1/N)\langle \sum_{n=1}^{N} (A[n] - \langle A \rangle)^2 \rangle \tag{45}$$

which can be expressed in terms of the autocorrelation function[82]

$$C(t) = (1/\chi)[\langle A(t)A(0) \rangle - \langle A \rangle^2] \tag{46}$$

where time is measured as a MCS. At long times, one can define a correlation time τ via $C(t) \sim e^{-t/\tau}$. (Here, τ is the number of MCS required to obtain two uncorrelated spin configurations.) If $N \gg \tau$, then a useful approximation for the error in $\bar{A}$ is given by

$$(\delta \bar{A})^2 = \frac{\chi}{N}(1 + 2\tau). \tag{47}$$

This expression is similar to the standard deviation but with an effective number of independent measurements given by

$$\frac{N}{1 + 2\tau}. \tag{48}$$

It is thus clear that critical slowing down can significantly reduce the number of independent measurements. MC simulations on large lattices at temperatures near T_N require N to be very large if high accuracy is required, as in the case of attempts to extract critical exponents. Unfortunately, it is very computer-time consuming to estimate τ at each lattice size for each thermodynamic quantity of interest. A simple qualitative estimate of errors can be made by repeating a MC run a number of times under the same conditions and examining the variation in $\bar{A}$.

Even more insidious is the possibility of *systematic* errors. These may arise due to the number of MCS being too small (so that the condition $N \gg \tau$ is not satisfied), or lattice sizes that are too small to reveal true critical behavior,[83] or even correlations within random numbers generated for the simulation.[84] Finite-size effects are important in the study of critical behavior since the correlation length ξ becomes comparable in length to a system dimension L. One can make use of the fact that ξ becomes very large at temperatures close to T_N and implement finite-size scaling analysis to study critical behavior.[15] The histogram technique can be particularly useful for this purpose.

Alternate MC algorithms have been proposed which can significantly reduce the effects of critical slowing down.[70] MC steps which involve generating new configurations from clusters of spins in the same state are used instead of updating individual spins as in the Metropolis method. This results in an ability to treat larger systems with a concomitant increase in the accuracy of critical-exponent estimates.[15]

Unfortunately, it appears that such methods are ineffective in the case of frustrated systems.[85]

The great advantage of the histogram method of analysis of MC data is that a run at a single temperature T_1 can be used to extract results for a continuous range of nearby temperatures.[86] Consider the partition function

$$Z(\beta_1) = \sum_{\{s\}} e^{-\beta_1 E[s]} = \sum_E W(E) e^{-\beta_1 E} \qquad (49)$$

where $W(E)$ is the density of states at energy E. Both the density of states and the histogram of energies, $H_1(E)$, are related to the probability that the system has energy E at β_1

$$P_E(\beta_1) = H_1(E)/N = W(E)[e^{-\beta_1 E}/Z(\beta_1)] \qquad (50)$$

for a run of N Monte Carlo steps. From this relation, one can extract an expression for the density of states

$$W(E) = a_1 H_1(E) e^{-\beta_1 E} \qquad (51)$$

where a_1 is a proportionality constant. Since $W(E)$ does not depend on temperature, the probability that the system has energy E at some temperature β can be related to the system at β_1 by

$$P_E(\beta) = \frac{H_1(E) e^{-(\beta-\beta_1)E}}{\sum_E H_1(E) e^{-(\beta-\beta_1)E}}. \qquad (52)$$

This is a powerful result. The continuous temperature dependence of any thermodynamic quantity can be estimated by

$$\langle A \rangle = \sum_E A(E) P_E(\beta). \qquad (53)$$

In practice the range of temperature over which $\langle A \rangle$ may be estimated from a single MC run at T_1 is limited by the range over which reliable statistics can be expected for $H(E)$. A rough guide is $T_a < T < T_b$, where T_a and T_b correspond to the energies $\langle E_a \rangle$ and $\langle E_b \rangle$ at which $H(E) \simeq \frac{1}{2} H_{max}$. Since $H(E)$ becomes more sharply peaked with system size, the valid temperature range becomes smaller as L increases. Multiple histograms made at a number of nearby temperatures may be combined to increase accuracy. The number of MCS is largerly a matter of trial and error, with some guidance provided by previous estimates for τ. Runs of 10^6 for smaller lattices to 10^7 at larger lattices are typical.[15]

A particularly useful application of this method is in the estimation of transition temperatures, which are characterized by extrema at $T_e(L)$, where $T_e(L \to \infty) \to T_N$, in various thermodynamic quantities as a function of T. For the purpose of locating T_e and estimating extrema A_e, it is desirable to have histogram data at a temperature close to T_c. Mapping-out phase diagrams, for example, as a function of magnetic-field and temperature, can be done more accurately for the same amount of computer time with the histogram method.

Recent finite-size scaling estimates of critical exponents in unfrustrated systems[15] appear to yield more accurate results than previous methods based on Fisher-type scaling relations using conventional MC data taken at a number of temperatures close to T_N (see, e.g., Ref.[65]). At a phase transition, thermodynamic quantities of interest include the specific heat C, susceptibility χ, and various logarithmic derivatives of the order parameter M[15] such as

$$V = \frac{\partial}{\partial\beta}ln\langle M\rangle = \frac{\langle ME\rangle}{\langle M\rangle} - \langle E\rangle. \tag{54}$$

At a continuous transition, the maxima in these quantities scale with linear dimension L as

$$C_{max} = C_1 + C_2 L^{\alpha/\nu}, \ \chi_{max} \sim L^{\gamma/\nu}, \ V_{max} \sim L^{1/\nu}. \tag{55}$$

The order parameter evaluated *at* the critical temperature scales as

$$M_c \sim L^{-\beta/\nu}. \tag{56}$$

Values of these exponent ratios for the universality classes discussed previously are given in the following table.

universality class	α/ν	β/ν	γ/ν	$1/\nu$
n=2	-0.018	0.519	1.960	1.49
n=3	-0.177	0.518	1.958	1.41
n=4	-0.297	0.527	1.986	1.35
n=2 chiral	0.63	0.47	2.09	1.85
n=3 chiral	0.41	0.51	1.98	1.69
mean-field	0	1	2	2
mean-field tricritical	1	$\frac{1}{2}$	2	2

In fact, all the scaling relations are also obeyed by the thermodynamic quantities evaluated *at* T_N. Although earlier studies involved finite-size scaling of maxima determined by the histogram method (including our own), it is becoming increasing clear that scaling at T_N is a more profitable approach. This is due to the fact that the extremum in each thermodynamic quantity occurs at a different temperature, which are not necessarily very close to each other. Thus it is often necessary to construct histograms at more than one temperature, which can become very computer-time expensive. If the critical temperature can be estimated accurately, however, then MC runs can be made at T_N only, to obtain $C_c, \chi_c, V_c, M_c, \dots$ at a number of system sizes L. A method to accurately estimate T_N, proposed by Binder,[87] involving the fourth-order cumulant

$$U_M(T) = 1 - \frac{\langle M^4 \rangle}{3 \langle M^2 \rangle^2}, \tag{57}$$

has proven to be very useful[15] when used in conjunction with the histogram technique.

Much progress has been made in recent years in the development of methods to determine by MC simulations if a transition is first order, even if it is only very weak (see, e.g., Ref.[65] for a review, and also Ref.[83]). In this case, the thermodynamic quantities discussed above (except M_c) should scale with system volume L^D. Even though $D = 3$ is considerably different from the critical-exponent ratios listed above, volume-dependent scaling may not be evident from MC data on smaller lattices where *pseudo-critical exponent* values may be observed.[66,83] Another useful indicator is the asymptotic behavior of the energy cumulant[88]

$$U(T) = 1 - \frac{\langle E^4 \rangle}{3 \langle E^2 \rangle^2}, \tag{58}$$

which exhibits a minimum at a phase transition, with the property $U_{min} \to U^* \to \frac{2}{3}$ in the infinite-lattice limit for the case of a continuous transition. For a first-order transition, U^* achieves a smaller value. (Note that $U_c = U(T_N)$ exhibits the same behavior). Of course, a precise value of $\frac{2}{3}$ is never achieved from MC simulations so that some care must be taken in evaluating results. Comparison with U^*-values of exactly solvable models which are known to exhibit weak first-order transitions (e.g., Potts models) is instructive. In addition, U^* is a useful quantity for estimating the latent heat from MC data.[50,88]

4. Simple hexagonal antiferromagnet: magnetic phase diagrams and multicritical points

The results of mean-field analysis and MC simulations (and to a limited extent, renormalization-group and scaling results) of magnetic phase diagrams are presented in this section for a variety of models on a stacked triangular (simple hexagonal) antiferromagnet. In particular, the influence of a magnetic field, and next-nearest-neighbor (NNN) interactions, on the ordered states which occur in the planar model, as well as in the Heisenberg model with weak planar or axial anisotropy, are examined. Special attention is given to the order of a transition line and to regions where three or more phases coexist, so-called multicritical points. Critical phenomena associated with symmetry-breaking perturbations are of interest since they can stabilize states with different symmetry and induce a *crossover* between univerality classes.[89] In the region near a multicritical point, the shape of phase boundary lines is governed by one or more crossover exponents ϕ. In cases where the multicritical point occurs at T_N, such studies can reveal useful information regarding the nature of the phase transitions discussed in the Sect. 2. Mean-field models are adequate to describe general features in most cases; however, several notable exceptions are demonstrated below.

It is instructive to illustrate these concepts in several cases of unfrustrated anti-ferromagnets. Such examples will also serve to demonstrate the effects of frustration by comparison with later results. The Landau-type free energy developed in the Sect. 3 is useful for this purpose. In the presence of a magnetic field $\mathbf{H}$, it is necessary to add to the spin density (4) a uniform component[5] $\mathbf{m}$. Using this new expression in (19) yields the result (compare (24))

$$F = A_Q S^2 + D S_z^2 + \tfrac{1}{2} A_0' m^2 + \tfrac{1}{2} D m_z^2$$
$$+ B_0 [\tfrac{1}{4} m^4 + S^4 + 2m^2 S^2 + 2(\mathbf{m} \cdot \mathbf{S})^2] - \mathbf{m} \cdot \mathbf{H} \qquad (59)$$

where the fourth-order anisotropy terms have beeen omitted, and

$$A_0' = a(T - T_0'). \qquad (60)$$

Ordered phases are characterized by the polarization vector $\mathbf{S}$. Since the term $(\mathbf{m} \cdot \mathbf{S})^2$ has a positive coefficient, it is clear that an applied magnetic field tends to favor a configuration $\mathbf{S} \perp \mathbf{H}$. The competition between this tendency and the anisotropy term can lead to multicritical points. Following Shapiro,[89] we present in Fig. 4 schematic phase diagrams in the $H^2 - T$ plane for the three cases of the planar (with $\mathbf{H} \| \hat{\mathbf{x}}$), Heisenberg, and weak-axial Heisenberg (with $\mathbf{H} \| \hat{\mathbf{z}}$) models.

By including the physically inaccessible region of the phase diagram $H^2 < 0$, the true structure of multicritical points which occur at $H = 0$ can be revealed (as well as *virtual* multicritical points which can occur at $H^2 < 0$). Note that for $H^2 < 0$, the term $(\mathbf{m} \cdot \mathbf{S})^2$ is negative so that a configuration $\mathbf{S} \| \mathbf{H}$ is favored. Since the system is unfrustrated, the number of spin components n fully characterizes the universality classes in these $3D$ systems. Each multicritical point shown here at (T_m, H_m) is associated with a spin-flop transition (*degenerate*, in the the cases (a) and (b) where $H_m = 0$) and, since they are characterized by two emerging critical lines, these are examples of *bicritical* points. Each has a value of n_m which is the sum of the values of n characterizing the critical lines. Renormalization group and scaling analyses demonstrate that the two critical lines emerge tangentially to the spin-flop line (giving an *umbilicus* structure) and are governed by the same crossover exponent

$$H^2 \sim (T_m - T)^\phi \qquad (61)$$

with $\phi \simeq 1.25$ for $n_m = 3$, and $\phi \simeq 1.18$ for $n_m = 2$. These values differ from the mean-field result $\phi = 1$. Support for these predictions in the case of weak axial anisotropy (Fig. 4 (c), where $H_m \sim \sqrt{DJ}$) has been found experimentally[89] and in MC simulations.[90]

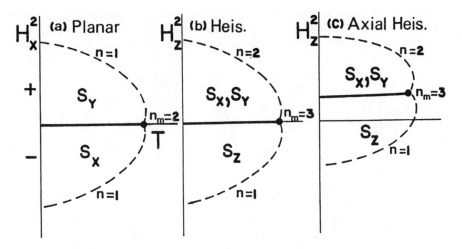

FIG. 4. Schematic phase diagrams of an unfrustrated lattice, revealing multicritical-point structures in three cases. Broken and solid curves denote continuous and first-order transitions, respectively. The region outside the curves is the paramagnetic state.

The remainder of this section is organized as follows. In Sect. 4.1, results are presented for $H - T$ phase diagrams of the planar model, as well as the Heisenberg model with weak planar anisotropy, for the case of AF coupling between frustrated triangular layers. Corresponding results for the planar model in the case of ferromagnetic interlayer interactions are given in Sect. 4.2. $H - T$ phase diagrams for the weak axial Heisenberg model are presented in Sect. 4.3. The effects of NNN in-plane coupling are reviewed in Sect. 4.4.

4.1. Planar model with antiferromagnetic interlayer coupling

The utility of the analysis of a phenomenological Landau-type free energy in reproducing experimental results for magnetic phase diagrams is nicely illustrated in the case of the quasi-1D hexagonal AF $CsMnBr_3$, which exhibits the 120° spin structure as shown in Fig. 1. This generalized mean-field theory can reproduce the gross quantitative features in addition to providing a qualitative understanding of the symmetry of each phase and the terms responsible for their stability. This material has a relatively large planar anisotropy, and the planar model, with the assumption $S\perp\hat{c}$, has proven adequate. In this case, the appropriate free energy is given by[91]

$$F = A_Q S^2 + \tfrac{1}{2}A'_0 m^2 + B_1 S^4 + \tfrac{1}{2}B_2 \,|\, \mathbf{S} \cdot \mathbf{S} \,|^2$$
$$+ \tfrac{1}{4}B_3 m^4 + 2B_4 \,|\, \mathbf{m} \cdot \mathbf{S} \,|^2 + B_5 m^2 S^2 - \mathbf{m} \cdot \mathbf{H}. \qquad (62)$$

The resulting phase diagram for the case of a magnetic field applied in the triangular plane, e.g. $\mathbf{H} \| \hat{\mathbf{x}}$, is shown in Fig. 5a together with data from a neutron diffraction study[92] on $CsMnBr_3$. Similar behavior observed in Ho[93] has not been analysed in detail.

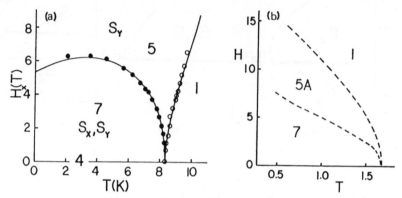

FIG. 5. (a) Magnetic phase diagram of $CsMnBr_3$ from Ref.[91] where full lines represent a fit of the phenomenological Landau free energy analysis to the experimental data (circles) of Ref.[92]. (b) Corresponding results from molecular-field theory with $J_{\|} = 1$ and $J_{\perp} = 1$.

Phases are labeled following a convention established in Ref.[75] and characterized by the nonzero components of the polarization vector $\mathbf{S} = \mathbf{S}_a + i\mathbf{S}_b$, as described in the table below, where $L \Rightarrow Linear$, $E \Rightarrow Elliptical$, and $H \Rightarrow Helical$. Phase 1 denotes the paramagnetic state.

1	2	3	4	5	6	7	8	9	9'
$S=0$			S_{ax}	S_{ay}	S_{ax}	S_{ax}	S_{ax}	S_{ax}	S_{ax}
	S_{az}	S_{az}					S_{az}	S_{az}	S_{ay}
		S_{by}	S_{by}			S_{by}	S_{by}		
	L	E	H	L	L	E	E	L	L

As the field is increased, the B_4-term reduces the component of $\mathbf{S}$ parallel to the field, S_x, and distorts the helix (stabilized by the B_2-term) into an elliptical configuration. At sufficiently high fields, the B_4-term dominates and a linear (non-chiral) S_y configuration is stabilized. The difference between this phase diagram and that of Fig. 4a is due to the extra degrees of freedom in this frustrated system: Both S_a and S_b are required to describe the magnetic order. Analytic expressions for the phase-boundary lines are given in Ref.[91]. In order to account for the increase in temperature of the $1 - 5$ boundary with increasing field, a value $B_5 < 0$ was required for the fit. This point is discussed further below.

It is seen from these results that T_N is a multicritical point, but it is only by examination of the region $H^2 < 0$ that two additional critical lines are also found to merge at T_N and the true *tetracritical* point structure is revealed.[89,94] Although phase 5 is a linear state, the $1 - 5$ transition does *not* belong to the Ising universality class.[95] This is due to the presence of both *cos* and *sin* contributions to the spin density, as discussed in Sect. 2.1, so that (S_a, S_b) forms a two-component variable and the transition is the standard XY $(SO(2))$ universality. At the $5 - 7$ transition, the single component S_y is involved and this transition is of simple Ising (Z_2) universality. Thus, the application of a magnetic field reveals the $V = Z_2 \times SO(2)$ symmetry present at T_N. Scaling analysis[96] of this system demonstrates that all four lines merging at the (chiral) tetracritical point are governed by the same crossover exponent, with a value estimated from a $4 - \epsilon$ expansion to be $\phi \simeq 1.04$.

A detailed analysis has also been made of the phase diagrams based on the corresponding exchange Hamiltonian for the planar model[97]

$$\mathcal{H} = J_\| \sum_{\langle ij \rangle} \mathbf{s}_i \cdot \mathbf{s}_j + J_\perp \sum_{\langle kl \rangle} \mathbf{s}_k \cdot \mathbf{s}_l - \mathbf{H} \cdot \sum_i \mathbf{s}_i \qquad (63)$$

where $\mathbf{s}_i \perp \hat{\mathbf{c}}$ and $J_\|, J_\perp > 0$, with $\langle i, j \rangle$ and $\langle k, l \rangle$ summed over NN sites along the c axis and in the basal plane, respectively. The assumption (4) of a spin density characterized by a single Fourier component is oversimplified. With only NN couplings, the magnetic order can have a periodicity of 2 (or less) along the c axis and a periodicity of 3 (or less) in the basal plane. Expression (4) can thus be generalized for this system as

$$\mathbf{s}(\mathbf{r}) = \mathbf{m} + \sum_{n=1}^{3} (\mathbf{S}_n e^{i\mathbf{Q}_n \cdot \mathbf{r}} + \mathbf{S}_n^* e^{-i\mathbf{Q}_n \cdot \mathbf{r}}) \qquad (64)$$

where

$$\mathbf{Q}_1 = \tfrac{1}{2}\mathbf{G}_\| + \tfrac{1}{3}\mathbf{G}_\perp, \quad \mathbf{Q}_2 = \tfrac{1}{2}\mathbf{G}_\|, \quad \mathbf{Q}_3 = \tfrac{1}{3}\mathbf{G}_\perp, \qquad (65)$$

and $\mathbf{Q}_1$ is the principal wave vector. In this study, we set $J_\| \equiv 1$ and examined the effects of varying $J_\perp$.

Ground-state (T=0) $H - J_\perp$ phase diagrams revealed three types of linear phases 5, in addition to the elliptical phase 7. These can be distinguished by the nonzero components of

$$\mathbf{S}_n = \mathbf{S}_{na} + i\mathbf{S}_{nb}, \qquad (66)$$

for a field $\mathbf{H}\|\hat{\mathbf{x}}$, as

phase $5A$: $S_{1a}^y,\ S_2^y,\ S_{3a}^x$

phase $5B$: $S_{1a}^y,\ S_{1b}^y,\ S_2^y,\ S_{3a}^x,\ S_{3b}^x$

phase $5C$: $S_{1b}^y,\ S_{3a}^x$

phase 7 : $S_{1a}^y,\ S_{1b}^x,\ S_2^y,\ S_{3a}^x,\ S_{3b}^y$

The three phases 5 are primarily characterized by S_{1a}^y and S_{1b}^y, which may be described in terms of a phase angle ϕ by writing

$$S_{1a}^y = \mid S_1^y \mid \cos\phi, \quad S_{1b}^y = \mid S_1^y \mid \sin\phi. \tag{67}$$

In state $5A$ the phase angle takes values $\phi = m\pi/3$, and state $5C$ is characterized by $\phi = (2m+1)\pi/6$, with intermediate values (e.g., $0 < \phi < \pi/6$) for phase $5B$. Transitions between phases $7 - 5B$ or $7 - 5C$ involve a spin flop $\mathbf{S}_{1b}\|\hat{\mathbf{x}}$ to $\mathbf{S}_{1b}\|\hat{\mathbf{y}}$ so that these are necessarily first order. Phase 5 which appears in Fig. 5a has the symmetry of $5A$.

A Landau-type free energy based on a molecular-field treatment of the Hamiltonian (63), expanded to order s^6, was analysed to obtain phase diagrams for a number of values of $J_\perp$. The incipient relationship between the primary and secondary Fourier components is revealed by the results

$$\mathbf{S}_2 + \mathbf{S}_2^* \sim \tfrac{1}{2}\mathbf{S}_1(\mathbf{S}_1 \cdot \mathbf{S}_1) + \mathbf{S}_1(\mathbf{m} \cdot \mathbf{S}_3^*) + c.c. \tag{68}$$

$$\mathbf{S}_3 \sim \tfrac{1}{2}\mathbf{m}(\mathbf{S}_1 \cdot \mathbf{S}_1) + \mathbf{S}_1(\mathbf{m} \cdot \mathbf{S}_1). \tag{69}$$

Note that $S_2 \sim S_1^3 \cos\phi$ is zero in phase 5C and that S_3 is nonzero only in the presence of a magnetic field. Since $S_1^2 \sim (T_N - T)$, it is clear that these additional contributions to the spin density are small in regions not far from T_N (of interest here). The phase diagram for the case $J_\perp = 1$ shown in Fig. 5b is similar in structure to results from the phenomenological model. A notable difference is the decrease in temperature of the $1 - 5$ boundary line with increasing field. A rather different phase diagram is obtained in the case of $J_\perp = 1.5$, as shown in Fig. 6. All three phases 5 appear ($5B$ occupies a very narrow region between $5A$ and $5C$). Differences between the three phases 5 arise from terms in the free energy that are effectively sixth order in S_1 and thus are not relevant to the universality class at the paramagnetic phase boundary. The structure of the tetracritical point at T_N remains intact. This is not the case, however, for quasi-2D coupling $J_\perp = 10$, as shown in Fig. 7a, where the $5C - 7$ transition is first order. This picture is changed if effects due to critical fluctuations are included, as discussed below. The quasi-1D case $CsMnBr_3$ using parameters $J_\perp/J_\| = 0.002$ was also considered. The resulting phase diagram is similar in structure to Fig. 5b, with the 1 − 5 boundary line also curving to the left.

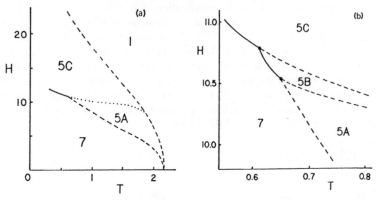

FIG. 6. Molecular-field theory of Ref.[97] with $J_\perp = 1.5$. Solid and broken curves denote first-order and continuous transition lines, respectively. The dotted curve in (a) represents two continuous transition lines as shown in (b).

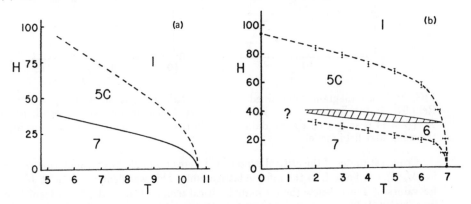

FIG. 7. (a) As in Fig. 6 with $J_\perp = 10$. (b) Corresponding results from MC simulations reported in Ref.[100]. Broken curves serve as guides to the eye. The nature of the region between phases 6 and 5C, as well as the low-T struture, is undetermined.

Monte Carlo simulations were also performed on the Hamiltonian (63) in an effort to examine the effects of critical fluctuations.[97] The standard Metropolis algorithm was used with periodic boundary conditions on $L \times L \times L$ lattices with L=12, 18, and 24. Fourier components of the spin density defined by[90]

$$M_a(\mathbf{q}) = [\frac{1}{L^3}]\langle[\sum_i s_{ia}e^{-i\mathbf{q}\cdot\mathbf{R}_i}]^2\rangle^{\frac{1}{2}} \tag{70}$$

were calculated, where $a = x, y$ and $\mathbf{q} = 0, \mathbf{Q}_n$ (n=1-3), given by (65). Averaging was done over 2-8 runs with random initial configurations and 10^4 MC steps, after discarding the initial 4×10^3 for thermalization. Phase boundaries were determined by the location of the point of inflection as a function of temperature or field in the primary component $M_a(\mathbf{Q}_1)$ for $L = 12$. The precise symmetry of the ordered phase was established by examination of the other Fourier components. Results shown in Fig. 8 for $J_\perp = 1$ are in qualitative agreement with the mean-field phase diagram of Fig. 5b. The results displayed in Fig. 8b near $T_N = 1.45(2)$ demonstrate that the predicted crossover exponent $\phi \simeq 1$ is consistent with the MC results, but it would be nearly impossible to obtain sufficiently accurate results to distinguish between the renormalization-group and mean-field values.

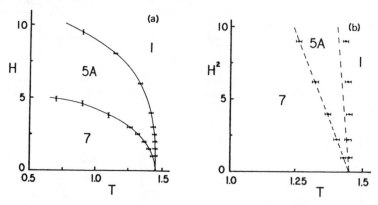

FIG. 8. Phase diagram determined by Monte Carlo simulations with $J_\perp = 1$ from Ref.[97]. In (a), lines serve as a guide to the eye. Linear temperature dependence of H_c^2 ($\phi = 1$) is suggested by the data shown in (b), where the broken lines are from the molecular-field model.

Crude estimates of the critical exponent β were also made by extrapolating results for $M_a(\mathbf{Q}_1)$ at finite L to the infinite lattice limit. This was done for data at $H = 3$, for values of T just below the estimated critical temperature of the $1 - 5$ transition, $T_c = 1.430(5)$. The result $\beta = 0.34(2)$ is consistent with the expected XY universality. For the $5 - 7$ transition at T=0.9, field scaling of the form $M_x \sim (h^2)^\beta$ (where $h^2 = (H_c^2 - H^2)/H_c^2$) was assumed. The estimate $\beta = 0.32(2)$ is consistent with the anticipated Ising universality. A variety of experimental estimates for the critical exponents associated with the phase transitions in $CsMnBr_3$ have been reported (see, e.g., Refs.[5, 9, 92]).

None of the phase diagrams discussed above, except for the phenomenological model, has shown a tendency for the $1 - 5$ boundary line to curve to the right at "low" field values as seen in the experimental results on $CsMnBr_3$. This feature was reproduced, however, by MC simulations[98] performed with quasi-1D exchange parameters appropriate for this material. It appears that the combined effects of critical fluctuations and the quasi-1D nature of the interactions are responsible for this behavior. An initial increase in the Néel temperature with field strength is also found in models of quasi-1D antiferromagnets where single-chain thermodynamics is treated exactly and interchain coupling is approximated by mean-field theory.[99]

MC simulation results for the phase diagram in the quasi-2D case[100] $J_\perp = 10$ shown in Fig. 7b can be compared with the corresponding mean-field theory of Fig. 7a. There is dramatic failure of the molecular-field approximation for this system. Phase 6 in Fig. 7b is a linear state and it is expected that the $1-6$ and $6-7$ phase transitions belong to XY and Ising universality classes, respectively. These results demonstrate that the $Z_2 \times SO(2)$ tetracritical behavior near T_N is restored by critical fluctuations.

Finally we consider briefly some results based on a phenomenological Landau-type free energy for the case of the Heisenberg model with weak planar anisotropy.[75] Schematic results are shown in Fig. 9. The structure in the absence of any fourth-order anisotropy effects (see, e.g., Eq.(19)) shown in Fig. 9a indicates a spin-flop transition from phase 7 characterized by S_x, S_y to phase 3 characterized by S_y, S_z for $\mathbf{H}\|\hat{\mathbf{x}}$. The B_4-term in the free energy (62) is responsible for stabilizing the high-field configuration $\mathbf{S}\perp\mathbf{H}$. (In the cases of strong planar anisotropy described above, long-range order is destroyed before such a spin-flop can occur.) Recent experimental results[101] suggest that such a phase diagram is realized by $RbMnBr_3$. Some possible effects of fourth-order anisotropy are indicated in Fig. 9b, showing that the spin-flop transition can be replaced by two lines of continuous transitions with an intermediate phase 8.

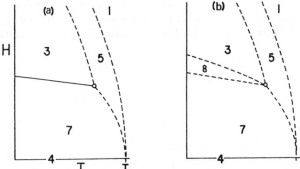

FIG. 9. Example phase diagrams for the case of weak planar anisotropy from the phenomenological model of Ref.[75]. Broken and solid curves represent continuous and first-order transition lines, respectively. Case (b) can occur if effects of fourth-order anisotropy are important.

4.2. Planar model with ferromagnetic interlayer coupling

Magnetic field phase diagrams with rather different structure can occur in the case of ferromagnetic interlayer coupling, $J_\| < 0$ in the Hamiltonian (63). Results from

Ref.[50] are included in this section of analysis by phenomenological, and molecular-field based, Landau-type free energies, as well as from numerical histogram MC simulations. Differences from the case of $J_{\parallel} > 0$ arise due to the fact that the primary ordering wavevector is now $\mathbf{Q}_3$, given by (65), so that $3\mathbf{Q} = \mathbf{G}$. (There are no secondary Fourier components in this case, only $\mathbf{m}$ induced by the magnetic field.) From this relation, it may be understood that in the evaluation of integrals as in (19), fourth-order terms which are *cubic* in $\mathbf{S}_3 \equiv \mathbf{S}$ and linear in $\mathbf{m}$ are allowed. Specifically, there is an additional contribution

$$F_3 = B_6[(\mathbf{m} \cdot \mathbf{S})\mathbf{S} \cdot \mathbf{S} + c.c.] \tag{71}$$

to the free energy (62). This term tends to favor a linearly polarized state with $\mathbf{S} \| \mathbf{m}$, in contrast with the B_4-term. For such a phase, one can write $\mathbf{S} = S_r e^{i\phi}$ so that $F_3 \sim mS_r^3 \cos(3\phi)$. If a transition from the paramagnetic phase to an ordered state of this type occurs within mean-field theory, it should be first order due to this term cubic in the order parameter. The phase angle can take three inequivalent values $\phi = n\pi/3$ so that the ordered state should have the symmetry of the 3-state Potts model. It has recently been demonstrated by MC simulations that the *discrete* 3-state Potts model in 3D exhibits a very weak first-order transition (see, e.g., Ref.[50]). (Since the transition is only weakly first order, it is possible to obtain pseudo critical exponents.[83,102] Special care must be taken in analysing MC data in such cases.) This is in contrast with the 2D case where the transition is continuous, with a unique set of critical exponents, and where mean-field theory (being independent of D) predicts a first-order transition. Higher-order terms in the free energy can stabilize a phase angle with values $\phi = (2n+1)\pi/6$, in which case the contribution F_3 is zero.

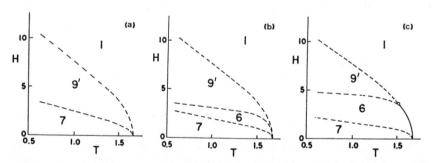

FIG. 10. Example phase diagrams for the case of ferromagnetic interplane coupling from an analysis the Landau-type free energy where phase 6 is characterized by S_{ax} and phase 9' by S_{ax}, S_{ay}: (a) represents results from molecular-field theory with fourth-order coefficients $B_i = bT$; (b) is obtained by changing only one coefficient $B_4 = 1$; (c) results from using $B_4 = 0.1$, where the 1 − 6 transition is first order.

The results of the molecular-field model ($B_i = bT$) are presented in Fig. 10a with $\mathbf{H} \| \hat{\mathbf{x}}$. The phase diagram is similar to those found for the case of $J_{\parallel} > 0$, but with a linear phase 9′ having both x and y components (also see Ref.[103]). This structure changes significantly if one or more of the fourth-order coefficients B_i is set to be different from the others, as in the phenomenological model. A new linear phase 6, with $\mathbf{S}\|\mathbf{H}$, appears in Figs. 10b and 10c, which were constructed with values $B_4 \equiv 1$ and $B_4 \equiv 0.1$, respectively. (These values are not so different from bT since $b = \frac{9}{5}$). This state appears to be accidentally excluded from the molecular-field treatment ($B_4 = bT$). The multicritical-point structure at T_N is seen to change dramatically. Phase 6 has the symmetry of the 3-state Potts model and the first-order nature of the $1 - 6$ transition within this mean-field theory is confirmed. Fig. 10c is seen to have features in common with the quasi-2D case $J_\perp = 10$ of Fig. 7b.

Standard Metropolis MC simulations were used to determine the effects of critical fluctuations on the structure of the phase diagram by the method described in Sect. 4.1 on lattices with L=12-24. The results shown in Fig. 11a are similar to those of Fig. 10c and notably different from those of the molecular-field theory, Fig. 10a. It is evident that the phenomenological model mimics some effects due to critical fluctuations. Similar MC results were obtained some ten years ago by Lee *et al.*[31] for the 2D triangular AF. This is not surprising since the (ferromagnetic) coupling between layers in the present case is relatively unimportant for the magnetic structure in the presence of a magnetic field; both 2D and 3D systems are characterized by $3\mathbf{Q} = \mathbf{G}$.

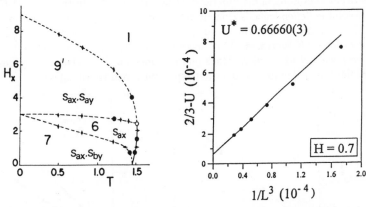

FIG. 11. From Ref. [50]. (a) Phase diagram with $J_\| = -1$ determined by standard MC simulations. Solid circles at H=0.7, 1.5, 2.7, and 4.0 indicate regions where histogram analysis was made. The $1-6$ boundary is a line of first-order transitions and all other boundaries are continuous transitions. (b) Scaling of the energy-cumulant minima with volume at the $1-6$ transition. Data at L= 12 and 15 have been omitted to allow for an expanded scale. The straight line is fitted to the four largest lattice sizes.

The criticality of each transition line was investigated by means of finite-size scaling through the histogram technique, as indicated by the solid circles on Fig. 11a. In all but one case, scaling of *extrema* in the thermodynamic functions $U, C, \chi, V, and M$, as described in Sect. 3.3, were examined. For the 6 − 9' boundary line at H=2.7, it was not possible to obtain reliable estimates of extrema as a function of temperature because the slope of this line is very small; thermal fluctuations between the two phases were very large. A crude estimate of the critical temperature $T_c \simeq 1.230$ was determined, however, based on these results so that scaling could be performed at the critical point. Lattice sizes $L = 12 - 33$ were used and thermodynamic averages were typically made from 1×10^6 MC steps for the smaller lattices and up to 3×10^6 for larger L. In most cases, $2-3 \times 10^5$ initial MC steps were discarded for thermalization. For the 6 − 9' boundary line at H=2.7, lattices $L = 12 - 27$ were examined using up to 7×10^6 MC steps. Special care was taken to determine if a transition was first order by examination of the energy cumulant U and checking for volume dependence of all the thermodynamic functions calculated. This was of particular importance regarding the 1 − 6 boundary line since it was not known if the present *continuous* version of the 3-state Potts model would also exhibit a first-order transition, possibly very weak at small values of the magnetic field.

Scaling of the energy-cumulant minima with volume at H=0.7 for the 1 − 6 transition is shown in Fig. 11b. The extrapolated value for $L \to \infty$ is $U^* = 0.66660(3)$, very close to the value expected for a continuous transition, 2/3. That this transition is, however, very weakly first order is confirmed by the volume-dependent scaling of the other thermodynamic functions, shown in Fig. 12. In the cases of U, C, and V, this volume dependence is evident only at larger lattice sizes. Corresponding results at H=1.5 indicate that the 1 − 6 transition becomes more strongly (but still weak) first-order as H increases. This suggests that the transition at $H = 0$ is continuous or *very* weakly first order. Since phase 6 does not have the same symmetry as the zero-field helical state, T_N does not represent a real tricritical point.

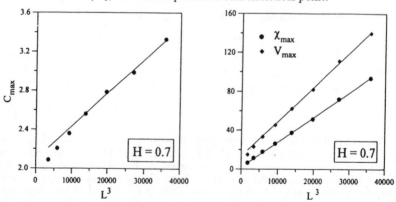

FIG. 12. From Ref. [50]. Scaling of the thermodynamic-function maxima with volume at the 1 − 6 transition.

It is not clear if the field-induced cubic contributions like F_3 would be relevant at T_N regarding the critical behavior. Following an idea suggested by Heinonen and

Petschek[104] in a related context, one can imagine, however, a scenario in which such systems at H=0 may generate coupling between the primary Fourier component of the spin density and the less stable $q = 0$ contribution. For example, if $J_\perp$ is small, then there is little energy difference between $q = Q_3$ and $q = 0$ states so that short range m-order may exist at temperatures above T_N at $H = 0$. In the case of $J_\parallel > 0$, a phase diagram of the strucure in Fig. 11a occurs in the presence of a *staggered* (along the c axis) field. In that case, cubic terms are generated by coupling bewteen $q = Q_1$ and Q_2 states. Such concepts might be important in the case of $CsMnBr_3$ where short-range AF order along the c axis is well established at temperatures near T_N. The small values observed for the exponent β may be indicating that the transition is in fact very weakly first order.

Histogram finite-size scaling of the other points indicated in Fig. 11a are consistent with the expected critical behavior based on symmetry: The $6-7$ and $6-9'$ transitions should exhibit Ising universality and the $1 - 9'$ transition should belong to the XY universality class. The possibility that the latter transition was weakly first order was also examined since phase $9'$ has a nonzero contribution from F_3. Within mean-field theory, however, $\phi \to \pi/6$ as $S \to 0$ so that F_3 vanishes rapidly at the transition and is not relevant to the criticality.

4.3. Heisenberg model with axial anisotropy

As discussed in Sect. 3.1, the Heisenberg model with weak axial anisotropy exhibits two successive phase transitions at zero field, $T_{N1} > T_{N2}$, where a linear phase 2 characterized by S_z first appears as T is lowered, followed by an elliptical state 3 with nonzero S_x, S_z. (Low-temperature thermodynamics and order by disorder in the related exchange-anisotropy model have recently been examined.[105]) An appropriate phenomenological free energy for the study of magnetic-field effects (in the case of AF interplane coupling) is given by[78]

$$F = A_Q S^2 + A_z \mid S_z \mid^2 + \tfrac{1}{2} A_{z0} m_z^2 + \tfrac{1}{2} A_0' m^2 + B_1 S^4$$
$$+ \tfrac{1}{2} B_2 \mid \mathbf{S} \cdot \mathbf{S} \mid^2 + \tfrac{1}{4} B_3 m^4 + 2 B_4 \mid \mathbf{m} \cdot \mathbf{S} \mid^2 + B_5 m^2 S^2 - \mathbf{m} \cdot \mathbf{H}, \qquad (72)$$

which is simply Eq. (62) with anisotropy added. The study of the phase diagram for this system with $\mathbf{H} \| \hat{\mathbf{c}}$ was motivated by the experimental results of Johnson *et al.*[106] on the quasi-1D system $CsNiCl_3$. A fit of the results from the phenomenological model provides good agreement with these data based on susceptibility measurements, as shown in Fig. 13a. A value $B_5 < 0$ was necessary to reproduce the increase in temperature with increasing field of the $1 - 4$ boundary line (as in the case of the $1-5$ line for $CsMnBr_3$). This phase diagram may be compared with the unfrustrated case of Fig. 4c. The multicritical point is characterized by a first-order spin-flop line together with three lines of continuous transitions. It is the first example of such a multicritical point in magnetic systems. The high-field phase 4 is characterized by

$S_x = S_y$ and is equivalent to the helical 120° spin stucture occurring at zero field in the planar model (and $CsMnBr_3$). Such a state is stabilized by the B_2 and B_4 terms.

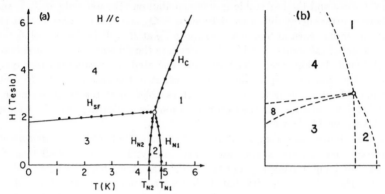

FIG. 13. (a) Magnetic phase diagram of $CsNiCl_3$ from Ref.[78] where lines represent a fit of the phenomenological Landau model to the experimental data (circles) of Ref.[106]. (b) Corresponding results from Ref.[75] showing some possible effects of fourth-order anisotropy.

Relevant experimental results on a number of ABX_3 compounds are reviewed in Refs. [5] and [107]. A notable feature of these data is the lack of clear evidence that the $3 - 4$ transition is first order. This aspect has been examined recently[108] and is likely due to the quasi-1D nature of the exchange interactions. The possibility of an intermediate phase 8, with nonzero spin components S_x, S_y, S_z as shown in Fig. 13b was also considered. In this case, the spin-flop line is replaced by two lines of continuous transitions and the multicritical point represents the convergence of *five* phases. A number of other possible phase diagrams resulting from the phenomenological model are examined in Ref.[75].

Detailed analyses of the ground state phase diagram[109] and molecular field model[5] of the axial Heisenberg Hamiltonian (anisotropy added to Eq. (63))

$$\mathcal{H} = J_{\parallel} \sum_{\langle ij \rangle} \mathbf{s}_i \cdot \mathbf{s}_j + J_{\perp} \sum_{\langle kl \rangle} \mathbf{s}_k \cdot \mathbf{s}_l + D \sum_i (s_{iz})^2 - \mathbf{H} \cdot \sum_i \mathbf{s}_i, \tag{73}$$

with varying $D < 0$ and $\mathbf{H} \| \hat{c} \| \hat{z}$ have been made. The derived Landau-type free energy was expanded to sixth order in s and all Fourier components as in Eq. (64) were included. As in the case of the planar model, this allows for distinction of linear and elliptical states by the appropriate phase angle, which are labeled A and B:

phase $2A$: S_{1b}^z, S_{3a}^z phase $2B$: S_{1a}^z, S_2^z, S_{3a}^z

phase $3A$: S_{1b}^z, S_{1a}^x, S_2^x, S_{3a}^z, S_{3b}^x phase $3B$: S_{1a}^z, S_{1b}^x, S_2^z, S_{3a}^z, S_{3b}^x

Example results using $J_{\parallel} = J_{\perp} = 1$ showing the effects of increasing anisotropy are depicted in Fig. 14. Since A and B states have the same symmetry, transitions between these phases are necessarily first order. The phase diagram of Fig. 14a with $D = -0.2$ demonstrates that the main features of Fig. 13a are captured by the molecular-field approximation. Smaller values of $| D |$ yield qualitatively the same results. The structure of the phase diagram changes considerably for larger values of $| D |$. Fig. 14b shows that for $D = -0.6$, the $2 - 3$ phase boundary moves away from the multicritical point and terminates at a *critical end point*.[110] Analysis of the free energy suggests that coupling of the primary order parameter to S_3 induces an effective fourth-order anisotropy of the form $\sim m^2 \mid S_z \mid^4$ which can break the symmetry at the multicritical point where second-order terms vanish. Results for $D = -1.0$ shown in Fig. 14c exhibit a number of new features. The $1 - 4$ phase boundary terminates at a critical-end point, there now appears a tricritical point on the $1 - 2$ line, and both $2 - 3$ boundaries terminate at critical end-points.

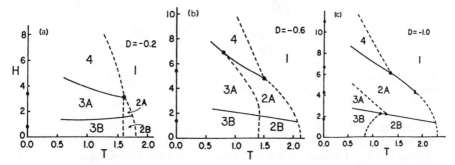

FIG. 14. Results of a molecular-field treatment of the axial Heisenberg model with $J_\parallel = J_\perp = 1$ and varying the anisotropy D. Solid circles on the $T = 0$ axis represent locations of phase boundaries from an exact ground-state calculation.[109] Solid and broken curves represent continuous and first-order boundary lines, respectively.

A signature of the difference between phases $2A$ and $2B$ is the appearance of a non-zero secondary component S_2^z in the latter case.[76] Neutron diffraction results[111] on $CsMnI_3$ indicate that the intensity associated with the (111) reflection (related to Q_2 by a reciprocal lattice vector) develops at a temperature between T_{N1} and T_{N2} (see Fig. 4 of that work). This suggests that a sequence of transitions $3B - 2B - 2A - 1$ occurs with increasing temperature at zero field. Although this possibility is not realized within the microscopically derived molecular-field theory, as seen from Fig. 14, it can result from a phenomenological version of this approach. Second-order isotropic contributions to the free energy appear as

$$F_2 = \tfrac{1}{2}a(T - T_0')m^2 + \sum_n a(T - T_{Q_n})S_n^2 \qquad (74)$$

where $\mathbf{Q}_n$ is defined as in (64)-(65), and (see (39)) $T_q = -J_q/a$:

$$T_0' = -2(J_\parallel + 3J_\perp)/a, \quad T_{Q_1} = 2(J_\parallel + 3J_\perp)/a,$$
$$T_{Q_2} = 2(J_\parallel - 3J_\perp)/a, \quad T_{Q_3} = 2(-J_\parallel + 3J_\perp)/a. \qquad (75)$$

Instead of determining the T_q by the values of $J_{\parallel}$ and $J_{\perp}$, we can treat these as phenomenological parameters. The results of Fig. 15 were obtained in this manner in an effort to consider possible phase diagrams which may be relevant to $CsMnI_3$. There has been no experimental effort to discriminate between A and B phases at finite field.

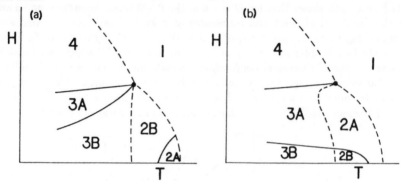

FIG. 15. Results of a phenomenological molecular-field treatment of the axial Heisenberg model in an effort to explore possible phase diagrams relevant to $CsMnI_3$, using $D = 0.1$, $T_0' = -7.2$, $T_{Q_1} = 0.8$, $T_{Q_2} = -3$ with (a) $T_{Q_3} = -2$ and (b) $T_{Q_3} = 0$.

Referring to the $CsNiCl_3$-type phase diagram of Fig. 13a, symmetry analysis[96] suggests that both $1 - 2$ and $2 - 3$ transitions belong to the XY universality class and that the $1 - 4$ transition is $n = 2$ chiral, within the framework of Kawamura's proposals. Only some of the experimental estimations of critical exponents at zero field are in support of this results.[96,112] Recent specific-heat measurements along the $1 - 4$ boundary at a number of increasing field strengths yield values for the exponent α apparently consistent with crossover from $n = 3$ to $n = 2$ chiral universality.[113] Scaling arguments indicate that all three critical lines near the multicritical point are governed by the same crossover exponent $\phi \simeq 1.06$, estimated from $4 - \epsilon$ expansions. Experimental results are at least consistent with this expectation.[114]

Monte Carlo simulations were performed on hexagonal systems governed by the Hamiltonian (73) with $J_{\parallel} = J_{\perp} = 1$, and for two values of anisotropy, $D = -1.0$ and -0.2 and using $12 \times 12 \times 12$ lattices. Results for $D = -1.0$ shown in Fig. 16a were determined using the standard Metropolis method to evaluate the relevant order parameter for each transition.[109] The structure of the phase diagram is in qualitative agreement with the corresponding molecular-field results of Fig. 14c, but no effort was made to determine boundary lines between phases A and B. The more detailed results for the case $D = -0.2$ shown in Fig. 16b were obtained using the histogram method, where phase boundaries were determined by the location of maxima in the staggered susceptibility.[107] Typically, 1×10^6 MC steps were used to calculate thermodynamic averages. The region near the multicritical point was of special interest and for this reason the study was numerically challenging due to large fluctuations between the many nearly degenerate phases. There is clear indication, however, that both the

$1-2$ and $1-4$ boundary lines merge at the multicritical point tangential to the $3-4$ spin-flop line. It is less clear that the $2-3$ boundary exhibits this expected behavior. In marked contrast with the usual bicritical point of unfrustrated systems (Fig. 4c), the upper critical line $(1-4)$ in the present case shows an initial curvature to the left. There is no requirement from scaling or renormalization-group arguments that the initial slopes of the critical lines have the same sign. Much more detailed results would be necessary to obtain an estimate of the crossover exponent ϕ. Comparison of these results with the corresponding molecular-field phase diagram, Fig. 14a, indicates qualitative agreement with the overall structure but that critical-fluctuation effects are pronounced near the multicritical point.

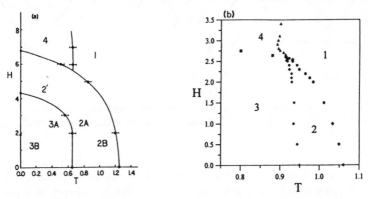

FIG. 16. Monte-Carlo results for the axial Heisenberg model using $J_\| = J_\perp = 1$ from Refs. [107] and [109] with (a) $D = -1.0$ (where lines serve as guides to the eye) and (b) $D = -0.2$.

4.4. Next-nearest-neighbor interactions

Some effects of including NNN in-plane interactions on the (stacked) triangular lattice are considered here. Recent work[115] has focused on the ground-state and low-temperature magnetic structures. For convenience, we label the first- and second-neighbor in-plane interactions by $J_1 > 0$ and J_2, respectively, and define $a = J_2/J_1$. For $a < \frac{1}{8}$, the helical 120° spin structure remains stable at $T = 0$. For $\frac{1}{8} < a < 1$, a peculiar and highly degenerate state occurs. It is defined by the four spins at the corners of a parallelogram constructed from two adjacent triangles and the only requirement is that $\mathbf{S}_1 + \mathbf{S}_2 + \mathbf{S}_3 + \mathbf{S}_4 = 0$. For $a > 1$, an incommensurate (IC), three-fold degenerate, phase is stabilized. The high degeneracy of the intermediate state is removed by thermal fluctuations in an order-by-disorder process. The resulting state has a simple period-2 structure along two basal-plane axes and is ferromagnetic along the third. There are three ways to construct such a configuration, which correspond to the three wavevectors[16]

$$\mathbf{Q}_a = \frac{\pi}{b}\hat{\mathbf{y}}, \quad \mathbf{Q}_b = \frac{\pi}{a}\hat{\mathbf{x}} + \frac{\pi}{2b}\hat{\mathbf{y}}, \quad \mathbf{Q}_c = \frac{\pi}{a}\hat{\mathbf{x}} - \frac{\pi}{2b}\hat{\mathbf{y}}, \tag{76}$$

where $b = \sqrt{3}/2$ and $\hat{\mathbf{x}}$ and $\hat{\mathbf{y}}$ are defined as in Fig. 1. These relations also satisfy $\mathbf{Q} = \frac{1}{2}\mathbf{G}$. The three degenerate linearly polarized spin configurations are shown in Fig. 17, where the spins are free to globally rotate.

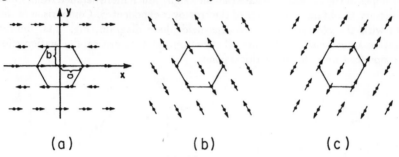

(a)   (b)   (c)

FIG. 17. Three degenerate linearly (L) polarized states corresponding to the wave vectors of Eq. (76) (see Ref. [16]) stabilized at intermediate values of J_2. Spins are free to globally rotate.

This state, as well as the 120° and IC phases mentioned above, is realized at the correct values of a by simple minimization of the exchange function J_Q. A similar three-fold degenerate linearly polarized phase can also be induced by dipole-dipole interactions if the coupling between triangular layers is ferromagnetic and the system is quasi-1D, as realized by $CsNiF_3$.[16] A possibly significant difference, however, is that dipole terms couple to the lattice and configurations $\mathbf{S}\|\hat{\mathbf{a}}$ are selected; there is no global rotational degeneracy.

Monte Carlo simulations on the XY and Heisenberg models with NNN interactions using lattice sizes $L = 12 - 36$ have been used to determine the $J_2 - T$ phase diagram shown in Fig. 18.[36,116] The general behavior expected from the ground-state and low-T analysis is confirmed. Several points are of interest. The boundary between the period-3 H and period-2 L phases is first order, as expected between two (commensurate) states not related by symmetry; however, the transition between L and paramagnetic phases is also first order, although only weakly so at values of J_2 near $\frac{1}{8}$ and near 1. The latter result is surprising and is not found within mean field theory. A possible explanation lies in the similarity to the 3-state Potts model, but in the present case there is no contribution to the free energy which is cubic in the order parameter. The 3-fold degeneracy is hidden in the coefficient of the quadratic term, J_Q. The $IC - P$ transition appears to be continuous but no attempt has been made to estimate critical exponents. In addition to a 3-fold degeneracy involving relative spin angles, the discrete 2-fold chiral degeneracy is also present and it is not clear to which universality class this transition may belong. Results similar to those shown in Fig. 18 were also found for the XY model.

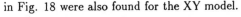

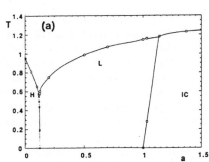
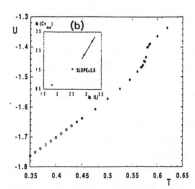

FIG. 18. Results from MC simulations of the Heisenberg model with NNN interactions $a = J_2/J_1$ after Ref.[116]. (a) Phase diagram where P, H, L, and IC denote paramagnetic, period-3 helical, period-2 linear and incommensurate phases, respectively. $H - L$ and $P - L$ transitions are first order. (b) Evidence that the $P - L$ transition at $a = 0.120$ is first order from the discontinuity in the internal energy U and near-volume dependent scaling of the specific heat maxima C_{max} (inset).

5. Other problems of interest

The preceding sections contain an overview of results for simple models on simple lattices. Although they are relevant for many real materials, it is more often the case that additional interactions, which can arise from a distortion of the lattice, must be accounted for in attempts to explain experimental data. In addition to reviewing some randomly selected results from our work on more complicated models, a number of other manifestations of frustration are examined in this section.

The positioning of the magnetic ions in $RbFeBr_3$ and related materials[59,117] is obtained by starting from the simple hexagonal lattice and displacing two-thirds of the ions (A sites) along the c axis by an amount δc, where δ is an irrational number. The undisplaced B sites form a triangular lattice whereas the displaced A ions occupy the sites of a (bipartite) honeycomb lattice. The new hexagonal unit cell has lattice constants $a = \sqrt{3}a_0$, $c = c_0$, where a_0 and c_0 refer to the original undisplaced simple hexagonal lattice. The dominant exchange interactions are assumed to be between NN in the A plane, $J = J_{AA}$, and between NN A and B sites $J' = J_{AB}$. A mean-field analysis of the planar model assuming AF interactions on this lattice, as believed to be relevant for $RbFeBr_3$, yields a $T - J'$ phase diagram which is identical in structure to the $H^2 - T$ phase diagram of $CsMnBr_3$ discussed in Sec. 4.1. In particular, there is a tetracritical point where an elliptical (chiral) phase, two linearly polarized states, and the paramagnetic phase meet. In contrast with $CsMnBr_3$, where only two of the four lines are physically accessible, the possiblity exists to study the complete tetracritical structure in $RbFeBr_3$ by changing the ratio J'/J through the application of uniaxial stress or ionic substitution.

The XY antiferromagnet $RbMnBr_3$ exhibits an incommensurate 130° (approximately) spin structure. The origin of this state has been attributed to the existence of two types of NN exchange interactions in the basal plane, J along two of the crystallographic axes and J' along the third.[117-119] The mean-field phase diagram (which appears to be the same as found from MC simulations[120]) exhibits the IC helical state for $J'/J > 0.5$ and a period-2 linear phase (see Fig. 17) for $J'/J < 0.5$. The critical behavior of the paramagnetic-helical phase boundary appears not to be affected by the incommensurability and is thus expected to be the same as at the Néel point in $CsMnBr_3$.[118]

Another XY AF with interesting properties is TMMC,[121] which undergoes a very small monoclinic distortion of the high-temperature simple hexagonal structure. In addition to producing two basal-plane exchange constants, as in the case of $RbMnBr_3$, another effect of this perturbation of the triangular symmetry is believed to be the introduction of a small basal-plane anisotropy of the form $E\sum_i s_{xi}^2$. A mean-field treatment of this model reveals an $H-T$ phase diagram with a tetracritical point at a finite field, $H_m \sim \sqrt{EJ}$. Unlike $CsMnBr_3$ where $H_m = 0$, a study of TMMC offers the possibility to observe all four critical lines.

Symmetry arguments have recently been made which suggest that an applied *electric* field can remove the chiral degeneracy of the XY triangular AF.[122] This may be simply understood by noting that the $\pm Q$ degeneracy exists due to inversion symmetry, which is broken by the application of an electric field through magnetoelectric coupling. This coupling effectively introduces a Dzyaloshinskii-Moriya (DM) type interaction of the form $\sum_{ij} D_{ij} \cdot (s_i \times s_j)$, which is known to favor one helicity. In the present case, the coefficient D is proportional to the applied field. In addition to chirality selection, this term also drives the ordering to be slightly incommensurate and elliptical. The transition to such a state from the paramagnetic phase is expected to belong to the usual XY universality class since the symmetry of the order parameter is simply $V = SO(2)$. In this case, the set of chirality exponents may also be measured. However, an electric field also introduces a term of the form $|\mathbf{E} \cdot \mathbf{S}|^2$, which favors a linear polarization and hence commensurate (120°) structure since the DM term is zero in this case. The transition from the paramagnetic phase to this state can be either first order or continuous with XY universality. Preliminary neutron diffraction results on $CsMnBr_3$ suggest that the transition is first order at the values of electric field used in the experiment (1.1 and 1.5 KV/cm).[123]

An application of an XY spin-type model to the orientational ordering in the discotic liquid crystal HHTT has yielded some interesting results.[124] This material is composed of disc-like molecules which have themselves 3-fold symmetry and are arranged in weakly coupled helical columns in a triangular array. The novel feature of having both the objects of interest (discs) as well as the lattice with the same 3-fold symmetry allows for an additional contribution (G-term) to the planar-model type Hamiltonian:

$$\mathcal{H} = \sum_{ij} J_{ij} \cos(3\theta_i - 3\theta_j) + \sum_{ij} G_{ij} \sin(3\theta_i) \sin(3\theta_j). \tag{77}$$

Our work focused on the relative chirality of the columns and finite-temperature effects through mean-field theory. Note that the above Hamiltonian can be rewritten as

$$\mathcal{H} = \sum_{ij} J_{ij} \cos(3\theta_i) \cos(3\theta_j) + \sum_{ij} G'_{ij} \sin(3\theta_i) \sin(3\theta_j), \tag{78}$$

where $G' = J + G$. This model (with Heisenberg spins) has recently been examined by MC simulations on the 2D triangular lattice where a Kosterlitz-Thouless transition is expected for the case $G' = J$.[125]

An effect of frustration on the spin excitations (possibly relevant to the thermal phase transitions) in the Heisenberg AF on a simple hexagonal lattice is also of interest. Spin-1 quasi-1D AFs like $CsNiCl_3$ are candidates for the observation of effects due to the Haldane conjecture. Indeed, there is little doubt that spin excitations measured in this material in the *paramagnetic* phase at low temperatures are due to the existence of a singlet ground state and triplet first excited state. A phenomenological nonlinear sigma model formulated to incorporate the effects of interchain coupling, and hence explain the observed spin excitations in the *ordered* phase, however, is more controversial.[126,127] A significant feature of these results is the prediction of an independent longitudinal mode, which does not appear in standard spin-wave theory. Our approach has also been phenomenological, based on symmetry arguments and formulated within the framework of the Lagrange equations of motion.[128] A longitudinal mode also appears in this model, but it is parasitic to the transverse modes and *exists entirely due to frustration*. The importance of frustration in understanding the excitations in such systems has also been emphasized more recently.[129]

Finally, we mention the unusual superconductung state of hexagonal UPt_3. Due to coupling with the long-range magnetic order, the superconducting order parameter is a complex vector (ψ_x, ψ_y). The appropriate Landau-Ginzburg free energy is quite similar to Eq. (62) used to describe $CsMnBr_3$. The $H - T$ phase diagram of UPt_3 also exhibits a tetracritcal point, but at a finite value of H_m.[130]

6. Concluding remarks

The review made here of critical properties of frustrated spin systems has focused on our own and related work, from which we can make some general observations. Systems frustrated by lattice geometry are of interest due to the many types of ordered phases which can appear as a function of temperature and various applied fields, the consequently rich variety of multicritical point structures which occur in phase diagrams, and the novel types of critical phenomena which have been predicted. We have demonstrated the utility of mean-field theory, in particular the description of ordered states in terms of the spin density s(r) written as a Fourier expansion, and analysis of Landau-type free energies $F[s(r)]$. Such studies make clear the symmetries broken at phase transitions and can lead to very good agreement with experimental results on magnetic phase diagrams. Applications of standard renormalization-group expansions appear to yield results that are difficult to interpret, or justify, for frustrated systems. Apart from performing expansions in $4 - D$ to much higher order, more extensive Monte Carlo simulations offers some hope in resolving the Kawamura-Azaria *et al* controversy concerning the critical behavior of the simple stacked triangular AF and related systems. Three recent finite-size scaling analyses of histogram MC results for the Heisenberg case appear to give results for critical exponents in agreement with those found earlier by Kawamura using standard MC simulations.[35-37] These results thus support the existence of a new universality class. One can always argue, however, that the simulations were performed on systems which were too small or that the number of steps was too few (the correlation time may be larger). It would nevertheless be surprising that four rather different MC simulations should yield the same exponents if this latter scenario were correct. Several groups are currently investigating the case of the XY (planar) spin model. The alternative proposals of a new universality class or mean-field tricritical behavior will, however, be difficult to differentiate since the respective set of critical exponents are similar. MC simulations results should at the very least be useful in determining if the transition is weakly first order.

Spin models on frustrated lattices are useful in the description of a variety of types of non-magnetic systems. These include those mentioned in Sec. 2.1, the dipole phase of superfluid ^3He and Josephson-junction arrays in a transverse field, as well as the discotic liquid crystal $HHTT$ discussed in Sec. 5. It is of interest to mention that another large class of frustrated systems may be found among the helical chain polymers. An example is PFTE, which exhibits a pressure-temperature phase diagram containing four different ordered states.[131] Single helical polymer chains are often described by a chiral-clock model of the form

$$\mathcal{H} = -J_\| \sum_i \cos(\theta_i - \theta_{i-1} - \alpha) \tag{79}$$

where α is the natural (right- or left-handed) pitch of a helix determined by intermolecular interactions. Interchain interactions tend to favor commensurate modulations. Dynamic helicity-reversal of such chiral chains has also been examined.[132]

With such a wide variety of applications among magnetic and non-magnetic systems, a continuing interest in the study of frustrated spin systems is assured.

7. Acknowledgements

Our work reviewed here has benefited from collaborations and discussions with P. Azaria, A.N. Berker, A. Chubukov, B. Delamotte, A. Ferrenberg, B. Gaulin, A. Harrison, M. Hébert, O. Heinonen, K. Hood, H. Kawamura, S. Miyashita, M. Poirier, J. Reimers, Y. Trudeau, D. Visser, and G. Zumbach. We thank A. Johnson for a careful proofreading of the manuscript.

Note Added. Several articles of relevance to the discussion of Sect. 2 may also be found in the recent publication *Renormalization Group 91*, edited by Shirkov and Priezzhev (World Scientific, Singapore, 1992). In particular, the work by S. A. Antonenko and A. I. Sokolov shows the absence of a fixed point in three dimensions for $n = 3$. For a discussion on this work, see subsection 3.1 and section 5 of the chapter by Azaria and Delamotte in this book. We thank G. Zumbach, P. Azaria and B. Delamotte for bringing this work to our attention.

8. References

[*] Present address: INRS-EAU, 2800 rue Einstein, C.P. 7500, Sainte-Foy, Québec, Canada G1V 4C7.

1. G. Toulouse, Commun. Phys. **2** (1977) 115.
2. S. Miyashita, Prog. Theor. Phys. Suppl. **87** (1986) 112.
3. R.S. Gekht and V.I. Ponomarev, Phase Trans. **20** (1990) 27.
4. K. Hirakawa, Phase Trans. **28** (1990) 33.
5. M.L. Plumer and A. Caillé, J. Appl. Phys. **70** (1991) 5961.
6. A. Harrison, Ann. Reps. Prog. Chem. **87, 1990** (1992) 211.
7. A. Mailhot, Ph. D. Thesis, Université de Sherbrooke (1992).
8. H.T. Diep, in *Recent Advances in Magnetism of Transition Metal Compounds*, edited by A. Kotani and N. Suzuki (World Scientific, Singapore, 1993).
9. H. Kawamura, in *Recent Advances in Magnetism of Transition Metal Compounds*, edited by A. Kotani and N. Suzuki (World Scientific, Singapore, 1993). The possibility of new critical behavior in the stacked triangular antiferromagnet was first suggested by Kawamura in earlier work: J. Phys. Soc. Jpn. **54** (1985) 3220; **55** (1986) 2095; **56** (1987) 474. The $SO(3)$ symmetry of the Heisenberg case was revealed by Kawamura and Miyashita: J. Phys. Soc. Jpn. **53** (1984) 4138. Kawamura's most recent estimates of critical exponents from MC simulations are found in: J. Phys. Jpn. **61** (1992) 1299.
10. J. Villain, J. Phys. C **10** (1977) 1717.
11. B.D. Gaulin, J.N. Reimers, T.E. Mason, J.E. Greedan, and Z. Tun, Phys. Rev. Lett. **69** (1992) 3244; P. Chandra, P. Coleman, and I. Ritchey, J. Phys. I France **3** (1993) 591.
12. J. Yeomans, in *Solid State Physics*, edited by H. Ehrenreich and D. Turnbull (Academic, New York, 1988), Vol.41.

13. W. Selke, in *Phase Transitions*, edited by C. Domb and J.L. Lebowitz (Academic, New York, 1992), Vol.15.
14. R.B. Griffiths, Phys. Rev. Lett. **24** (1970) 1479.
15. P. Peczak, A.M. Ferrenberg, and D.P. Landau, Phys. Rev. B **43** (1991) 6087; A.M. Ferrenberg and D.P. Landau, Phys. Rev. B **44** (1991) 5081; C. Holm and W. Janke, Phys. Rev. B **48** (1993) 936; K. Chen, A.M. Ferrenberg, and D.P. Landau, Phys. Rev. B **48** (1993) 3249.
16. See, for example: M.L. Plumer and A. Caillé, Phys. Rev. B **37** (1988) 7712.
17. J.N. Reimers, A.J. Berlinsky, and A.-C. Shi, Phys. Rev. B **43** (1991) 865.
18. T. Nagamiya, in *Solid State Physics*, edited by F. Seitz, H. Ehrenreich, and D. Turnbull (Academic, New York, 1967), Vol.20.
19. H.T. Diep, Europhys. Lett. **7** (1988) 725; Phys. Rev. B **39** (1989) 397.
20. J. Villain, J. Phys. (Paris) **38** (1977) 385.
21. D. Mukamel and S. Krinsky, Phys. Rev. B **13** (1976) 5065, 5078; P. Bak and D. Mukamel, Phys. Rev. B **13** (1976) 5086.
22. T. Garel and P. Pfeuty, J. Phys. C **9** (1976) L245.
23. I.E. Dzyaloshinskii, JETP **45** (1977) 1014.
24. Z. Barak and M.B. Walker, Phys. Rev. B **25** (1982) 1969.
25. H. Kawamura, Phys. Rev. B **38** (1988) 4916; **42** (1990) 2610(E); J. Phys. Soc. Jpn. **55** (1986) 2157. To lowest order in ϵ, a new stable fixed point was found for large $n > 21.8 - 23.4\epsilon + O(\epsilon^2)$. Kawamura then suggests that higher-order terms would stabilize the fixed point for $n = 3$ and $n = 2$ at $D = 3$. Support for this scenario was also found in his $1/n$-expansion results. The possibility of a weak first-order transition was also indicated. In this case, *universal* pseudo-critical exponents would be expected.
26. M.C. Barbosa, Phys. Rev. B **42** (1990) 6363.
27. L. Saul, Phys. Rev. B **46** (1992) 13847.
28. D.R.T. Jones, A. Love, and M.A. Moore, J. Phys C **9** (1976) 743; D. Bailin, A. Love, and M.A. Moore J. Phys C **10** (1977) 1159.
29. E. Granato and J.M. Kosterlitz, Phys. Rev. Lett. **65** (1990) 1267.
30. M.Y. Choi and S. Doniach, Phys. Rev. B **31** (1985) 4516; M. Yosefin and E. Domany Phys. Rev. B **32** (1985) 1778.
31. D.H. Lee, J.D. Joannopoulos, J.W. Negele, and D.P. Landau, Phys. Rev. Lett. **52** (1984) 433; Phys. Rev. B **33** (1986) 450.
32. E. Granato, J.M. Kosterlitz, and J. Lee, Phys. Rev. Lett. **66** (1990) 1090; J. Lee, J.M. Kosterlitz, and E. Granato, Phys. Rev. B **43** (1991) 11531; J. Lee, E. Granato, and J.M. Kosterlitz, Phys. Rev. B **44** (1991) 4819.
33. J.C. Le Guillou and J. Zinn-Justin, J. Phys. Lett. (Paris) **46** (1985) L137.
34. P. Azaria, B. Delamotte, and Th. Jolicoeur, Phys. Rev. Lett. **64** (1990) 3175; J. Appl. Phys. **69** (1991) 6170; P. Azaria, B. Delamotte, F. Delduc, and Th. Jolicoeur, Nucl. Phys. B **408**, 485 (1993).
35. T. Bhattacharya, A. Billoire, R. Lacaze, and Th. Jolicoeur, J. Phys. I (France) **4**, (1994) 122.
36. D. Loison and H.T. Diep, unpublished.

37. A. Mailhot *et al.*, unpublished.
38. H. Kawamura, J. Phys. Soc. Jpn. **59** (1990) 2305.
39. A.V. Chubukov, Phys. Rev. B **44** (1991) 5362; H. Kawamura, J. Phys. Soc. Jpn. **60** (1991) 1839; Phys. Rev. B **47** (1993) 3415.
40. H. Kunz and G. Zumbach, J. Phys. A **26** (1993) 3121. Also see: G. Zumbach, Phys. Rev. Lett. **71** (1993) 2421; Nucl. Phys. B (to be published). A possibility raised by these latter works, which use the local potential approximation, is that the $SO(3)$ transition is "almost" second order, with a unique set of critical exponents.
41. M.L. Plumer, Phys. Rev. B **44** (1991) 12376.
42. D.A. Tindall, M.O. Steinitz, and T.M. Holden, J. Appl. Phys. **73** (1993) 6543.
43. M.K. Wilkinson, W.C. Koehler, E.O. Wollan, and J.W. Cable, J. Appl. Phys. Suppl. **32** (1961) 48S.
44. W.C. Koehler, H.R. Child, E.O. Wollan, and J.W. Cable, J. Appl. Phys. **34** (1963) 1335.
45. D.A. Tindall, M.O. Steinitz, and M.L. Plumer, J. Phys. F **7** (1977) L263.
46. S.W. Zochowski, D.A. Tindall, M. Kahrizi, J. Genossar, and M.O. Steinitz, J. Magn. Magn. Mat. **54-57** (1986) 707; H.U. Åström and G. Benediktsson, J. Phys. F **18** (1988) 2113.
47. K.D. Jayasuriya, A.M. Stewart, S.J. Campbell, and E.S.R. Gopal, J. Phys. F **14** (1984) 1725.
48. C.L. Henley, J. Appl. Phys. **61** (1987) 3962. Also see: M.T. Heinilä and A.S. Oja, Phys. Rev. B **48** (1993) 7227.
49. J. Villan, R. Bidaux, J.-P. Carton, and R. Conte, J. Phys. (Paris) **41** (1980) 1263.
50. M.L. Plumer, A. Mailhot, and A. Caillé, Phys. Rev. B **48** (1993) 3840 (a more detailed description is to be published in Phys. Rev. B (1994)). There is an error in the first paragraph of this work: Bhattacharya *et al.* (see Ref. 35) have estimated critical-exponent values which support the proposal by Kawamura of a new chiral universality class, and *not* $O(4)$ universality.
51. W. Minor and T.M. Giebultowicz, J. Phys. (France) **C8** (1988) 1551; H.T. Diep, and H. Kawamura, Phys. Rev. B **40** (1989) 7019; H.T. Diep, Phys. Rev. B **45** (1992) 2863; M.T. Heinilä and A.S. Oja, Phys. Rev. B **48** (1993) 16514.
52. H. Kadowaki, H. Kikuchi, and Y. Ajiro, J. Phys. **2** (1990) 4485.
53. E. Rastelli and A. Tassi, J. Phys. C **19** (1986) L423; J. Phys. C **21** (1988) 1003; I.M. Vitebskii and N.M. Lavrinenko, Low Temp. Phys. **19** (1993) 380.
54. S.S. Applesnin and R.S. Gekht, JETP **69** (1989) 1224.
55. F. Matsubara and S. Inawashiro, J. Phys. Soc. Jpn. **56** (1987) 4087.
56. J.N. Reimers and A.J. Berlinsky, Phys. Rev. B **48** (1993) 9539.
57. A.B. Harris, C. Kallin, and A.J. Berlinsky, Phys. Rev. B **45** (1992) 2899.
58. A.V. Chubukov, Phys. Rev. Lett. **69** (1992) 832.
59. M.L. Plumer, A. Caillé, and H. Kawamura, Phys. Rev. B **44** (1991) 4461.
60. J.T. Chalker, P.C.W. Holdsworth, and E.F. Shender, Phys. Rev. Lett. **68** (1992) 855.

61. A.P. Ramirez, G.P. Espinosa, and A.S. Cooper, Phys. Rev. Lett. **64** (1990) 2070; C. Broholm, G. Aeppli, G.P. Espinosa, and A.S. Cooper Phys. Rev. Lett. **65** (1990) 3173.
62. I. Ritchey, P. Chandra, and P. Coleman, Phys. Rev. B **47** (1993) 15342; D.L. Huber and W.Y. Ching, Phys. Rev. B **47** (1993) 3220; E.F. Shender, V.B. Cherepanov, P.C.W. Holdsworth, and A.J. Berlinsky, Phys. Rev. Lett. **70** (1993) 3812.
63. J.N. Reimers (private communication).
64. J.N. Reimers, Phys. Rev. B **45** (1992) 7287.
65. J.N. Reimers, J.E. Greedan, and M. Björgvinsson, Phys. Rev. B **45** (1992) 7295; A system with planar anisotropy has recently been studied by MC simulations, where a strong first-order transition has been found: S.T. Bramwell, M.J.P. Gingras, and J.N. Reimers, J. Appl. Phys. (to be published).
66. A. Mailhot and M.L. Plumer, Phys. Rev. B **48** (1993) 9881.
67. H.T. Diep, A. Ghazali, and P. Lallemand, J. Phys. C **18** (1985) 5881.
68. To our knowledge, this is the only published work where a value for γ less than unity has been found. R. Fisch (private communication).
69. R. Fisch (private communication).
70. J.M. Yeomans, *Statistical Mechanics of Phase Transitions* (Clarendon Press, Oxford, 1992); J.J. Binney, N.J. Dowrick, A.J. Fisher, and M.E.J. Newman, *The Theory of Critical Phenomena* (Clarendon Press, Oxford, 1992).
71. D.I. Uzunov, *Theory of Critical Phenomena* (World Scientific, Singapore, 1993).
72. J.-C. Tolédano and P. Tolédano, *The Landau Theory of Phase Transitions* (World-Scientific, Singapore, 1987).
73. C.J. Bradley and A.P. Cracknell *The Mathematical Theory of Symmetry in Solids* (Clarendon Press, Oxford 1972).
74. C. Kittel, *Introduction to Solid State Physics*, Fourth Edition (John Wiley, New York, 1971).
75. M.L. Plumer, A. Caillé, and K. Hood, Phys. Rev. B **39** (1989) 4489.
76. X. Zhu and M.B. Walker, Phys. Rev. B **36** (1987) 3830.
77. X. Zhu and M.B. Walker, Phys. Rev. B **34** (1986) 8064.
78. M.L. Plumer, K. Hood, and A. Caillé, Phys. Rev. Lett. **60** (1988) 45, 1885(E).
79. P. Bak and J. von Boehm, Phys. Rev. B **21** (1980) 5297.
80. N. Suzuki, J. Phys. Soc. Jpn. **52** (1983) 3199.
81. R.M. White, *Quantum Theroy of Magnetism* (Springer-Verlag, Berlin, 1983).
82. H. Müller-Krumbhaar and K. Binder, J. Stat. Phys. **8** (1973) 1; H. Gould and J. Tobochnik, Computers in Physics, Jul/Aug (1989) 82; A.M. Ferrenberg, D.P. Landau, and K. Binder, J. Stat. Phys. **63** (1991) 867.
83. P. Peczak and D.P. Landau, Phys. Rev. B **39** (1989) 11932; J. Lee and J.M. Kosterlitz, Phys. Rev. B **43** (1991) 3265; K. Binder *et al.*, Int. J. Mod. Phys. C **3** (1992) 1025; K. Vollmayr, J.D. Reger, M. Scheucher, and K. Binder, Z. Phys. B **91** (1993) 113; A. Billoire, T. Neuhaus, and B. Berg, Nucl. Phys. B **396** (1993) 779; G. Zumbach, Phys. Rev. Lett. **71** (1993) 2421; W. Janke,

Phys. Rev. B **47** (1993) 14757.

84. A.M. Ferrenberg, D.P. Landau, and Y. J. Wang, Phys. Rev. Lett. **69** (1992) 3382.

85. P.W. Leung and C.L. Henley, Phys. Rev. B **43** (1991) 752. However, also see Ref.[40] and: G. Zumbach and H. Kunz, Phys. Lett. A **165** (1992) 235; P.D. Coddington and L. Ham (preprint).

86. A.M. Ferrenberg and R.H. Swendsen, Phys. Rev. Lett. **61** (1988) 2635; *ibid* **63** (1989) 1195; Computers in Physics, Sept/Oct (1989) 101.

87. K. Binder, Phys. Rev. Lett. **47** (1981) 693.

88. M.S.S. Challa, D.P. Landau, and K. Binder, Phys. Rev. B **34** (1986) 1841; A. Billoire *et al*, Phys. Rev. B **42** (1990) 6743.

89. *Multicritical Phenomena*, edited by R. Pynn and A. Skjeltrop (Plenum, New York, 1984); Y. Shapira, in *Multicritical Phenomena* and J. Appl. Phys. **57** (1985) 3268.

90. D.P. Landau, and K. Binder, Phys. Rev. B **17** (1978) 2328.

91. M.L. Plumer and A. Caillé, Phys. Rev. B **41** (1990) 2543.

92. B.D. Gaulin, T.E. Mason, M.F. Collins, and J.Z. Larese, Phys. Rev. Lett. **62** (1989) 1380.

93. M.O. Steinitz, M. Kahrizi, and D.A. Tindall, Phys. Rev. B **36** (1987) 783; F. Willis, N. Ali, M.O. Steinitz, M. Kahrizi, and D.A. Tindall, J. Appl. Phys. **67** (1990) 5277.

94. M.L. Plumer, A. Caillé, and H. Kawamura, Phys. Rev. B **44** (1991) 4461.

95. D. Blankschtein, M. Ma, A.N. Berker, G.S. Grest, and C.M. Soukoulis, Phys. Rev. B **29** (1984) 5250.

96. H. Kawamura, A. Caillé, and M.L. Plumer, Phys. Rev. B **41** (1990) 4416.

97. M.L. Plumer and A. Caillé, Phys. Rev. B **42** (1990) 10388.

98. T.E. Mason, M.F. Collins, and B.D. Gaulin, J. Appl. Phys. **67** (1990) 5421.

99. J. Villain and J.M. Loveluck, J. Phys. Lett. (Paris) **38** (1977) L-77.

100. M.L. Plumer and A. Caillé, J. Appl. Phys. **69** (1991) 6161.

101. L. Heller, M.F. Collins, Y.S. Yang, and B. Collier, Phys. Rev. B (to be published).

102. G. Bhanot *et al.*, Phys. Rev. B **48** (1993) 6183.

103. D.H. Lee *et al.*, Phys. Rev. B **29** (1984) 2680.

104. O. Heinonen and R.G. Petschek, Phys. Rev. B **40** (1989) 9052; M.L. Plumer *et al.*, Phys. Rev. B **47** (1993) 14312; A. Bunker, B.D. Gaulin, and C. Kallin, Phys. Rev. B **48** (1993) 15861.

105. Q. Sheng and C.L. Henley, J. Phys. **4** (1992) 2937.

106. P.B. Johnson, J.A. Rayne, and S.A. Friedberg, J. Appl. Phys. **50** (1979) 1853.

107. A. Mailhot, M.L. Plumer, and A. Caillé, Phys. Rev. B **48** (1993) 15835.

108. Y. Trudeau, M.L. Plumer, M. Poirier, and A. Caillé, Phys. Rev. B **48** (1993) 12805.

109. A. Mailhot, M.L. Plumer, and A. Caillé, J. Appl. Phys **67** (1990) 5418.

110. M.C. Barbosa and M.E. Fisher, Phys. Rev. B **43** (1991) 10635.

111. A. Harrison, M.F. Collins, J. Abu-Dayyeh, and C.V. Stager, Phys. Rev. B **43** (1991) 679.

112. Y. Ajiro, T. Inami, and H. Kadowaki, J. Phys. Soc. Jpn. **59** (1990) 4142; H. Kadowaki *et al.*, J. Phys. Soc. Jpn. **60** (1991) 1708; D. Beckmann, J. Wosnitza, H. von Löhneysen, and D. Visser, J. Phys. **5** (1993) 6289; Y. Oohara, H. Kadowaki, and K. Iio, J. Phys. Soc. Jpn. **60** (1991) 393.

113. D. Beckmann, J. Wosnitza, H. v. Löhneysen, and D. Visser, Phys. Rev. Lett. **71** (1993) 2829.

114. M. Poirier, A. Caillé, and M.L. Plumer, Phys. Rev. B **41** (1990) 4869; H.A. Katori, T. Goto, and T. Ajiro, J. Phys. Soc. Jpn. **62** (1993) 743.

115. Th. Jolicoeur, E. Dagotto, E. Gagliano, and S. Bacci, Phys. Rev. B **42** (1990) 4800; A.V. Chubukov and Th. Jolicoeur, Phys. Rev. B **46** (1992) 11137; S.E. Korshunov, Phys. Rev. B **47** (1993) 6165.

116. D. Loison and H.T. Diep, J. Appl. Phys. **73** (1993) 5642.

117. W.-M. Zhang, W.M. Saslow, M. Gabay and M. Benakli, Phys. Rev. B **48** (1993) 10204.

118. H. Kawamura, Prog. Theor. Phys. Suppl. **101** (1990) 545.

119. W.-M. Zhang, W.M. Saslow, and M. Gabay, Phys. Rev. B **44** (1991) 5129.

120. W.M. Saslow, M. Gabay, and W.-M. Zhang Phys. Rev. Lett. **68** (1992) 3627.

121. M.L. Plumer and A. Caillé, Phys. Rev. B **45** (1992) 12326.

122. M.L. Plumer, H. Kawamura, and A. Caillé, Phys. Rev. B **43** (1991) 13786; Ferroelectrics (to be published).

123. D. Visser *et al.*, Ferroelectrics (to be published).

124. M.L. Plumer, A. Caillé, and O. Heinonen, Phys. Rev. B **47** (1993) 8479.

125. T. Horiguchi, D. Loison, H.T. Diep, and O. Nagai (unpublished).

126. I. Affleck, Phys. Rev. Lett. **62** (1989) 474; **65** (1990) 2477(E); **65** (1990) 2835(E).

127. I. Affleck ad G.F. Wellman, Phys. Rev. B **46** (1992) 8934. We comment that the model discussed in their Appendix, while apparently motivated by our work, does not have the same kinetic contribution. As an example, the spin-wave velocity of our Lagrangian is proportional to the long-range order parameter, and not inversely proportional as in their Appendix model.

128. M.L. Plumer and A. Caillé, Phys. Rev. Lett. **68** (1992) 1042.

129. T. Ohyama and H. Shiba, J. Phys. Soc. Jpn. **62** (1993) 3277.

130. R. Joynt, Europhys. Lett. **16** (1991) 289; Phys. Rev. Lett. **71** (1993) 3015.

131. O. Heinonen, Phys. Rev. B **47** (1993) 2661.

132. O. Heinonen and P.L. Taylor, Polymer **30** (1989) 585.

Magnetic System with Competing Interaction
Ed. H. T. Diep
©1994 World Scientific Publishing Co.

CHAPTER II

THE RENORMALIZATION GROUP APPROACH

TO FRUSTRATED HEISENBERG SPIN SYSTEMS

Patrick Azaria

*Laboratoire de Physique Théorique des Liquides, Université Pierre et Marie Curie,
4, Place Jussieu, 75230 Paris Cedex 05, France.*

and

Bertrand Delamotte

*Laboratoire de Physique Théorique et des Hautes Energies, Université Denis Diderot,
2, Place Jussieu, 75251 Paris Cedex 05, France.*

1. Introduction

The most striking feature with frustrated non colinear systems is that while the fluctuating field at the scale Λ^{-1} of the lattice spacing is a vector (the spin itself), the order parameter which is the relevant long wavelength fluctuating field is a rotation matrix R of $SO(3)$. It is typically the case of the Heisenberg antiferromagnet on the triangular lattice (AFT) where the spins display the famous 120 degree structure on each elementary cell at $T = 0$ because of the geometrical frustration. This is very different from colinear systems where the order parameter is a vector. Therefore, we expect that the low-lying excitation spectrums of frustrated systems are drastically different from those of non frustrated magnets. As we shall see in the following, this has dramatic consequences on the low energy, long distance physics. In particular, we shall see that it is not straightforward to write down an effective theory for these low energy modes.

There are two problems that we want to address in this chapter. The first one is the nature of the thermally activated phase transition of *classical* frustrated Heisenberg magnets. We would like to know if they belong to a new universality class different from the usual $O(N)/O(N - 1)$ Wilson-Fisher classes. This question has been the object of a considerable amount of work. However, both from the theoretical as well as from the experimental point of view, there is up to now no definitive conclusion. The present theoretical situation can be summarized as follows. It was first Garel and Pfeuty[1] and next Bailin, Love and Moore[2] who first studied the relevant effective Landau-Ginzburg-Wilson (LGW) model for this problem. They found no fixed point near four dimensions and concluded on a first order transition. On the other hand, results from Monte-Carlo simulations[3,4]*on the stacked AFT model in $D = 3$ are in favor of a second order phase transition while the experimantal results are not compatible among themselves[5,6,7,8,9,10] (see the chapter of Gaulin in this

---

*See the article of M.L. Plumer et al. in this book for an extensive review of the litterature on this subject

book) . It was only quite recently, that a new strategy was employed to investigate this problem. With F. Delduc and T. Jolicoeur, the authors have studied the relevant Non Linear Sigma (NLσ) model for frustrated Heisenberg models in a systematic $\epsilon = D - 2$ expansion[11,12]. We found a stable fixed point near two dimensions where the model becomes $SO(4)$ symmetric. Thus nonetheless we concluded in favor of a second order phase transition but we found no new universality class but the general phenomenon of increased symmetry at a fixed point. As seen, in contrast with colinear magnets, there is no agreement between the different theoretical approaches and with Monte Carlo simulations. It is precisely this mismatch that makes the critical physics of frustrated magnets non trivial.

In the second part of this chapter, we study the effect of *quantum* fluctuations on the low energy physics of frustrated *antiferromagnets*. The reason is that these fluctuations play an important role when both the quantum spin number S and the space dimension D is small, and particularly in the important case $S = 1/2$, $D = 2$[13,14]. In fact it is possible that the semiclassical Néel ground state is destabilized by quantum fluctutions in favor of a quantum disordered one, a spin liquid state. Apart from evident theoretical interest, such spin fluid phases are thought to be relevant in some theories of high T_c superconductors[15].

At $T = 0$, as one varies the magnitude S of the spin, quantum antiferromagnets can undergo a *quantum* phase transition between semiclassical Néel order at large S and quantum disorder at small S[14]. The natural question is to know how one can describe such a quantum phase transition and what is its nature compared to ordinary thermally activated transitions. In addition, one would like to know the nature of the low energy excitations in both phases. To this end one needs an effective theory for the low energy physics which enables us to give quantitative as well as qualitative predictions that descriminate between semiclassical order and quantum disorder. Such a theory is of importance when one wants to understand both experiments and numerical results. In a beautiful work, Chakravarty, Halperin and Nelson[16,14] have studied the relevant theory for colinear antiferromagnets. They established the existence of a critical value S_c above which the semiclassical Néel state is stable. In this regime, the spectrum consists into two relativistic spin wave modes and the low energy physics is well controlled by the infrared fixed point at $S = \infty$ so that one is able to give quantitative as well as qualitative tests for the existence of long range order. As a result, together with experimental and numerical results, there is no doubt that colinear antiferromagnets, such as the 2D Heisenberg antiferromagnetic model on the square lattice, displays long-range order in the ground state, even for $S = 1/2$. When frustration is taken into acount, the situation is less clear. Since one expects frustration to enhance disorder, it is natural to think that frustrated quantum antiferromagnets are promising candidates to exhibit quantum disordered ground states[17]. This is why they are actively studied. As in the unfrustrated case, we want an effective theory for the low energy physics and we want to understand the nature of the quantum phase transition in the presence of frustration[18,19]. It is the goal of the second part of this chapter to derive and study this effective theory.

There is in fact a strong analogy between the classical and quantum problems. It turns out that when *the net magnetization is zero* the effective theory that governs the thermally activated phase transition in the *classical* systems in D dimensions is formally identical to the one that governs *quantum* fluctuations in $D - 1$ dimensions as the spin S is varied. As an immediate consequence, we see that the nature of both types of transitions should be identical and that we shall have the same problem to conclude about universality. Let us comment a little further on this problem. In the Renormalization Group (RG) ideology, the effective theory governing the low energy, long wavelength fluctuations, in which only a small number of relevant and marginal operators survive, should be obtained by integrating out the high energy degrees of freedom. In the RG language this means this theory lives in the neighborhood of a fixed point. In practice, of course, such a decimation procedure is impossible to carry out and one has to have recourse to perturbation theory. It is worth stressing that, as soon as we are doing perturbation theory, we have to make an ansatz on the nature of the relevant low energy excitations. In general, we are constrained by symmetry considerations, renormalizability and simplicity of the effective theory. There are two such theories that we expect to control perturbatively in the neighborhood of dimensions $D = 2$ and $D = 4$. They are the NLσ and the LGW or ϕ^4-like models. The NLσ model is the theory of the interacting spin waves (or Goldstone bosons) resulting from the breakdown of the (continuous) symmetry of the problem. It accounts well for the low energy physics of the symmetry broken phase, i.e. when $S > S_c$ or $T \leq T_c$, since in this regime it is controlled by the infrared fixed point at $S^* = \infty$ or $T^* = 0$. In the critical regime, or in the disordered phase, the spin waves are stongly interacting but one can still control the physics at sufficiently high energy if there exists a non trivial ultraviolet fixed point at $S^* = 0(1/(D - 2))$ or $T^* = 0(D - 2)$. In the LGW theory, one makes an ansatz on the nature of the massive spectrum which is justified by the existence of an infrared fixed point at $S^* = 0(1/(4 - D))$ or $T^* = 0(4 - D)$. In the simplest cases, as it is the case for colinear antiferromagnets, both theories are consistent and one has a good understanding of the low energy physics in both phases. In the frustrated case, as will be shown below, one does not find a stable infrared fixed point in the LGW theory. While one cannot question the existence of a well defined spin wave effective theory in the symmetry broken phase, the existence of a universal effective theory of frustration near T_c or S_c can be seriously questioned. As seen, the question of universality of the phase transition in classical frustrated Heisenberg models is deeply related to the nature of the low lying spectrum of quantum frustrated magnets.

We certainly do not expect to answer to all these important questions in the present chapter but rather we rather try to present the theoretical situation at the time being. We are mainly concerned with the RG point of view and mostly by our own works. We refer to complementary works each time it is possible. This chapter is organized as follows.

The second section is concerned with the derivation of the effective theories for frustrated magnets. We obtain both LGW and NLσ models. A short comment on the large N approaches is also made. Finally, we focus at the end of this section on the particular case of the stacked triangular antiferromagnetic (STA) model.

The third section is entirely devoted to the renormalization properties of the LGW models.

In the fourth section we find and solve the RG equations of the NLσ model. This part of the work is quite technical but we find useful to take some time to explain it as pedagogically as possible since it can be generalized to other problems and that a comprehensive review for non specialist on the subject, is still lacking in our opinion.

In the fith section, we quickly expose the different possibilities for interpolating the results between $D = 2$ and $D = 4$.

We investigate the possible effects of topological excitations (Z_2 vortices) in classical two dimensional systems in the sixth section. Our principal aim is to calculate the low T behavior of the physical observables, such as the correlation length and the spin stiffnesses, that should be observed if vortices were irrelevant at sufficiently low temperatures. Results of Monte-Carlo simulations made on the two dimensional classical AFT Heisenberg model are finally discussed.

The seventh section is devoted to the study of 2D quantum frustrated antiferromagnets. We derive the relevant effective low energy theory, and prove the existence of a critical value of the spin S above which a stable semiclassical Néel state exists. We also explain the mechanism of spontaneous symmetry breaking in quantum antiferromagnets and show which constraints the low lying spectrum of a *finite* system has to fulfil to be ordered in the thermodynamical limit. We also compute the low temperature behavior of physical observables in the symmetry broken phase. Finally we comment and refer to complementary works, more suitable for the study of the quantum disordered phase, such as the large N Schwinger boson theory.

Concluding remarks are finally made in the eigth section.

2.   Effective action for frustrated models

As well known, the critical behavior of a statistical system is governed by its long distance behavior. Therefore, to study the critical behavior, one needs an effective long distance theory that describes the low energy collective excitations. In principle, to obtain such an effective theory from a given microscopic Hamiltonian, one should integrate out the rapid fluctuations until one reaches a regime where all the irrelevant operators have disappeared from the theory and where one is left with an effective Hamiltonian that depends only on a finite number of relevant and marginal operators[20]. In practice, this procedure is extremely cumbersome or even impossible to carry out and one has to have recourse to perturbation theory. There exist different approaches which have their own advantages and shortcomings, and the most popular are the $\epsilon = 4 - D$ and $\epsilon = D - 2$ expansions together with the $1/N$ expansion (N being the number of spin components)[21]. Needless to say that in general, the physically relevant case, $N = 3$ and $D = 3$, is out of the domain of validity of the perturbative approaches and one has to compare the results obtained by the different perturbative approaches to draw a consistent picture of the physics in $D = 3$ and $N = 3$. Such a procedure has been successfully achieved for $O(N)$ models for which both $\epsilon = 4 - D$ and $\epsilon = D - 2$ expansions predict a second order phase transition[22,23], a fact which is consistent with the large N expansion. As we shall see this is *not the case* for Frustated Heisenberg models[11,24], and one must keep in mind that results obtained from pertubation theory may not be correct.

The ϵ-expansions allow to treat consistently, only a finite number of relevant and marginal operators. Depending on which dimension we expand around, two or four, these relevant operators are different and so will be the relevant effective long distance theories. They are respectively near $D = 2$ and $D = 4$ the Non Linear Sigma (NLσ) model and the Landau-Ginzburg-Wilson (LGW) model. In the present section we derive these relevant effective low energy theories for frustrated Heisenberg models. As a warm-up example, we first review the unfrustated case. Then both NLσ and LGW effective actions will be explicitly obtained for the frustrated case. The AFT model will be finally considered with particular emphasis on symmetry properties.

2.1. Non frustrated Heisenberg models.

Let us first review how the effective long distance theories can be derived for the non frustrated magnets. We start with the Heisenberg model which Hamiltonian reads:

$$H = - \sum_{<ij>} J_{ij} \mathbf{S}_i . \mathbf{S}_j. \tag{1}$$

In this equation the vectors $\mathbf{S}_i$ are classical Heisenberg spins with fixed unit length. In the ferromagnetic case, i.e. when J_{ij} is positive, the continuum limit is achieved by letting the spins fluctuate around their common expectation value $\mathbf{S}_0$. We then replace in (1) $\mathbf{S}_i . \mathbf{S}_j$ by $(\nabla \mathbf{S}(x))^2$ to obtain the continuum action[22]:

$$S = \frac{1}{2T} \int d^D x \, (\partial \mathbf{S})^2 \tag{2}$$

with

$$\mathbf{S}^2 = 1 \tag{3}$$

In Eq.(2) we have dropped the higher order derivative terms since they can be shown to be irrelevant near both four and two dimensions. If not for the constraint $\mathbf{S}^2 = 1$, the effective action (2) would be that of a free field theory. The different perturbative approaches, ϵ or $1/N$ expansions, correspond to different ways of taking into account of the constraint (3):

• one can parametrize (3) by $\mathbf{S}(\mathbf{x}) = \sigma(x) \mathbf{u} + \pi(x)$ with $\sigma^2 + \pi^2 = 1$ and $\mathbf{u} \perp \pi(x)$, and integrate exactly on $\sigma(x)$. This is the NLσ model that depends only on $\pi(x)$ and their derivatives. The $\pi(x)$ modes are still constrained by $\pi^2(x) < 1$ which cannot be taken into account in perturbation theory. The basic approximation is then to let the fields $\pi(x)$ fluctuate in the range $] - \infty, +\infty[$. As it can be readily shown, this is equivalent to neglecting terms of order $\exp(-1/T)$. In the double expansion in $\epsilon = D - 2$ and in the temperature T these terms are always irrelevant and the model is renormalizable. It is worth stressing that the confidence we can have in the results thus obtained relies on the existence of a fixed point that survives in three dimensions and that terms of order $\exp(-1/T)$ are inessential to the physics at long distance.

• one can also approximate (3) by a potential

$$\delta(\mathbf{S}^2 - 1) \to e^{-V}$$

$$V = \frac{1}{2}m^2\mathbf{S}^2 + \frac{1}{4}\Lambda^{4-D}u\mathbf{S}^4 \tag{4}$$

The resulting action is that of the well known LGW, $\mathbf{S}^4$ theory. It is renormalizable in a double expansion in $\epsilon = 4 - D$ and u. The basic assumption behind (4) is the existence of a fixed point at $u* \sim \epsilon$ that guarantees that the neglected terms when truncating the delta by a potential are indeed irrelevant.

In both expansions, we obviously control the perturbative results only in the vicinity of the two critical dimensions. In particular $D = 3$ is in the non perturbative regime of both models. To control the role of the dimension we need another expansion which is non perturbative in D. This is the large N expansion. Both the NLσ and the LGW models can be studied by this technique. In the NLσ model, the constraint is rewritten as:

$$\delta(\mathbf{S}^2 - 1) \to \int_{-i\infty}^{+i\infty} d\lambda e^{\lambda(\mathbf{S}^2 - 1)} \tag{5}$$

while in the LGW model the $\mathbf{S}^4$ term is decoupled via a Hubbard-Stratonovitch transformation:

$$e^{-g\mathbf{S}^4} \to \int_{-i\infty}^{+i\infty} d\lambda e^{\frac{1}{2}\lambda^2 + \sqrt{2g}\lambda\mathbf{S}^2} \tag{6}$$

Then, in both cases $\mathbf{S}(x)$ can be integrated out since the functional integral is now quadratic in $\mathbf{S}$. The N-dependence in the effective action on λ comes from the fact that $\mathbf{S}$ is taken as a N-component vector and that the gaussian integral on $\mathbf{S}$ gives a factor N, one for each component. Then $1/N$ orders naturally the perturbation theory around the saddle point in λ which is exact in the limit $N \to \infty$. Once the spatial integrals are performed, the dimension D is just a parameter in perturbation theory in contrast to the ϵ expansions. A second order phase transition governed by a $O(N)/O(N-1)$ fixed point is found in the large N expansion of both NLσ and LGW models that are therefore consistent. The expression for the critical exponents are the same order by order in $1/N$. Moreover, it can be shown[21] that the scaling relations verified by the correlation functions are identical in the critical domain of both models. In addition the critical exponents obtained so far agree with that obtained from both $D = 2 + \epsilon$ and $D = 4 - \epsilon$ expansions of NLσ or LGW models. Thus the natural conclusion for non frustrated systems is that the lattice system has a unique fixed point which is a function of N and D that can be reached in perturbation theory for any N in $D = 2 + \epsilon$ and in $D = 4 - \epsilon$ with respectively the NLσ model and the LGW model and for any D in the large N expansion of either the NLσ or LGW model. This gives a completely consistent picture of the critical physics of non frustrated models.

2.2. Frustrated Heisenberg models

Let us see how this scheme works in the case of frustrated models. The Heisenberg hamiltonian for frustrated systems is still given by Eq.(1) but now, the J_{ij}'s are such that the ground state is *canted*. A well known example is the Heisenberg antiferromagnetic model on the triangular lattice (AFT) wich displays the 120 degree structure of Fig.(1).

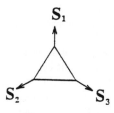

Fig. 1. The 120 degree structure of the AFT model.

For the time being we shall keep our discussion as general as possible since the nature of the order parameter alone allows to draw many conclusions about physical properties of frustrated models. A particular attention on the AFT model will be paid in the next subsection.

The general idea to perform the continuum limit is first to identify the order parameter and then to write down the most general renormalizable hamiltonian (relevant and marginal couplings) built with the order parameter that reproduces the symmetry breaking pattern that occurs on the lattice system. This gives a prominent role to the symmetries of the system and more particularly to the symmetry breaking scheme. The crucial difference between non frustrated and frustrated systems is that since the former has a canted ground state, the rotation group is completely broken in the low temperature phase, contrary to the colinear systems where the $O(2)$ rotation group around the (staggered) magnetization is left unbroken. As we shall see in the following, this will imply that the ordre parameter is a rotation matrix $R \in SO(3)$ instead of a vector pointing on the sphere $S^2 = SO(3)/SO(2)$. This fact turns out to have relevant dynamical consequences. As a result of Goldstone theorem, there are *three* spin waves or Goldstone modes in the low temperature phase of frustrated models and only two in colinear systems. Let us now identify the order parameter and derive the effective action of frustrated systems.

When the interaction distribution $\{J_{ij}\}$ leads to a canted ground state the continuum limit is not obvious since neighboring spins do not fluctuate around the same mean expectation value. To overcome this difficulty, one has to consider the magnetic cell with n sublattices $(\mathbf{S}^1, \ldots, \mathbf{S}^n)$ as the basis of a new superlattice where the continuum limit is taken. Practically, this procedure depends on the detailed microscopic model: lattice symmetry, ground state structure and interaction parameters. We shall however give qualitative arguments valid for many canted models.

Let us define in each elementary cell an orthonormal basis $\{\mathbf{e}_a(x)\}$:

$$\mathbf{e}_a(x).\mathbf{e}_b(x) = \delta_{ab} \quad ; \quad a = 1, 2, 3 \quad , \tag{7}$$

where x is a superlattice index. We may parametrize our n sublattice spins $\mathbf{S}^\alpha(x), \alpha = 1, .., n$ as:

$$\mathbf{S}^\alpha(x) = \sum_a C_a^\alpha(x)\mathbf{e}_a(x). \tag{8}$$

In the ground state, all the $\mathbf{S}^\alpha$ are in general *not* independent. There is in fact a maximum of three of them which are independent. Equivalently, there is a minimum of $n-3$ linear combinations of the $\mathbf{S}^\alpha(x)$ which have zero expectation value in the ground state. Such combinations cannot be part of an order parameter. They correspond to relative motions of the spins within each unit cell. They are massive modes with short range correlations and are thus irrelevant to the critical behavior. We ignore them by imposing the constraints that *locally*, i.e. within each unit cell, the spins are in the ground state configuration. We call this requirement "local rigidity"[12]. Thus, up to an appropriate field redefinition, the order parameter of canted magnets will be the orthonormal basis $\{\mathbf{e}_a(x)\}$ defined on each site of the superlattice. As a consequence canted magnets are equivalent in the critical region to a system of interacting solid rigid bodies. The $\{\mathbf{e}_a(x)\}$ can be gathered in a 3×3 matrix R:

$$R = (\mathbf{e}_1, \mathbf{e}_2, \mathbf{e}_3) \tag{9}$$

that belongs to $SO(3)$ since the $\{\mathbf{e}_a\}$ are orthonormal. R is the expected order parameter.

The continuum effective action, is now obtained through the standard gradient expansion of the $\{\mathbf{e}_a(x)\}$ as in ferromagnets:

$$S = \frac{1}{2}\Lambda^{D-2} \int d^D x \sum_{a=1}^{3} p_a \left(\partial\mathbf{e}_a\right)^2, \tag{10}$$

where the $\mathbf{e}_a$ fields are subject to the constraints:

$$\mathbf{e}_a(x).\mathbf{e}_b(x) = \delta_{ab} \tag{11}$$

The partition function reads[25,11,18]:

$$Z = \int D\mathbf{e}_{1,2,3}(x) \left(\prod_{ab} \delta(\mathbf{e}_a(x).\mathbf{e}_b(x) - \delta_{ab})\right) e^{-S}. \tag{12}$$

Action (10) can be rewritten in terms of the rotation matrix R as:

$$S = \frac{1}{2}\Lambda^{D-2} \int d^D x \, Tr\left(P(\partial R^{-1})(\partial R)\right), \tag{13}$$

where P is the diagonal matrix of coupling constants: $P = \mathrm{diag}(p_1, p_2, p_3)$. In this language, the partition function reads:

The Renormalization Group Approach to ... 59

$$Z = \int DR(x) e^{-S} \tag{14}$$

As seen, on Eq.(13) and contrary to the non frustrated case, we need three coupling constants to specify the low energy physics of frustrated models. The couplings p_i can easily be related to the stiffnesses η_a's associated to relative rotations around the top axis $\mathbf{e_a}$. It is not difficult to show that to an infinitesimal rotation of angle $\delta\theta$ around $\mathbf{e_a}$, the energy increment is $\delta E = (p_b + p_c)(\delta\theta)^2$ where $a \neq b \neq c$. This defines the three stiffnesses of our problem: $\eta_1 = p_2 + p_3, \eta_2 = p_3 + p_1$ and $\eta_3 = p_1 + p_2$. The relations between theses stiffnesses are connected to the symmetry properties of the model that we now discuss in some details since they play an important role in this chapter.

The original rotational symmetry of the Heisenberg hamiltonian is implemented on S as the right action of $SO(3)$:

$$R(x) \to UR(x), \quad U \in SO(3). \tag{15}$$

These are the usual rotations of the vectors $\mathbf{e_a}$ with respect to space.

Apart from the rotational invariance, the symmetry group of the hamiltonian contains, in general, a discrete group of transformations, such as the point group of the lattice in the AFT model, that mixes together the sublattices $\mathbf{S}^\alpha$ or equivalently the $\mathbf{e_a}$. These are rotations around the top axis and are implemented on R as the right action:

$$R(x) \to R(x)V, \quad V \in SO(3). \tag{16}$$

As readily seen, the action Eq.(13) is invariant under the group of right transformations V such that $[P, V] = 0$. This shows how the relative values of the couplings p_a are related to the lattice symmetry. We have to distinguish between three cases:

- when two couplings are equal, say $p_1 = p_2$, then the right invariance group is parametrized by a block matrix:

$$V = \begin{pmatrix} A & 0 \\ 0 & 1 \end{pmatrix} \quad , \quad A \in SO(2) \tag{17}$$

since V commutes with P. The complete symmetry of the high temperature phase is therefore $G = SO(3) \otimes SO(2)$. We shall see in the following that this is precisely the case of the triangular model and that the right $SO(2)$ symmetry is reminiscent of the C_{3v} point group of the triangular lattice that implies $p_1 = p_2$.

- when all the couplings are equal $p_1 = p_2 = p_3$, P is proportional to the identity matrix and the right symmetry is thus the whole $SO(3)$ group: $G = SO(3) \otimes SO(3)$.

- finally, when none of the p_i's are equal no right symmetry exists.

To summarize, depending on the P matrix, actions (13) or (10) are invariant under $G = SO(3) \otimes SO(p)$ implemented as a left and a right rotation group and where $SO(p)$ is the subgroup of $SO(3)$ that commutes with the P matrix. To complete our discussion, we have to characterize the symmetry breaking pattern. To this end, we need the subgroup H of G that leaves the ground state invariant. The ground state R_0 of (13) or (10) is obtained when all the tops are parallel. It is not difficult to show that the conjugate left and right transformations

$$R_0 \to U R_0 U^{-1}, \quad U \in SO(p) \tag{18}$$

leaves the ground state invariant and is the isotropy subgroup H. The symmetry breaking pattern is therefore for all $p = 1, 2, 3$, $G = SO(3) \otimes SO(p) \to H = SO(p)$. This shows that whatever the right symmetry is, there exist always three Goldstone modes: the additional $SO(p)$ right invariance, when it exists, has no dynamical consequence, and simply accounts for relations between the three spin stiffnesses of our problem.

We come now to an important point which is the connection between the nature of the order parameter and the characterization of the relevant effective low energy theory. As readily seen, if not for the constraints of Eq.(11), the action of Eq.(10) would be that of a free field theory for the e_a. In fact, one may still think of Eq.(10) as a free theory, but defined in terms of constrained fields leaving on a manifold defined by the set of equations Eq.(11). This manifold is nothing but the order parameter space which is the space of parameters such that to any point corresponds a value of the order parameter. This space is also conveniently defined as the set of transformations of a reference value of the order parameter, the ground state, into a generic one[26]. Therefore, given both the symmetry groups of the symmetric phase G and the isotropy subgroup H of the ground state, the order parameter manifold is just the (homogeneous) coset space $V = G/H$. As an example, in the case of the ferromagnetic N-component vector model it is the set of rotations of a fixed reference vector $\mathbf{S}^0$ onto a generic value of $\mathbf{S}(x)$. For a given value of $\mathbf{S}$ there are many such rotations since if R is such a rotation so is $R^h = R.h$ where h is any rotation around the reference vector $\mathbf{S}^0$, $h\mathbf{S}^0 = \mathbf{S}^0$. The order parameter space is thus the set of rotations of $SO(3)$ up to the rotations of $SO(2)$ around the reference vector $\mathbf{S}^0$ and therefore $V = SO(3)/SO(2)$. An element of this space is the equivalent class of all the rotations that rotate the reference vector onto $\mathbf{S}(x)$, and is determined by two angles. Considered as a manifold, $SO(3)/SO(2)$ is the unit sphere of the three dimensional space. A point of this sphere is clearly associated to a unique value of the order parameter $\mathbf{S}$.

In the frustrated case, the order parameter is a rotation matrix $R(x) \in SO(3)$. The symmetry group G of the high temperature phase as well as the isotropy subgroup H depends on the interaction scheme defined by the p_a. The possible order parameter spaces that one can define are therefore the homogeneous coset spaces: $V(p) = SO(3) \otimes SO(p)/SO(p)$, $p = 1, 2, 3$. In frustrated magnets, the fact that the order parameter is a rotation matrix does not completely specify the order parameter space. However, the $V(p)$ are all topologically equivalent to $V(3) = SO(3) \otimes SO(3)/SO(3) \sim SO(4)/O(3) \otimes Z_2$. The spaces $V(p)$ thus differ only by their metric properties and more precisely by the isometries that can be implemented on the metric.

2.2.1. The Non Linear Sigma model

As in the non frustrated case, one can solve the constraints Eq.(11) in terms of the Goldstone modes. There are infinitely many different ways to parametrize the solutions, as many as there are different parametrizations of the $SO(3)$ manifold. We choose first to eliminate the $\mathbf{e}_3$ field withthanks to the condition: $\mathbf{e}_3 = \mathbf{e}_1 \wedge \mathbf{e}_2$. Then the remaining constraints can be rewritten in terms of

$$\Phi = (\mathbf{e}_1, \mathbf{e}_2) \tag{19}$$

as:

$$^t\Phi\Phi = 1_2 \tag{20}$$

If we parametrize Φ as

$$\Phi = \begin{pmatrix} \Pi \\ A \end{pmatrix} \tag{21}$$

with Π a 1×2 matrix and A a 2×2 matrix, we find that the general solution of (20) is

$$A = \omega(x)\sqrt{1_2 - {}^t\Pi\Pi} \quad , \quad \omega \in O(2) \tag{22}$$

Π together with ω contains the 3 Goldstone modes of our problem. Let us call π^i, $i = 1, 2, 3$ these fields and σ^a the three remaining massive modes. We shall see how the σ^a's are related to the $\mathbf{e}_a$'s in the next section. At this stage, it is important to notice that we have chosen a particular way of parametrizing the orthonormality constraints by choosing the fields Π, by orthogonal projection. Without changing the physical predictions of the model, we could perfectly have chosen another projection, for instance the stereographic one, to parametrize the spheres described by the vectors $\mathbf{e}_{1,2}$. Therefore the choice of parametrization of the fields is irrelevant (although it can be convenient for practical calculations to choose a particular parametrization), only the order parameter manifold is relevant. The set of the 3 independent components of the π_a's is a coordinate system on the manifold $SO(3)$ whose equations in the 6 dimensional embedding Euclidean space are the three orthonormality constraints, Eq.(20). It is now not difficult to write the order parameter $R(x)$ in terms of the π_a:

$$R(\pi(x), \omega(x)) = \begin{pmatrix} \sqrt{1 - \pi\,{}^t\pi} & \pi \\ -\omega(x){}^t\pi & \omega(x)\sqrt{1_2 - {}^t\pi\pi} \end{pmatrix}. \tag{23}$$

The matrix $R(\pi(x), \omega(x))$ parametrizes the manifold $SO(3)$. Plugging this result into Eq.(13) we obtain:

$$\mathcal{S} = \frac{1}{2}\int d^D x \; g_{ij}(\pi)\partial_\mu\pi^i(x)\partial_\mu\pi^j(x) \tag{24}$$

where $g_{ij}(\pi)$ is a non trivial function of the π_a and the p_a. The form of this action is dictated by the form of $\mathcal{S}$, Eq.(10), which is quadratic in ∂_μ and does not involve a potential. The interpretation of Eq.(24) is that $\mathcal{S}$ in Eq.(10) is formally identical to the line element on the order parameter space and therefore $g_{ij}(\pi)$ is the metric on

$SO(3)$ viewed as a metric space. This result generalizes to arbitrary manifold of the form G/H.

We saw above that depending on the P matrix, S is invariant under $G = SO(3) \otimes SO(p)$. Note that as can be seen on Eq.(20), the π_a's span a *non-linear* representation[†] of G, a fact from which follows the name Non Linear Sigma model. In addition, one can show, in the same way, that the action of $H = SO(p)$, the isotropy subgoup of G, still acts linearly on the the the π_a's. The symmetry properties of S we have discussed above have the following geometrical interpretation. While all the manifolds $V(p)$ can be parametrized by a unique coordinate system (they are topologically equivalent), as metric manifold they are different: depending on p, the metrics have different isometries, the $SO(p)$ right invariance.

As in the ferromagnetic case, the π_a's are still constrained with $\pi_a{}^2 < 1$ and we still have to let them fluctuate in the range $]-\infty, +\infty[$ to perform perturbative calculations. Thus the basic approximation, in the double $\epsilon = D - 2$ and T expansion, is simply to forget about $\exp(-1/T)$ terms that originate from global properties of the manifold. This fact is of considerable consequence since, as we shall see in section **4**, it implies that the dynamical properties of the model are solely determined by the geometry of the order parameter space which is itself determined by the group structures of G and H. Otherwise, we should have to take into account the topological structure of the manifold as well. Therefore we can already guess that a completely geometric and group theoretical language instead of a conventional field theoretical one will be powerful for the study of the NLσ model. In particular, this geometric language has at least one fundamental advantage on the other formulations: it allows to worry only about intrinsic, i.e. coordinate independent, quantities.

2.2.2.   The Landau-Ginzburg-Wilson models

As in the non frustrated case, the LGW models can be deduced from Eqs.(13, 10). This is done by replacing the constraints Eq.(11) by an appropriate potential. There are two ways to do it.

• *The two-vector model*

First of all, one can use the constraint $\mathbf{e}_3 = \mathbf{e}_1 \wedge \mathbf{e}_2$ to integrate out exactly this field. The resulting action is therefore a function only of $\mathbf{e}_1$ and $\mathbf{e}_2$. Then the remaining constraint can be replaced by a potential :

$$V_1(\mathbf{e}_1, \mathbf{e}_2) = \frac{1}{2}m^2(\mathbf{e}_1^2 + \mathbf{e}_2^2) + \frac{\Lambda^{4-D}}{4}u_1(\mathbf{e}_1^2 + \mathbf{e}_2^2)^2 + \frac{\Lambda^{4-D}}{2}u_2(\mathbf{e}_1{}^2\mathbf{e}_2{}^2 - (\mathbf{e}_1.\mathbf{e}_2)^2) \quad (25)$$

wich disfavors field configurations not satisfying Eq.(11).[‡] If we restrict ourselves to cases where there is an additional $O(2)$ right invariance we obtain the LGW action

---

[†]The representation spanned by the π^i's must be non linear since the orthogonal projection (for instance) of the $\mathbf{e}_i$'s is a non linear operation on these fields.

[‡]To obtain Eq.(25) we have used $(\mathbf{e}_1 \wedge \mathbf{e}_2)^2 = \mathbf{e}_1{}^2\mathbf{e}_2{}^2 - (\mathbf{e}_1.\mathbf{e}_2)^2$

first obtained by Garel and Pfeuty[1]:

$$S_1 = \int d^D x \, \frac{1}{2} \left((\partial e_1)^2 + (\partial e_2)^2 \right) + V_1(e_1, e_2) \tag{26}$$

In the last equation we have omitted an $SO(2)$ current term which comes from the kinetic term $(\partial e_3)^2$ of the e_3 field:

$$(e_1 \partial e_2 - e_2 \partial e_1)^2 \tag{27}$$

which, in four dimensions is of dimension six and therefore is non renormalizable and irrelevant. The symmetry group $G = SO(3) \otimes SO(2)$ is implemented on Eq.(26) as follows. The left $SO(3)$ group of space rotations acts as usual on e_1 and e_2 and the right $SO(2)$ mixes the e_i together. However, contrary to the NLσ model, this model can never be invariant under a right $SO(3)$ group since it would have a *non linear* representation on e_1 and e_2: while a right rotation of axis e_3 acts linearly on e_1 and e_2, a right rotation of axis e_1 or e_2 generates both a current term Eq.(27) and monomial terms up to order 8 in e_1 and e_2 which are irrelevant terms[§] We shall see that this has dramatic consequences since, as we shall show in the next section, the $2 + \epsilon$ analysis of the NLσ model predicts a fixed point which is $SO(3) \otimes SO(3)$ symmetric. What we would like is therefore a linear theory that takes into account (at least partially) of terms like the current term of Eq.(27).

• *The three-vector model*

A possible way to do it is to keep the e_3 field. To this end, we replace the constraints Eq.(11) by another potential[12]:

$$V_2(e_1, e_2, e_3) = \frac{1}{2}m^2(e_1^2 + e_2^2 + e_3^2) + \frac{1}{4}\Lambda^{4-D}u_1\left(e_1^2 + e_2^2\right)^2 + \frac{1}{2}\Lambda^{4-D}u_2\left(e_1^2 e_2^2 - (e_1.e_2)^2\right)$$
$$+ \frac{1}{4}\Lambda^{4-D}u_5(e_3^2)^2 + \frac{1}{2}\Lambda^{4-D}u_4 e_3^2\left(e_1^2 + e_2^2\right) - \frac{1}{2}\Lambda^{4-D}u_3\left((e_1.e_3)^2 + (e_2.e_3)^2\right) \tag{28}$$

In a system which has either an $SO(2)$ or an $SO(3)$ right invariance we obtain a different LGW model which is also suitable for the study of frustrated model :

$$S_2 = \int d^D x \, \frac{1}{2} \left((\nabla e_1)^2 + (\nabla e_2)^2 + (\nabla e_3)^2 \right) + V_2(e_1, e_2, e_3) \tag{29}$$

Note that now the $SO(3)$ right group can be implemented linearly on Eq.(29). In particular, when $u_5 = u_1$, $u_2 = u_3$ and $u_4 = u_1 + u_2$, S_2 is $SO(3) \otimes SO(3)$ invariant. Indeed, the order parameter e_1, e_2, e_3 span the 9-dimensional *tensor* representation of $SO(4)$.

As seen, there is no unique way to write an effective LGW model for frustrated models. Both actions Eqs.(26,29) are, from standard arguments, good candidates

---

[§]while the action of the $SO(2)$ right rotation of axis e_3 acts linearly on e_1 and e_2, it is clear that a rotation of axis e_1 or e_2 has a non linear realization on e_1 and e_2 since $e_3 = e_1 \wedge e_2$

to account for the physics near $D = 4$. They differ only by the number of massive modes and by the representation of the symmetry group G spanned by the order parameter. However, the choice of one potential rather than the other one leads to different predictions near a non trivial fixed point. When such a fixed point exists, we expect that the NLσ model and the LGW models describe the same physics. There is a priori no guarantee that it is the case in general. We shall see in the following that precisely, it is the non existence of a fixed point in the LGW models, that makes the whole analysis of the frustrated models non trivial.

2.2.3. The large-N approaches

As in the unfrustrated case one can generalize both the LGW and NLσ actions to fields e_a having N components[27]. There are many way to generalize our model to N component fields. This fact is not particular to frustrated models since, as well known, even in the ferromagnetic case, there are two natural large N generalizations of $O(3)/O(2)$ which are the $O(N)/O(N-1)$ and the $Cp_{N-1} = SU(N)/SU(N-1) \otimes U(1)$ models. In the following, we restrict ourselves to models having for any N an $SO(2)$ right invariance. These are for the LGW model action S_1 of Eq.(26) with N component e_1 and e_2 fields. For the NLσ we take as Lagrangian density:

$$\mathcal{L} = \frac{1}{2}\Lambda^{D-2}\left(g_1\left(\partial e_1^2 + \partial e_2^2\right) + g_2\left(e_1\partial e_2 - e_2\partial e_1\right)^2\right), \qquad (30)$$

where e_1 and e_2 are two orthonomal N component fields. To obtain Eq.(30) from Eq.(10) we integrate over the e_3 field subject to the constraint $e_3 = e_1 \wedge e_2$. The kinetic term for the e_3 field yields in particular the current term Eq.(27). We then generalize to N component fields. The resulting model, is the NLσ model defined on the coset space $SO(N) \otimes SO(2)/SO(N-2) \otimes SO(2)$ which is associated to the linear action S_1 of Eq.(26).

The large N expansions of both the LGW and NLσ models arise when either:

- the three fourth order terms in V_1: $e_1{}^2e_2{}^2$, $(e_1e_2)^2$ and $(e_1{}^2+e_2{}^2)^2$ are decoupled via a Hubbard-Stratonovitch transformation. The resulting action is then expressed in terms of three Lagrange multipliers $\lambda_1, \lambda_2, \lambda_3$,

- the constraints in the NLσ model are rewritten as:

$$\delta(e_1{}^2 - 1)\delta(e_1{}^2 - 1)\delta(e_1.e_2) \rightarrow \int_{-i\infty}^{+i\infty} d\lambda_{1,2,3}\ e^{\lambda_1(e_1{}^2-1)+\lambda_2(e_2{}^2-1)+\lambda_3 e_1.e_2} \qquad (31)$$

In both large N expansions, a current decoupling gauge field A_μ has also to be introduced.

As seen, the large N expansion of frustrated models is considerably more complicated than that of ferromagnetic systems. We shall not get into the details in this chapter and refer the interested reader to reference[27].

2.2.4. Topological considerations

Let us conclude this subsection by topological considerations. Perturbation theory, of both NLσ and LGW models, is only sensitive to non singular field configurations of the order parameter. In some cases, there may exist singular field configurations that may be essential for the physics as it is the case for the $2D$ XY model. The existence of such singular configurations depends on the topological properties of the order parameter space and can be classified thanks to homotopy theory. In the case of frustrated systems, there exist both point defects in two dimensions and line defects in three dimensions which cannot be continuously deformed into the ground state (i.e. which are stable against spin wave interaction). This result stems from the fact that $\Pi_1(V(p)) = \Pi_1(SO(4)/O(3) \otimes Z_2) = Z_2$[28]. These defects are therefore labelled by an integer that can take two values $0, 1$. Field configurations whith charge 0 belong to the trivial class and can be continuously deformed to the ground state in contrast with field configurations whith charge 1 which constitute the relevant vortices. Note that, contrary to the XY case, vortices are their own anti-vortices since $1 + 1 = 0$ in Z_2. In all the following, the effect of vortices will be assumed to be irrelevant for the critical behavior. This fact is absolutely non trivial. Indeed, nonetheless a vortex anti-vortex unbinding process has been observed in the numerical simulations made on the $2D$ AFT model, but also recent works made on the nematic-paramagnetic transition in $D = 3$ have concluded that vortex lines are relevant for the critical behavior in this system[29]. Our strategy in all this chapter is to inquire all what can be said about the critical behavior of frustrated spin systems with no recourse to topological properties, hoping that there exist systems where vortices are indeed irrelevant. As we shall show, even without vortices, the critical behavior of frustrated systems is already highly non-trivial.

2.3. The Stacked Triangular Heisenberg Antiferromagnet

In the preceding subsection, we have identified, on general grounds, the order parameter of canted systems as a rotation matrix R of $SO(3)$ and we have written down the most general renormalizable action with such an order parameter. We found that it depends, in general on three couplings p_a, $a = 1, 2, 3$ and that relations between them define the isometries of the order parameter space. When dealing with a particular model such as the AFT model, we would like to relate both the order parameter R to the original spins of the lattice and also the isometries of the order parameter space to the symmetry properties of the lattice hamiltonian (1). This was done directly by taking the continuum limit from the lattice by Dombre and Read[25], who have found that $p_1 = \sqrt{3}J/4T$ and $p_3 = 0$. We find particulary illuminating to show how a group theoretical analysis enables to characterize the complete symmetry breaking pattern of the AFT model. In this case, the hamiltonian (1) is invariant under the space group of the triangular lattice and under the $SO(3)$ group of spin rotations. The effective coarse grained models must be symmetric under these two groups if we want to capture correctly the critical physics associated with the symmetry breaking scheme. The $SO(3)$ symmetry can be trivially implemented in the effective models and the real problem comes from the lattice symmetry.

Let us consider the elementary plaquette which is a triangle. The point group of the equilateral triangle is C_{3V} that has six elements: the identity, two rotations of angle $2\pi/3$ and $4\pi/3$ and the three reflections in the three bisectrices. This group has three irreducible representations, two of dimension one and one of dimension two. In the high temperature phase where the symmetry is not broken the spins of a plaquette are permuted under the action of C_{3V} so that the partition function has a C_{3V} invariance. The three spins in the elementary plaquette span a representation of C_{3V} which turns out to be reducible. The linear combinations of $\mathbf{S}_1, \mathbf{S}_2, \mathbf{S}_3$ that span respectively the unit trivial and the dimension two representations of C_{3V} are:

$$\Sigma = \mathbf{S}_1 + \mathbf{S}_2 + \mathbf{S}_3 \qquad (32)$$

$$\begin{pmatrix} \mathbf{e}_1 \\ \mathbf{e}_2 \end{pmatrix} \propto \begin{pmatrix} -\frac{\sqrt{3}+1}{2}\mathbf{S}_1 + \frac{\sqrt{3}-1}{2}\mathbf{S}_2 + \mathbf{S}_3 \\ \frac{\sqrt{3}-1}{2}\mathbf{S}_1 - \frac{\sqrt{3}+1}{2}\mathbf{S}_2 + \mathbf{S}_3 \end{pmatrix} \qquad (33)$$

In the ground state configuration Σ, is vanishing and has therefore a vanishing expectation value at any temperature. It cannot be an order parameter. It actually corresponds to relative motion of the spins inside a plaquette and represents modes which are never critical (massive modes). We ignore them by imposing the constraints that *locally*, i.e. within each unit cell, the spins are in the 120^0 ground state configuration:

$$\Sigma(x) = <\Sigma(x)> = 0. \qquad (34)$$

This is the "local rigidity" constraint introduced above. It allows fluctuations of the spins between cells but not within the cells. Since Σ spans the unit representation of C_{3V}, this constraint is invariant and, therefore, can be consistently imposed. On the other hand, it can be readily checked that $\mathbf{e}_1$ and $\mathbf{e}_2$ have a non vanishing value in the ground state so that they are the sought order parameter which is therefore a (rectangular) matrix and not only a vector. Let us emphasize that the rigidity constraint is *not* specific to the canted systems. In the colinear case, the elementary cell consists of two neighboring spins (S_i, S_{i+1}) and the local order parameter is the staggered magnetization $\phi(x_i) = S_i - S_{i+1}$. The analog of the mode Σ is $S_i + S_{i+1}$. The local rigidity constraint consists in imposing that this latter mode is zero, i.e. to forget it. This is the reason why the (classical) antiferromagnetic model is identical to the ferromagnetic one.

As we did in Eq.(19) we define:

$$\Phi = (\mathbf{e}_1, \mathbf{e}_2) \qquad (35)$$

After imposing the local rigidity constraint, we find that $\mathbf{e}_1$ and $\mathbf{e}_2$ are orthogonal unit vectors. We also define the vector $\mathbf{e}_3$ as the cross product of the two others $\mathbf{e}_3 = \mathbf{e}_1(x) \wedge \mathbf{e}_2(x)$. $(\mathbf{e}_1, \mathbf{e}_2, \mathbf{e}_3)$ form, on each plaquette, an orthonormal basis and therefore a rotation matrix which is the order parameter. It is also straightforward to see that since all the spins belonging to a given sublattice are colinear in the ground state and thus have an effective ferromagnetic-like interaction, the interactions between the axis of two neighboring orthonormal basis are also ferromagnetic: in the ground state all the vectors $e_1(x)$ for instance are parallel. *Thus we conclude that the critical behavior of the AFT is identical to that of a system of rigid bodies (tops)*[12]. This is different from the non frustrated case where the order parameter is the staggered magnetization and the critical behavior is equivalent to that of a system of *rigid rotators*. Depending on the lattice model, these tops can be either spheric, symmetric or asymmetric. Here again our group theoretical analysis will tell us which kind of top model we get. The implementation of the group action on (e_1, e_2) is straightforward since, by construction, C_{3V} is represented on (e_1, e_2) by the dimension two representation. For instance, the representations of the $2\pi/3$ or $4\pi/3$ rotations on e_1, e_2, e_3 are given by a right transformation identical to those of Eq.(17):

$$(e_1, e_2, e_3) \to (e_1, e_2, e_3) \begin{pmatrix} \cos\theta & \sin\theta & 0 \\ -\sin\theta & \cos\theta & 0 \\ 0 & 0 & 1 \end{pmatrix}, \tag{36}$$

and one of the reflection by:

$$(e_1, e_2, e_3) \to (e_1, e_2, e_3) \begin{pmatrix} 0 & 1 & 0 \\ 1 & 0 & 0 \\ 0 & 0 & -1 \end{pmatrix}. \tag{37}$$

Note that the e_3 vector spans the non trivial one dimensional representation of C_{3V}. We conclude that the two axis e_1 and e_2 are exchanged under C_{3V}, i.e. they must play an identical role in our effective models, and that the third axis is different from the two others. This means that our tops are symmetric. To summarize, once the local rigidity constraint is imposed, the AFT is equivalent to a system of symmetric tops that we can define on the center of each elementary plaquette and that interact ferromagnetically with the neighboring tops, see Fig.(2).

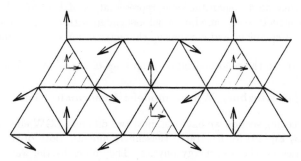

Fig. 2. The tops are drawn at the center of each elementary plaquette of the triangular lattice. The e_3 vectors are chosen perpendicular to the plane of the figure.

The symmetry group of this model is the usual rotation group times the symmetry group of the symmetric top, Eq.(36,37). This group is actually larger than C_{3V} since any left rotation like (36) leaves the top invariant. Thus the symmetry group of the high temperature phase of the top model is $G = O(3) \otimes O(2)$. To completely characterize the critical behavior we also need to know the symmetry breaking pattern of the AFT model.

In the low temperature phase, the $O(3)$ and the triangular lattice symmetries are spontaneously broken and to identify the remaining unbroken symmetry we have to choose one particular ground state and find out the subgroup of $O(3) \otimes O(2)$ that leaves this ground state configuration invariant. We choose:

$$\mathbf{e}_1(x) = \mathbf{e}_1^0 = (0, 1, 0) \quad , \quad \mathbf{e}_2(x) = \mathbf{e}_2^0 = (0, 0, 1) \tag{38}$$

The set of transformations that leave invariant this ground state configuration is given by the combined action of a left $O(2) \subset O(3)$ rotation and a right $O(2)$ with the opposite angle:

$$\begin{pmatrix} 0 & 0 & 0 \\ 0 & \cos\theta & -\sin\theta \\ 0 & \sin\theta & \cos\theta \end{pmatrix} \begin{pmatrix} 0 & 0 \\ 1 & 0 \\ 0 & 1 \end{pmatrix} \begin{pmatrix} \cos\theta & \sin\theta \\ -\sin\theta & \cos\theta \end{pmatrix} = \begin{pmatrix} 0 & 0 \\ 1 & 0 \\ 0 & 1 \end{pmatrix}. \tag{39}$$

This group is the diagonal group of the subgroup $SO(2)$ in $SO(3)$ generated by J_x and the right $SO(2)$.[¶] Thus the rotation group is completely broken and the symmetry breaking scheme is $G = SO(3) \otimes SO(2) \to H = SO(2)_{diag}$. It would be straightforward to generalize the previous analysis to the case where $G = O(3) \otimes O(p)$ with the results that the symmetry breaking scheme is always:

$$G = O(3) \otimes O(p) \to H = O(p)_{diag} \tag{40}$$

as already said in the previous section. The Goldstone modes of the triangular model span a representation of the unbroken $SO(2)_{diag}$ group: two of them form of vector of $SO(2)$ and the other one a scalar. The number of Goldstone modes together with the fact that they do not span the same representation of $H = SO(2)_{diag}$ ultimately determine the structure of the NLσ model associated with the AFT model.
We then have a complete and convenient characterisation of the symmetries of the AFT model
We now turn to the renormalization properties of the LGW model.

3. The Landau-Ginzburg-Wilson models: Renormalization

In the preceding section we have obtained two different LGW effective actions $\mathcal{S}_1$ Eq.(26) and $\mathcal{S}_2$ Eq.(29) for frustrated Heisenberg models. Both of them are natural effective theories for the low energy physics. They differ by the way the constraint

---

[¶]On the lattice, the low temperature symmetry is the diagonal C_{3v} that leaves the 120^0 structure invariant.

Eq.(11) are handled in perturbation theory and most importantly on the way one takes into account of operators like $(e_1 \partial e_2 - e_2 \partial e_1)^2$. The question of the relevance for the physics of such a term goes beyond the perturbation theory, and the best one can do is to investigate RG properties of both models to see if, and how, their critical behavior differ.

3.1. The two vector model

Action S_1 of Eq.(26), has been first extensively studied by Garel and Pfeuty[1] and then by Bailin, Love and Moore[2] in the late seventies. This action is invariant under the action of the left $SO(N)$ group that acts on the components of the e_a's and the action of the right $SO(2)$ group that mixes e_1 and e_2. The potential is bounded from below when $u_1 > 0$ and $2u_1 + u_2 > 0$ which defines the region of physical interest. When $u_2 = 0$, our model degenerates onto the standard $2N$ component $(e^2)^2$ theory relevant for non frustrated magnets. When $u_2 \neq 0$, the action S_1 is not trivial as we now show. Let us first perform the mean-field analysis. To do so we have to search for the minimum of the potential for constant values of the fields:

$$\begin{cases} \dfrac{\partial V}{\partial e_1} = 0 \\[2mm] \dfrac{\partial V}{\partial e_2} = 0 \end{cases} \tag{41}$$

which reads explicitly:

$$\begin{cases} \left[m^2 + 4u_1(e_1^2 + e_2^2) + 2u_2 e_2^2 \right] e_1 = 2u_2(e_1.e_2)e_2 \,, \\[2mm] \left[m^2 + 4u_1(e_1^2 + e_2^2) + 2u_2 e_1^2 \right] e_2 = 2u_2(e_1.e_2)e_1 \,. \end{cases} \tag{42}$$

The latter equations have solutions that depend on both the sign of m^2 and u_2. The case $u_2 > 0$ corresponds to a spin density wave model since, as it can be readily shown, Eqs.(42) are satisfied when e_1 and e_2 are parallel and is not relevant for our problem. In the following, we then restrict ourselves to the case where $u_2 < 0$. There are now two possibilities:

• $m^2 > 0$, and we are in the symmetric phase with:

$$\begin{cases} e_1 = e_2 = 0, \\[2mm] u_2 < 0. \end{cases} \tag{43}$$

- $m^2 < 0$, and the symmetry is spontaneously broken with:

$$\begin{cases} \mathbf{e}_1.\mathbf{e}_2 = 0, \\[2mm] |\mathbf{e}_1|^2 = |\mathbf{e}_2|^2 = \dfrac{-m^2}{8u_1 + 2u_2}, \\[2mm] u_2 < 0. \end{cases} \tag{44}$$

When $m^2 < 0$, we choose a ground state with $\mathbf{e}_1{}^0 = (e, 0, ..., 0)$ and $\mathbf{e}_2{}^0 = (0, e, ..., 0)$ and where e is given by Eq.(44). The spectrum in the symmetry broken phase is then obtained by diagonalizing the mass matrix:

$$\frac{\delta^2 V_1}{\delta e_a{}^0 \delta e_b{}^0} \tag{45}$$

and consists in $2N - 3$ Goldstone modes π and 3 massive modes σ_0, σ_1 and σ_2 that can be expressed in terms of the original e_a fields as:

$$\begin{cases} \sigma_0 = \dfrac{1}{\sqrt{2}}(e_{1x} + e_{2y}), \\[3mm] \sigma_1 = \dfrac{1}{\sqrt{2}}(e_{1x} - e_{2y}), \\[3mm] \sigma_2 = \dfrac{1}{\sqrt{2}}(e_{2x} + e_{1y}), \\[3mm] \pi_0 = \dfrac{1}{\sqrt{2}}(e_{2x} - e_{1y}), \\[3mm] \pi_{a,\alpha} = e_a{}^\alpha, \quad \alpha > 2, a = 1, 2. \end{cases} \tag{46}$$

The corresponding masses for these σ modes are:

$$\begin{cases} m_0{}^2 = -2m^2, \\[2mm] m_1{}^2 = m_2{}^2 = 2m^2 \dfrac{u_2}{4u_1 + u_2} \end{cases} \tag{47}$$

thus, at the mean-field level, the action S_1 describes a second order transition between a paramagnetic and a helical phase as $m^2 \sim T - T_c$ is varied. Corrections to mean-field behavior can be calculated, as in the standard $\mathbf{e}^4$ theory, in a double expansion in u_1, u_2 and $\epsilon = 4 - D$. The two loop recursion relations for u_1 and u_2 are:

$$\begin{cases} \dfrac{\partial u_1}{\partial l} = \epsilon u_1 - C\left[2(N+4){u_1}^2 + 2(N-1)u_1 u_2 + (N-1){u_2}^2\right] \\[2mm] \quad -\dfrac{1}{2}C^2\left[-10(N-1){u_2}^3 - 39(N-1)u_1{u_2}^2 - 44(N-1){u_1}^2 u_2 - 12(3N+7){u_1}^3\right] \\[3mm] \dfrac{\partial u_2}{\partial l} = \epsilon u_2 + C\left[(N-6){u_2}^2 - 12 u_1 u_2\right] \\[2mm] \quad +\dfrac{1}{2}C^2\left[-(9N-29){u_2}^3 - 4(N-31)u_1{u_2}^2 + 4(5N+41){u_1}^2 u_2\right] \end{cases}$$

$$(48)$$

where $C = 2^{1-D}/[\pi^{D/2}\Gamma(D/2)]$.

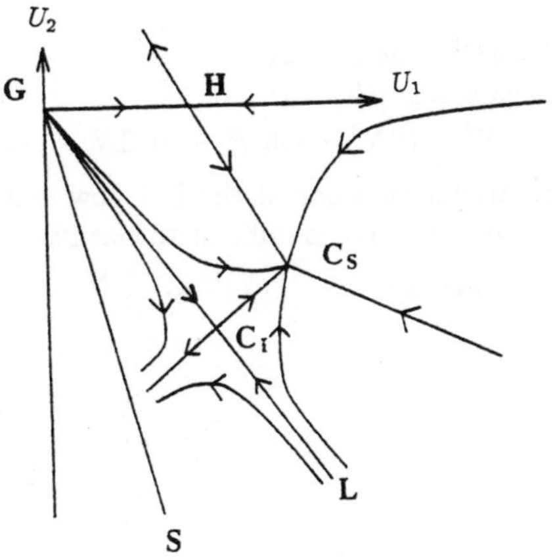

Fig. 3. Renormalization group flow near $D = 4$ for the two-vector LGW model for $N > N_c$.

Apart from the trivial gaussian fixed point G at $u_1{}^* = u_2{}^* = 0$, Eqs.(48) admit an $O(2N)$ fixed point H at $u_2{}^* = 0$ as expected. This fixed point has one direction of instability, which means that the u_2 symmetry breaking term is relevant. Now, depending on the value of N, i.e. when $N > N_c$, there may exist additional fixed points $C_I(N, \epsilon)$ and $C_S(N, \epsilon)$. While the former is always unstable, the latter is stable and thus governs a second order phase transition. A detailed analysis shows that N_c is a function of ϵ:

$$N_c(\epsilon) = 21.8 - 23.4\epsilon + O(\epsilon^2) \qquad (49)$$

The behavior of the renormalization group flow thus depends on both D and N as follows:

• when $N < N_c(\epsilon)$ there is no fixed point for $u_2 \neq 0$ and the transition is expected to be first order

• when $N > N_c(\epsilon)$ there are still two different regions in the portion of the parameter space where the potential is stable. One is the second order region. It lies above the line L joining the origin to C_I and is the basin of attraction of the fixed point C_I. The other one lies between L and the stability line S of the potential: $u_2 = -2u_1$. It is a region of runaway and is expected to correspond to a first order region. One can further show that as N tends to infinity, L tends to S, as in the spherical model. We show in Fig.(3) the flow diagram corresponding to Eqs.(48).

At the second order transition the critical exponents ν and η are:

$$
\begin{cases}
\nu = \dfrac{1}{2} + \dfrac{1}{8}A_N[N(N^2 + N + 48) + B_N(N + 4)(N - 3)]\epsilon \\
\eta = \dfrac{1}{8}A_N\left[C_N + (N + 4)(N - 3)B_N{}^3\right]\epsilon^2
\end{cases}
\tag{50}
$$

with:

$$
\begin{cases}
A_N = 1/N^3 + 4N^2 - 24N + 144, \\
B_N = (N^2 - 24N + 48)^{1/2}, \\
C_N = 5N^2 - 3N^4 - 16N^3 - 656N^2 + 3072N - 1152.
\end{cases}
\tag{51}
$$

As a consequence of the non trivial behavior of the RG flow, the situation of the physically interesting $N = 3$ and $\epsilon = 1$ case is difficult to clarify.

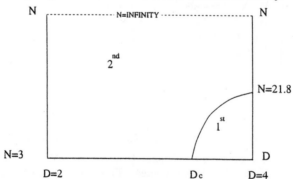

Fig. 4. The critical curve $N_C(D)$ separating a first order region at small N from second order one at large N in the (N, D) plane.

As suggested by Eq.(49) we expect N_c to be a function of the dimension D that separates a first order region from a second order one governed by the fixed point $C_S(N, D)$. Of course, the determination of the curve $N_c(D)$ goes far beyond perturbation theory and one may only conjecture two different qualitative behaviors. First, if the curve $N_c(D)$ is always above the $N = 3$ axis, then the transition is always of first order for any D. The interesting possibility, is that there exists a critical dimension D_c such that $N_c(D_c) = 3$. As seen on Fig.(4), for all dimensions $D < D_c$ the transition should be of second order and governed by the fixed point $C_S(3, D)$. Naivley, one would expect that the fixed point $C_S(3, D)$ can be obtained with the two vector LGW model by some adequate resumation procedure. If this is the case, it should be $SO(3) \otimes SO(2)$ symmetric and therefore should correspond to a new universality class. This is the scenario proposed by Kawamura[3] to account for the continuous transition found in his Monte-Carlo simulation of the $3D$ stacked AFT model. Now the β-function of the two vector model has been calculated to *three loop* order in $D = 3$ by Antonenko et Sokolov[30]. By using adequate Padé-Borel resumation of their results, they found no fixed point for $N = 3$. This result is in agreement with an other calculation by Zumbach[31] who found no fixed point in the full Wilsonian[20] RG approach within the local potential approximation. These two results together strongly suggest that if it is likely that a stable fixed point $C_S(3,3)$ exists when $D = N = 3$ it cannot be obtained pertubatively with the two vector LGW model and are in disfavour of a new universality class.

Since as already anticipated, a stable fixed point exists near $D = 2$ which is $SO(3) \otimes SO(3) \sim SO(4)$ symmetric, the absence of a fixed point in S_1 could be the consequence that it cannot be $SO(4)$ symmetric except if one has recourse to irrelevant terms.

3.2. The three vector model

We now turn to the study of action S_2 of Eq.(29). Let us first generalize S_2 to N-component fields having N components. There is a notable difference between S_1 and S_2 when $N > 3$. While S_2 is still left $SO(N)$ symmetric, the set of N-component vectors $(\mathbf{e}_1, \mathbf{e}_2, \mathbf{e}_3)$ decompose, under the action of the right $SO(2)$ symmetry, into two irreducible representations, one of dimension 2 spanned by $(\mathbf{e}_1, \mathbf{e}_2)$ and a one dimensional representation spanned by $\mathbf{e}_3$. Therefore, the LGW model S_2 couples two irreducible representations of the right $SO(2)$ symmetry. The mean field analysis can be done in the same way as that for S_1, and for suitable values of the u_i's one finds a symmetry breaking phase where $(\mathbf{e}_1, \mathbf{e}_2, \mathbf{e}_3)$ are orthogonal vectors. This ground state is left invariant under the action of the subgroup $SO(N-3) \otimes SO(2)$, which means that the symmetry breaking pattern is $SO(N) \otimes SO(2) \rightarrow SO(N-3) \otimes SO(2)$. As a consequence of Goldstone theorem, the spectrum in the symmetry broken phase consists in $3N - 6$ Goldstone modes together with 6 massive modes. Both models, even for $N = 3$, differ by their number of massive modes. Let us finally emphasize that, as it was the case for $N = 3$, when $u_1 = u_5$, $u_2 = u_3$ and $u_4 = u_1 + u_2$ the symmetry of the action S_2 is enlarged to $SO(N) \otimes SO(3)$ so that the symmetry breaking pattern becomes therefore $SO(N) \otimes SO(3) \rightarrow SO(N-3) \otimes SO(3)$.

We have obtained the one loop recursion relations for the couplings entering in Eq.(29):

$$\begin{cases} \dfrac{\partial u_1}{\partial l} = \epsilon u_1 - C\left[2(N+4)u_1{}^2 + (N-1)u_2{}^2 + 2(N-1)u_1u_2 + Nu_4{}^2 - 2u_3u_4 + u_3{}^2\right] \\[2.5ex] \dfrac{\partial u_2}{\partial l} = \epsilon u_2 + C\left[(N-6)u_2{}^2 - 12u_1u_2 + u_3{}^2\right] \\[2.5ex] \dfrac{\partial u_3}{\partial l} = \epsilon u_3 + C\left[(N+2)u_3{}^2 + u_2u_3 - 2u_1u_3 - 8u_3u_4 - 2u_3u_5\right] \\[2.5ex] \dfrac{\partial u_4}{\partial l} = \epsilon u_4 - C\left[4u_4{}^2 + u_3{}^2 + 2(N+1)u_1u_4 - 2u_1u_3 - u_2u_3 \right. \\[1ex] \qquad\qquad \left. +(N-1)u_2u_4 + (N+2)u_4u_5 - u_3u_5\right] \\[2.5ex] \dfrac{\partial u_5}{\partial l} = \epsilon u_5 - C\left[(N+8)u_5{}^2 + 2Nu_4{}^2 + 2u_3{}^2 - 4u_3u_4\right] \end{cases}$$

$$(52)$$

The analysis of the latter RG equations is as follows:

• when $u_3 = u_4 = 0$, Eqs.(52) become identical to Eqs.(48) and the discussion of the RG scaling proceeds as before.

• when $u_{3,4} \neq 0$, Eqs.(52) admit, apart from the gaussian fixed point, an $SO(3N)$ fixed point which is again unstable. Like in the preceding model, there exists a critical value N_{c2} below which no stable fixed point exists. When $N > N_{c2}$, there exists a fixed point at $u_1{}^* = u_5{}^*$, $u_2{}^* = u_3{}^*$ and $u_4{}^* = u_1{}^* + u_2{}^*$ where the symmetry is *enlarged for any* N to $SO(N) \otimes SO(3)$. To leading order we find $N_{c2} \sim 32.5$ which is larger than N_{c1} and we again have the same problem of extrapolating our results to $N = 3$ and $D = 3$. Focusing on the latter fixed point, we are led, in the same spirit as for the the preceding LGW model, to conjecture the existence of a critical dimension D_{c2} for which $N_{c2}(D_{c2}) = 3$. Again, when $N = 3$, the transition will be of second order for all $D < D_{c2}$ but this time it can be $SO(3) \otimes SO(3) \sim SO(4)$ symmetric. This is very different from what happens in the two vector model S_1, i.e. when $u_{3,4} = 0$, which can never be $SO(4)$ symmetric for $N = 3$ except if one relies on irrelevant terms such as the current term of Eq.(27). It is important to stress that perturbatively one cannot take into account of current-like terms, like that of Eq.(27), since they are not renormalizable, or equivalently irrelevant, near the gaussian fixed point. What we have seen, when studying our three vector model, is that in fact, one can take care in a renormalizable way of such terms (at least partially), by introducing an additional $\mathbf{e}_3$ field and by coupling it to the original $\mathbf{e}_1$, $\mathbf{e}_2$ fields via a quartic potential. In a certain sense, one may think of the $\mathbf{e}_3$ field as a decoupling, Hubbard-Stratonovich-like, field for the current term.

While the three vector LGW model has the suitable symmetry properties, one is still faced with the problem of the existence of a stable fixed point at $D = 3$ and for $N = 3$. In the simplest hypothesis, one may hope to see it by using adequate Padé-Borel resumation of the perturbation series, as for example what was done with the two vector model[30]. Such results are still lacking, and one must wait before drawing any definitive conclusions. Let finally stress that the problem may be even worse, and that other operators than the current term may be relevant for the physics in $D = 3$ so that even the three vector model may be not the relevant low energy effective action for frustrated magnets.

3.3. Concluding remarks

To summarize, we have found that for both LGW models we have studied, there exists a critical $N_{c,1,2}$ below which no stable fixed point exists. The simplest hypothesis, is that the transition is always of first order between $D = 4$ and $D = 2$.

However, as suggested above, there exists the possibility that a second order transition occurs in both models. When $N = 3$ and $D = 3$, we found that the two vector model, which is unable to account for the $SO(4)$ symmetry, does not possess a stable fixed point even by using more elaborate calculations than the ϵ expansion. On the other hand, we found that, when $N > N_{c2}$ the three vector model has an enlarged symmetry at the fixed point which is $SO(N) \otimes SO(2) \rightarrow SO(N) \otimes SO(3)$ and allows for the $SO(4)$ symmetry when $N = 3$. Although it looks quite unnatural to couple two irreducible representations[32], the three vector model is a more suitable effective theory than two vector one, from the symmetry point of view. However one still has no indication that its fixed point persists down to $D = 3$ and $N = 3$.

Both models differ in the way the constraint of Eq.(11) are taken into account or equivalently, by the nature of the low energy excitations in the symmetric phase they describe: an $SO(2)$ right doublet $[e_1, e_2]$ in the two vector model and an $SO(2)$ right doublet together with an $SO(2)$ right scalar $[(e_1, e_2), e_3]$ in the three vector model.

In the absence of any physical insight on the nature of these relevant low energy excitations one has no other choice than making an ansatz, which should be justified by the existence of a non trivial infrared fixed point. We saw that no fixed point is seen in perturbation theory in both models when $N = 3$. Although one has still to wait for more improved calculations with the three vector model, one may already suspect that the nature of the low energy fluctuations of frustrated models can be non trivial. As discussed in section 2., there is an alternative effective field theory which is the NLσ model. It describes the physics of the Goldstone modes resulting from the breaking of the $SO(3)$ group and no ansatz is required for the spectrum in the symmetric phase. Moreover, the current term of Eq.(27) is found to be renormalizable in a double expansion in $\epsilon = D - 2$ and temperature. We shall see in the following how the repeatedly anticipated $SO(4)$ symmetry will emerge from the RG analysis.

Before turning to the RG properties of the NLσ model, let us first review rapidly how the LGW and NLσ models are related in a more precise way than in Eq.(4) where the constraints were simply replaced by a potential. There are different methods to proceed. One possible method is to show that the correlations coincide in a $1/N$ expansion. This shows, at least for sufficiently large N that either model can be used. Unfortunately, we already know that the qualitative behavior of the correlations change drastically with N in our case because of the existence of a N_c. The other method is to show directly on the partition function the equivalence in the low temperature phase of the two models in the large mass limit of the massive sigma modes. The idea is to decompose the $e_i(x)$ fields of the LGW models on a orthonormal basis at x which are the fields of the NLσ model. Let us call $\rho_i^a(x)$ the components of $e_i(x)$ on the orthonormal basis. They are the massive σ modes contrary to the fields of the orthonormal basis. To relate the LGW models with the NLσ model we have to integrate on the $\rho_i^a(x)$. The integration over these massive fields leads to an effective action for the remaining massless modes that can be expanded in terms of local terms as long as we are interested in distances L much larger than the inverse mass of the ρ_i^a fields. For T close to T_c, this means $L \gg \xi \sim (T - T_c)^{-\nu}$. This effective action is the action of the NLσ model plus terms involving four derivatives that are irrelevant in any dimensions between two and four. Therefore, for very large distances the two models are equivalent. We are of course interested in the equivalence of the two models in a much more problematic region, namely the critical domain where $a \ll L \ll \xi$. In this domain we can only compare the results of the two different perturbative approaches and see if they are consistent. We shall see in the following that they are not always.

4. The non linear sigma model: Renormalization

As we have already said the NLσ model is obtained from (12) by integrating out as much modes as possible and by relaxing the modulus constraints on the three remaining modes. The action on the remaining unconstrained Goldstone modes reads (see Eq.(24)):

$$S = \frac{1}{2} \int d^D x g_{ij}(\pi) \partial_\mu \pi^i(x) \partial_\mu \pi^j(x) \tag{53}$$

where $g_{ij}(\pi)$ is the *metric* on G/H. This is is of considerable consequence since it implies that the dynamical properties of the model are completely determined by the geometry of the order parameter space which is itself determined by the group structures of G and H, contrary to the LGW model. Therefore we can already guess that a completely geometric and group theoretical language instead of a conventional field theoretical one will be powerful for the study of the NLσ model. In particular, this geometric language has at least one fundamental advantage on the other formulations: it allows to worry only about intrinsic, i.e. coordinate independent, quantities.

Let us now study in detail the renormalization of this model.

4.1. Renormalization of the NLσ model in $D = 2 + \epsilon$

The NLσ model is just renormalizable in two dimensions and the parameter expansion is the temperature. It is thus necessary to perform a double expansion in the temperature and in $\epsilon = D - 2$ if we want to study the three dimensional system.

The renormalizability in $D = 2 + \epsilon$ of Non Linear Sigma (NLσ) models defined on general homogeneous coset spaces G/H was studied by D.H.Friedan[33] and by Becchi et al.[34] some years ago. They showed that the renormalization properties depend only on geometrical quantities of the manifold G/H viewed as a metric space. In particular, the β function, which gives the evolution of the coupling constants as a function of the scale, is given, at two loop order, in terms of the Ricci and Riemann tensors of the manifold. As a consequence, renormalization depends only on *local* properties of G/H and is insensitive to its global structure (topological properties).

Considerations based on covariance show that the only possible couterterms resulting from the loopwise expansion can only involve tensors. Friedan has obtained so the first two terms of the β function for the bare metric in dimension $D = 2 + \epsilon$:

$$\beta_{ij} = -\epsilon g_{ij} + \frac{1}{2\pi}R_{ij} + \frac{1}{8\pi^2}R_i{}^{klm}R_{jklm} \tag{54}$$

where R_{ij} and R_{jklm} are the Ricci and Riemann tensors and indices are raised with used of the inverse metric g^{ij}.

The important point, is that contrary to the $4 - \epsilon$ expansion, the β function, and therefore the mere existence of fixed points, depends only on the geometry of G/H and thus is independent of the representation r of the group G spanned by the order parameter. This is not the case for the field renormalization Z which depends explicitly on r.

All these renormalization properties, when formulated within the framework of differential geometry, can be further shown to be entirely determined by the Lie algebras of G and H and more precisely by their structure constants defined by the following commutation rules :

$$\begin{aligned}
[T_a, T_I] &= f_{aI}{}^J T_J, \\
[T_a, T_b] &= f_{ab}{}^c T_c, \\
[T_I, T_J] &= f_{IJ}{}^K T_K + f_{IJ}{}^a T_a,
\end{aligned} \tag{55}$$

where $T_a \in \text{Lie}(H)$ and $T_I \in \text{Lie}(G)\text{-Lie}(H)$.

It is therefore extremely convenient to work directly on geometric quantities, i.e. to get rid of any dependence on the coordinate system π^i. This is done by going from the manifold itself to its tangent space. The point is that it is possible to find a basis in the tangent space (called the vielbein basis) where both Riemann and Ricci tensors are functions only of the structure constants of $\text{Lie}(G)$ and where the tangent space metric η_{IJ} is constant. In tangent space, Eq.(54) becomes:

$$\beta_{IJ} = -\epsilon\eta_{IJ} + \frac{1}{2\pi}R_{IJ} + \frac{1}{8\pi^2}R_I^{PQR}R_{JPQR} \tag{56}$$

where the Riemann tensor in tangent space can be expressed as :

$$R_{IJKL} = f_{IJ}{}^a f_{aKL} + \frac{1}{2}f_{IJ}{}^M \left(f_{MKL} + f_{LMK} - f_{KLM}\right)$$

$$+\frac{1}{4}\left(f_{IKM}+f_{MIK}-f_{KMI}\right)\left(f_{J}{}^{M}{}_{L}+f_{LJ}{}^{M}-f^{M}{}_{LJ}\right)$$

$$-\frac{1}{4}\left(f_{JKM}+f_{MJK}-f_{KMJ}\right)\left(f_{I}{}^{M}{}_{L}+f_{LI}{}^{M}-f^{M}{}_{LI}\right). \tag{57}$$

The indices a and $\{I, J \ldots\}$ refer to H and and $G-H$ respectively. $G-H$ indices are raised and lowered by means of η^{IJ} and η_{IJ} and repeated indices are summed over.

Eq.(57) is the central perturbative result of the renormalization of general NLσ models defined on compact homogeneous coset spaces. As a most important point , it shows, that there *always* exists, in the vinicity of dimension two, a non trivial infrared fixed point that governs a second order phase transition from ordered to disordered phases. However, it is worth stressing that this result is based on the asumption that at sufficiently low temperature, the physics, in perturbation theory, is entirely controlled by spin wave excitations or Golstone modes. In particular, it does not take into account the global structure of the manifold and its topological properties. Therefore, one must take care when one tries to extend results obtained so far at finite distance from two dimensions. Indeed, this is precisely the case of frustrated Heisenberg models since as dicussed in the preceding section, they have non trivial topological properties and since one knows, that at least in the neighbourhood of four dimensions, the massive modes may have non trivial effects on the phase transition.

Like any perturbation theory, the $\epsilon = D - 2$ expansion has it own possible source of failure, and one has to compare its predictions with other approaches to get a complete picture of the physical properties of a given system and it is worth stressing that in the case of frustrated systems, it allows to make qualitative as well as quantitative predictions that, contrary to the $4 - \epsilon$ expansion, can be compare with numerical data as well as experimental results. In addition, this formalism offers a very nice framework in which scaling properties depend only on abstract symmetry properties and on the dimension D. Although this approach is very powerful and esthetically appealing it may obscur, for people not familiar with differential geometry and abstract group theory, the physical mechanism of renormalization. This is why we shall consider, in the first subsection, the renormalization of general frustrated Heisenberg models in a more concrete and physical approach using what field theorists call the background field method. In the second subsection, generalization to arbitrary spin components as well as two loop results will be presented with use of the formalism described above. Finally, in the third subsection we give the results obtained from the large N expansion.

4.2. The $SO(3)$ model

In this section, we consider the one loop renormalization of general NLσ models with an order parameter in $SO(3)$. These are $O(3) \otimes O(p)/O(p)$ models with arbitrary diagonal P matrix as defined in the preceding section. We derive the basic RG functions for these models paying particular attention to the computation. Although the results presented in this subsection are limited to the one loop order, they shall enable us to draw general conclusions about the physics of frustrated Heisenberg models.

The aim of the following calculation is to show how one loop calculations can be easily performed in the NLσ model and with a very intuitive approach: the background field method (see Polyakov[35]). The idea is to separate the field into a "classical" part obeying the (classical) equations of motion and a fluctuating part and to compute the modification of the mean field action when one integrates over the fluctuating fields representing the thermal (or quantum, in the quantum case) fluctuations around the classical configuration. This modification amounts to the divergent renormalization of the metric, i.e. of the coupling constants. We can therefore obtain the beta function by this method.

The Lagrangian density of the NLσ model is:

$$\mathcal{L}(R) = -\frac{1}{2}\Lambda^{d-2}Tr[P(R(x)^{-1}\partial_\mu R(x))^2] \tag{58}$$

with $\mathcal{S}(R) = \int d^D x \mathcal{L}(R)$ the action. Consider a finite system with boundary Σ and impose the following boundary condition:

$$R(x) = R_\Sigma, \quad x\epsilon\Sigma. \tag{59}$$

We focus on the physical quantity:

$$\Xi(R_\Sigma) = \int_{R(x\epsilon\Sigma)=R_\Sigma} DR(x) \ e^{-\mathcal{S}(R)}, \tag{60}$$

The functional $\Xi(R_\Sigma)$ gives the free energy in a finite system as a function of the boundary conditions. The classical or zero temperature limit corresponds to taking the minimum of the action with the boundary condition (59). Inclusion of quantum or thermal fluctuations results from the decomposition $R(x) = R_{cl}(x)h(x)$ where $R_{cl}(x)$ is solution of the classical equation of motion with the boundary condition (59):

$$\partial_\mu \frac{\delta\mathcal{L}}{\delta(\partial_\mu R)} = \frac{\delta\mathcal{L}}{\delta R}. \tag{61}$$

We can rewrite Eq.(60) as:

$$\Xi(R_\Sigma) = \Xi(R_{cl}(x)) = \int_{h(x\epsilon\Sigma)=I} Dh(x) \ e^{-\mathcal{S}(R_{cl}h)}, \tag{62}$$

where $h(x)$ is now subject to the boundary condition:

$$h(x) = I, \quad x\epsilon\Sigma, \tag{63}$$

I being the identity matrix. Note that the Ξ functional is really a functional of $R_{cl}(x)$ since there is a one to one correspondance between $R_{cl}(x)$ and R_Σ because of the Dirichlet theorem. The systematic loopwise expansion arises when one writes:

$$h(x) = e^{\varphi(x)} \sim 1 + \varphi(x) + \frac{1}{2}\varphi(x)^2 + \dots \tag{64}$$

where $\varphi(x)$ belongs to the Lie algebra of $SO(3)$. The nice thing with (62) is that we do not have to bother with the precise solution of (61) as far as we are concerned with the renormalization properties.

4.2.1. The coupling constant renormalization

To proceed further, we need the currents associated with the $SO(3)$ symmetry. For any matrix $R(x)$ belonging to $SO(3)$, $R(x)^{-1}\partial_\mu R(x)$ belongs to the Lie algebra of $SO(3)$ so one may write:

$$R(x)^{-1}\partial_\mu R(x) = \omega_\mu = \omega_\mu^a T_a \tag{65}$$

where the T_a's are generators of Lie$(SO(3))$ satisfying the well known commutation relations, $[T_a, T_b] = \epsilon_{ab}{}^c T_c$ with $\epsilon_{12}{}^3 = 1$, and are normalized such that $Tr(T_a T_b) = -2\delta_{ab}$. With these currents one can express the action $\mathcal{S}(R)$ as:

$$S(R) = \frac{\Lambda^{D-2}}{2} \int d^D x \ \eta_{ab} \ \omega_\mu^a \omega_\mu^b, \tag{66}$$

where η_{ab} is the metric in the vielbein basis:

$$\eta_{ab} = -\text{Tr}(PT_a T_b) = \eta_a \delta_{ab}, \tag{67}$$

with $\eta_1 = p_2 + p_3$, $\eta_2 = p_3 + p_1$ and $\eta_3 = p_2 + p_1$. To compute the one loop correction to the classical action we need to keep only the first terms in (64). Substituting (64) into (62) we get:

$$\Xi(\omega_{\mu,cl}) = \exp -\mathcal{S}(R_{cl}) \int D\varphi \ \exp -\Lambda^{D-2} \int d^D x \left[\frac{1}{2}\eta_{ab} \ \partial_\mu\varphi^a \partial_\mu\varphi^b + \delta L_1 + \delta L_2 \right]. \tag{68}$$

where

$$\delta\mathcal{L}_1 = \frac{1}{2} \left[f_{abc}(\varphi^a \partial^\mu \varphi^b - \varphi^b \partial_\mu \varphi^a) - g_{abc}(\varphi^a \partial_\mu \varphi^b + \varphi^b \partial_\mu \varphi^a) \right] \ \omega_{\mu,cl}^c,$$

$$\delta\mathcal{L}_2 = \frac{1}{2} V_{abcd} \ \varphi^a \varphi^b \ \omega_{\mu,cl}^c \omega_{\mu,cl}^d, \tag{69}$$

and

$$f_{abc} = \frac{1}{2} \ (\epsilon_{ab}{}^d \eta_{dc} + \epsilon_{ca}{}^d \eta_{db} + \epsilon_{bc}{}^d \eta_{da}),$$

$$g_{abc} = \frac{1}{2} \ (\epsilon_{ac}{}^d \eta_{db} + \epsilon_{bc}{}^d \eta_{da}), \tag{70}$$

$$V_{abcd} = \frac{1}{2} \ Tr[\ (2T_a P T_b - PT_a T_b - T_a T_b P)T_c T_d \ ].$$

The contributions $\delta\mathcal{L}_1$ and $\delta\mathcal{L}_2$ depend on the symmetry properties of the model. When $P \propto I$, there is only one coupling, the model is $O(3) \otimes O(3)$ symmetric and $\eta_{ab} = \eta\delta_{ab}$, so that both g_{abc} and V_{abcd} vanish. In all other cases, there are at least two different coupling constants and both terms contribute. In Eq.(68) the first term just gives the classical contribution to $\Xi(R_{cl})$. There are no terms linear in φ since they vanish as a consequence of the equation of motion Eq.(61). The remaining terms which are quadratic in φ give, after integration, a contribution which is quadratic in $\omega_{\mu,cl}$. From simple power counting, it is easy to see that this contribution is of order Λ^{D-2} in the limit $\Lambda \to \infty$ and therefore contributes to the renormalization of η_{ab}. There are two divergent terms resulting from integration over φ. We obtain:

$$\Xi(\omega_{\mu,cl}) = \exp - \int d^D x \frac{\Lambda^{D-2}}{2} \left[\eta_{ab} - \frac{R_{ab}}{2\pi\epsilon} \right] \omega^a_{\mu,cl}\omega^b_{\mu,cl} + \text{finite terms} \qquad (71)$$

where $\epsilon = D - 2$ and R_{ab} is the diagonal Ricci tensor in tangent space given by:

$$R_{11} = \frac{\eta_1^2 - (\eta_2 - \eta_3)^2}{2\eta_2\eta_3}$$

$$R_{22} = \frac{\eta_2^2 - (\eta_3 - \eta_1)^2}{2\eta_3\eta_1} \qquad (72)$$

$$R_{33} = \frac{\eta_3^2 - (\eta_1 - \eta_2)^2}{2\eta_1\eta_2}$$

This defines the relation between the bare and the renormalized metric at scale μ:

$$\mu^\epsilon \eta_{ab,r} = \Lambda^\epsilon \eta_{ab} - \frac{\Lambda^\epsilon}{2\pi\epsilon} R_{ab}, \qquad (73)$$

from which we deduce the β function:

$$\beta_{ab} = \Lambda \frac{\partial}{\partial\Lambda} \eta_{ab} = -\epsilon\eta_{ab} + \frac{1}{2\pi} R_{ab} \qquad (74)$$

This is the expected result for the beta function, Eq.(56). It can be explicitly checked that R_{ab} is the Ricci tensor of the manifold G/H. We now turn to the field renormalization that requires some more work.

4.2.2. The field renormalization

In general, in NLσ models a choice of coordinates π^i is not stable under renormalization. This means that since the canonical dimension of the field is zero in two dimensions and since the physical quantities are insentitive to the parametrization of the manifold, a field renormalization consists in general not only in a (divergent) amplitude redefinition of the fields but also in a reparametrization of the manifold, i.e. a non linear transformation of the coordinate system. Among all the possible coordinate systems, the ones that renormalize multiplicatively, i.e. that are stable under renormalization, are those that consist in the π^i field together with the massive modes σ: they build up a linear representation of the group G. The order parameter $R(x) = (\mathbf{e_1}(\mathbf{x}), \mathbf{e_2}(\mathbf{x}), \mathbf{e_3}(\mathbf{x}))$ itself, of course renormalizes multiplicatively since it transforms linearly under the action of the left $SO(3)$ group. However, when the coupling matrix P is arbitrary, there is no additional right symmetry, and the $\mathbf{e_a}(\mathbf{x})$ fields renormalize independently with different renormalization constants z_a. The relation between bare and renormalized fields can be written as:

$$\mathbf{e_a}(\mathbf{x}) = \mathbf{e_{ar}}(\mathbf{x})z_a^{\frac{1}{2}}, \tag{75}$$

which can be written into a matrix form as:

$$R(x) = R_r(x)Z^{\frac{1}{2}}, \tag{76}$$

where Z is the diagonal renormalization matrix $Z = \text{diag}(z_1, z_2, z_3)$.

To get the field renormalization we have to calculate the correlation function:

$$\Delta[K; x - y]_\Sigma = < Tr[KR^{-1}(x)R(y)] >_\Sigma, \tag{77}$$

where K is some arbitrary matrix. One can obtain from Eq.(77) all the two point correlation functions between the $\mathbf{e_a}(\mathbf{x})$'s with particular choices of K. In Eq.(77), the average value $< ... >_\Sigma$ is taken in a finite system with the boundary conditions (59). The calculation proceeds in the same spirit as that for Ξ. At leading order we obtain:

$$\Delta[K; x - y]_\Sigma = Tr(\Theta\Gamma(K; x - y)\Theta R_{cl}^{-1}(x)R_{cl}(y)) \tag{78}$$

where η^{ab} is the inverse tangent space metric and

$$\Gamma(K; x - y) = K - \Lambda^{-\epsilon}\eta^{ab} \ T_a K T_b \int \frac{d^D k}{(2\pi)^D} \frac{e^{ik(x-y)}}{k^2}$$

$$\Theta = I + \frac{1}{2}\Lambda^{-\epsilon}\eta^{ab} \ T_a T_b \int \frac{d^D k}{(2\pi)^D} \frac{1}{k^2} \tag{79}$$

In order to obtain Eq.(78), we have used that at this order:

$$< \varphi^a(x)\varphi^b(y) >_\Sigma = \Lambda^{-\epsilon}\eta^{ab} \int \frac{d^D k}{(2\pi)^D} \frac{e^{ik(x-y)}}{k^2}. \tag{80}$$

Using Eq.(76) we identify the Z ‖matrix as:

$$Z = \Theta^2 = I + (\Lambda^{-\epsilon}\eta^{ab})\frac{\Lambda^\epsilon}{2\pi\epsilon} \ T_a T_b. \tag{81}$$

‖At this order, $\Lambda^{-\epsilon}\eta^{ab} = \mu^{-\epsilon}\eta_r{}^{ab}$ and is kept fixed in the renormalization process.

This equation makes explicit that Z is indeed an operator which acts on the coordinates. This is a general result.[**]

When there is some additional right invariance as it is the case, for example, for $O(3) \otimes O(2)$ models, Eq.(81) can be written in a suggestive form:

$$Z = I + \frac{\Lambda^\epsilon}{2\pi\epsilon} \left(\frac{\Lambda^{-\epsilon}}{\eta_1} (\Sigma T_a^2) + \left(\frac{\Lambda^{-\epsilon}}{\eta_3} - \frac{\Lambda^{-\epsilon}}{\eta_1} \right) T_3^2 \right) \tag{83}$$

where the inverse metric is: $\eta^{ab} = (1/\eta_1, 1/\eta_1, 1/\eta_3)$ as a result of the $O(2)$ right invariance generated by T_3, and ΣT_a^2 and T_3^2 are the Casimir operators of both $SO(3)$ and $O(2)$ groups. This result generalizes to arbitray coset spaces of the form $G \otimes X/H \otimes X$ where X is the maximal subgroup of G that commutes with H. In this case, ΣT_a^2 and T_3^2 will be the Casimir operators of both G and X groups. We shall use this result in the next subsection when analyzing $O(N) \otimes O(2)/O(N-2) \otimes O(2)$ models.

Returning to our $SO(3)$ model we can obtain from Eq.(81) the γ Callan-Symanzyk matrix function:

$$\gamma = -\Lambda \frac{\partial}{\partial \Lambda} \log Z = -\frac{1}{2\pi} \eta^{ab} T_a T_b, \tag{84}$$

which is diagonal: $\gamma = \text{diag}(\gamma_1, \gamma_2, \gamma_3)$ with:

$$\gamma_1 = \frac{1}{2\pi} \left(\frac{1}{\eta_2} + \frac{1}{\eta_3} \right)$$

$$\gamma_2 = \frac{1}{2\pi} \left(\frac{1}{\eta_3} + \frac{1}{\eta_1} \right) \tag{85}$$

$$\gamma_3 = \frac{1}{2\pi} \left(\frac{1}{\eta_1} + \frac{1}{\eta_2} \right).$$

4.2.3. The renormalization group equations

Using the β and γ functions one can obtain the scaling equations for the bare correlation functions which are also the correlation functions of statistical mechanics:

$$\Delta_I(p, \eta_{ab}, \Lambda) = < e_I(0).e_I(p) > \tag{86}$$

---

[*]In the language of differential geometry, the factor Z is, at one loop order, given by the eigenvalue of the Laplace-Beltrami operator acting on the coordinates π^i. It can be shown that this operator is nothing but $g^{ij}\Gamma_{ij}^k$ where Γ_{ij}^k is the Christoffel connection on the metric manifold G/H:

$$Z\pi^k = \pi^k + \frac{\Lambda^{-\epsilon}}{2\pi\epsilon} (\Lambda^\epsilon g^{ij}) \Gamma_{ij}^k. \tag{82}$$

The only fields that renormalize multiplicatively are those that transform under a linear representation of the left $SO(3)$ group. The order parameter R itself transforms according to the representation of dimension 1 but we could have, as well, calculated the field renormalization of operators transforming under an other irreducible representation.

They satisfy the RG equation:

$$\left(\Lambda\frac{\partial}{\partial\Lambda} + \beta_{ab}\frac{\partial}{\partial\eta_{ab}} + \gamma_I\right)\;\Delta_I(p,\eta_{ab},\Lambda) = 0,\tag{87}$$

which can be solved by the method of characteristics by:

$$\Delta_I(p,\eta_{ab},\Lambda) = Z_I(\lambda)\;\Delta_I(p,\eta_{ab}(\lambda),\lambda\Lambda)\tag{88}$$

where:

$$\frac{\partial\eta_{ab}(\lambda)}{\partial\log\lambda} = \beta_{ab}\;,\quad \eta_{ab}(1) = \eta_{ab},$$

$$\frac{\partial\log Z_I(\lambda)}{\partial\log\lambda} = -\gamma_I\;,\quad Z_I(1) = 0,\tag{89}$$

$\eta_{ab}(1)$ being the metric at the lattice scale Λ^{-1}. These RG equations are valid, as usual, in the critical regime defined by, $\xi^{-1} \ll p \ll \Lambda$, where ξ is the correlation length satisfying the homogeneous equation :

$$\left(\Lambda\frac{\partial}{\partial\Lambda} + \beta_{ab}\frac{\partial}{\partial\eta_{ab}}\right)\;\xi = 0.\tag{90}$$

The phase diagram is obtained by solving the flow equations Eq.(89). Apart from the trivial zero temperature fixed point at $\eta^{ab*} = 0$, Eq.(89) admits a non trivial fixed point with $R_{ab}^* = 1/2\;\delta_{ab}$. At the latter fixed point, one has $\eta_{ab}^* = \dfrac{\delta_{ab}}{4\pi\epsilon}$ and $P^* = 1/8\pi\epsilon \mathrm{diag}(1,1,1)$ so that the model becomes $O(3) \otimes O(3)/O(3) \sim SO(4)/O(3)$ symmetric[11,12]. In the space η_1,η_2,η_3, there is a whole two dimensional critical surface that constitutes the basin of attraction of this $SO(4)$ fixed point. Furthermore, it has only one direction of instability and thus governs an ordinary second order phase transition. *What we have shown is that all models with an order parameter in $SO(3)$ belongs to the same $SO(4)$ Wilson-Fisher universality class.* Stated in the spin language, this result means, that at the fixed point, the system escapes in the space of spin dimension for freeing itself from strong local correlations induced by frustration. This is the very result of our analysis. We thus find no new universality class for frustrated Heisenberg models but meet the general phenomenon of increased symmetry at a fixed point.

From Eq.(88), one can obtain the asymptotic scaling of correlation functions near the $SO(4)$ fixed point:

$$\Delta_I(p,\eta_{ab}^*,\Lambda) \sim p^{-2+\eta},\tag{91}$$

where η is the anomalous dimension of the e_I field:

$$\eta = \gamma_I^* - \epsilon = 3\epsilon.\tag{92}$$

As a result of the increased symmetry at the $SO(4)$ fixed point, all the e_I fields have the same anomalous dimension $\eta = 3\epsilon$ which is different from that of a $N = 4$ unit vector in the $SO(4)$ model (which is $\eta' = \epsilon/2$). The reason for this is that the order parameter $R(x) = (e_1(x), e_2(x), e_3(x))$ spans the *tensor* representation of $SO(4)$. As a consequence, the exponent η of frustrated Heisenberg model is the anomalous dimension of a *composite* operator of the $N = 4$ vector model. To see this, we need the relationship between the $O(3)$ matrix R and a $SO(4)$ unit vector. It can be shown that to any unit 4-component vector:

$$\Psi = (\Psi_0, \Psi_i) \quad ; \quad \Psi_0^2 + \sum_i \Psi_i^2 = 1 \tag{93}$$

there exists a matrix R of $O(3)$ with components:

$$R_{ij} = 2(\Psi_i \Psi_j - \frac{1}{4}\delta_{ij}) + 2\epsilon_{ijk}\Psi_0\Psi_k + 2(\Psi_0^2 - \frac{1}{4})\delta_{ij} \tag{94}$$

Therefore, the expectation values of the vectors $< e_i(x) >, i = 1, 3$ are obtained from those of the *bilinear* forms $< (\Psi_i \Psi_j - \frac{1}{4}\delta_{ij}) >$.

To summarize, we have obtained that, in the vicinity of $D = 2$, frustrated Heisenberg models belongs to the well known $SO(4)$ universality class but in the tensor representation. In dimension $D = 3$, the exponent ν is very accurately known for the $N = 4$ model from $4 - \epsilon$ results: $\nu = 0.74$. However, the anomalous dimension of the composite operator $(\Psi_i \Psi_j - \frac{1}{4}\delta_{ij})$ is only known at the two loop order in $\epsilon = 4 - D$ and no accurate value is known at the time being.

4.3. $O(N) \otimes O(2)/O(N-2) \otimes O(2)$ models and two loop results

In this subsection we compute the two loop RG properties of general $O(N) \otimes O(2)/O(N-$ model as defined in the first section. We shall use the formalism described in () and show, on our concrete example, how abstract group theory provides a natural and powerfull framework for the renormalization of general NLσ models.

Let recall the action of our model Eq.(30):

$$S = \frac{\Lambda^{D-2}}{2} \int d^D x \left(g_1 \left(\partial e_1^2 + \partial e_2^2\right) + g_2 \left(e_1 \partial e_2 - e_2 \partial e_1\right)^2 \right), \tag{95}$$

where e_1 and e_2 are two orthonomal N component vectors that constitute the order parameter. As in the $SO(3)$ model, we want to write Eq.(95) in terms of the currents of the theory. Amongs all the currents associated to the $O(N)$ symmetry, only the $2N-3$ currents belonging to Lie$(O(N))$ -Lie$(O(N-2))$ are relevant. These are the one associated with the spin wave fluctuations and span the tangent space of the manifold $O(N) \otimes O(2)/O(N-2) \otimes O(2)$. Once the action Eq.(95) is written in terms of the currents and the metric coupling η_{ab} is identified one can obtain the Riemann tensor and therefore the β function by using Eqs.(56). To do so, it is convenient to define the rectangular matrix :

$$\Phi = (e_1, e_2). \tag{96}$$

The $O(N) \otimes O(2)$ transformations can be written:

$$^t\Phi' = {}^t r(x) \, {}^t\Phi \, {}^t R(x), \tag{97}$$

where $R \in O(N-2)$ and $r \in O(2)$. Once a ground state Φ^0 is chosen, one can parametrize all the fluctuations in terms of group elements of $O(N)$ and $O(N-2)$ with:

$$^t\Phi = {}^t r(x) \, {}^t\Phi^0 \, {}^t R(x), \tag{98}$$

However, since the ground state Φ^0 is invariant under the transformations of the little group $O(N-2) \otimes O(2)$:

$$^t\Phi^0 = h_1(x)^t\Phi^0 H(x), \tag{99}$$

with $h_1 \in O(2)$ and

$$H(x) = \begin{pmatrix} h_2(x) & 0 \\ 0 & h_1^{-1}(x) \end{pmatrix}, \qquad h_2 \in O(N-2) \tag{100}$$

the matrices $r(x)$ and $R(x)$ are defined up to the following local transformations:

$$\begin{cases} r(x) & \to & r(x)\, h_1(x) \\ R(x) & \to & R(x)\, H(x) \end{cases} \tag{101}$$

As a consequence, in the low temperature phase, we can rewrite Φ in terms of the $2N-3$ Goldstone modes as :

$$\Phi(x) = \begin{pmatrix} \pi(x) \\ \omega(x)\sqrt{1_2 - {}^t\pi\pi} \end{pmatrix}, \tag{102}$$

where π is a $(N-2) \times 2$ matrix and $\omega(x) \in O(2)$. The π_i^α, $i = 1, .., N, \alpha = 1, 2$ transform as two independent vectors under $O(N-2)$ and as a vector under $O(2)$. $\omega(x)\sqrt{1_2 - {}^t\pi\pi}$ represents one extra degree of freedom which is scalar under both $O(N-2)$ and $O(2)$. One can use the gauge freedom Eq.(101) to go from a general element $R(x) \otimes r(x)$ of $O(N) \otimes O(2)$ to the unique element in the same gauge orbit $L \otimes 1_2$:

$$L(\pi(x), \omega(x)) = \begin{pmatrix} \sqrt{1 - \pi \, {}^t\pi} & \pi \\ -\omega(x)^t\pi & \omega(x)\sqrt{1 - {}^t\pi\pi} \end{pmatrix}. \tag{103}$$

The matrix L thus parametrizes the coset space $O(N) \otimes O(2)/O(N-2) \otimes O(2)_{diag}$. Since $L^{-1}\partial_\mu L$ belongs to the Lie algebra of $O(N)$ we can write:

$$L^{-1}\partial_\mu L = \omega_\mu{}^I T_I + \omega_\mu{}^\alpha T_\alpha \tag{104}$$

where T_I belongs to Lie$(O(N))$ - Lie$(O(N-2))$ - Lie$(O(2))$ and T_α belongs to Lie$(O(2))$. The currents are then defined as:

$$\begin{aligned} \omega_\mu{}^I &= -\tfrac{1}{2}Tr(L^{-1}\partial_\mu L T_I), \\ \omega_\mu{}^\alpha &= -\tfrac{1}{2}Tr(L^{-1}\partial_\mu L T_\alpha). \end{aligned} \tag{105}$$

For example, the $O(2)$ current can be written in terms of the **e**'s fields as:

$$\omega_\mu{}^\alpha = -\frac{1}{2}(\mathbf{e_1}\partial_\mu\mathbf{e_2} - \mathbf{e_2}\partial_\mu\mathbf{e_1}) \qquad (106)$$

In terms of the currents one can write Eq.(95) as:

$$S = \frac{1}{2}\int d^D x \eta_1(\omega_\mu{}^I)^2 + \eta_2(\omega_\mu{}^\alpha)^2. \qquad (107)$$

The point is that the Lagrangian Eq.(95) becomes $O(2)$ left *gauge* invariant when $g_2 = -1/2g_1$. Indeed, when the latter condition is satisfied, our model becomes the Grassmanian model $O(N)/O(N-2)\otimes O(2)_{diag}$ which is symmetric. In this case, the metric is diagonal with $\eta_{IJ} = \eta_1\delta_{IJ}$, where I belongs to Lie$(O(N))$ - Lie$(O(N-2))$ - Lie$(O(2))$ and one can identify $\eta_1 = g_1$ from what follows $\eta_2 = 2g_1 + 4g_2$. We can thus write the metric as:

$$\eta_{ab} = \eta_1\delta_{aI}\delta_{bI} + \eta_2\delta_{a\alpha}\delta_{b\alpha}, \qquad (108)$$

where the α indices refers to Lie$(O(2))$. From the group theoretical point of view one can understand Eq.(107) by noticing that the tangent space of $O(N)/O(N-2)$ decomposes into two irreducible representations under the action of $O(N-2)\otimes O(2)$. $O(2)$ itself spans the adjoint representation of $O(2)$ and is a scalar under $O(N-2)$. $O(N) - O(N-2) - O(2)$ is irreducible because $O(N-2)\otimes O(2)$ is maximal in $O(N)\otimes O(2)$, stated otherwise $O(N)/O(N-2)\otimes O(2)_{diag}$ is a symmetric space. Thus, the two projected matrices $(L^{-1}\partial L)_{|O(N)-O(N-2)-O(2)}$ and $(L^{-1}\partial L)_{|O(2)}$ transform independently under the right action of the $O(N-2)\otimes O(2)$ group so that there are two independent couplings η_1 and η_2.

Using Eq.(56) we obtain the following two loop recursion relations valid for any $N \geq 3$:

$$\begin{cases} \dfrac{\partial\eta_1}{\partial l} = -\epsilon\eta_1 + N - 2 - \dfrac{1}{2}\dfrac{\eta_2}{\eta_1} + \dfrac{3N-4}{8}\dfrac{\eta_2^2}{\eta_1^3} + 3(1 - \dfrac{N}{2})\dfrac{\eta_2}{\eta_1^2} \\[2mm] \qquad\qquad + (3N - 8)\dfrac{1}{\eta_1} \\[4mm] \dfrac{\partial\eta_2}{\partial l} = -\epsilon\eta_2 + \dfrac{N-2}{2}\left(\dfrac{\eta_2}{\eta_1}\right)^2 + \dfrac{N-2}{8}\dfrac{\eta_2^3}{\eta_1^4} \end{cases} \qquad (109)$$

In the latter equations, a factor 2π has been reabsorbed in the definition of the coupling constants. Defining $T_{1,2} = 1/\eta_{1,2}$ we find that, apart from the trivial zero temperature line of fixed points: $T_1 = T_2 = 0$ with T_1/T_2 arbitrary there is one non trivial fixed point C_{NL} with coordinates:

$$\begin{cases} T_1^* = \dfrac{N-1}{(N-2)^2}\left(\epsilon - \dfrac{1}{2}\dfrac{3N^2 - 10N + 4}{(N-2)^3}\epsilon^2\right) + O(\epsilon^3) \\[4mm] T_2^* = \dfrac{1}{2}\dfrac{(N-1)^2}{(N-2)^3}\left(\epsilon - \dfrac{1}{2}\dfrac{5N^2 - 16N + 4}{(N-2)^3}\epsilon^2\right) + O(\epsilon^3) \end{cases} \qquad (110)$$

This fixed point has one direction of instability so that our model undergoes an ordinary second order phase transition with critical exponent ν:

$$\nu^{-1} = \epsilon + \frac{1}{2}\frac{6N^3 - 27N^2 + 32N - 12}{(N-2)^3(2N-3)}\epsilon^2 + O(\epsilon^3) \tag{111}$$

In order to complete our discussion, we have to specify the representation r of $O(N) \otimes O(2)$ spanned by the observable of the physical system under study. We are interested in the AFT model with N-component spins. In this case, the order parameter transforms under the vector representation of both $O(N)$ and $O(2)$, see Eq.(97). At one loop, it follows from Eq.(83) and Eq.(84) that the anomalous dimension η is:

$$\eta = \frac{3N^2 - 10N + 9}{2(N-2)^3}\epsilon + O(\epsilon^2) \tag{112}$$

4.4. The large-N results

To complete our RG study of frustrated models, we need, as discussed in section 2.2.1, the results of the large N expansion of both $O(N) \otimes O(2)/O(N-2) \otimes O(2)$ LGW and NLσ models. We shall not reproduce here the details of the calculations but shall rather give the results and refer the interested reader to the relevant references[36,3,27]. In both large N expansions, one finds a stable fixed point for all dimensions $2 < D \leq 4$, a fact that is consistent with the existence of a critical $N_c(D)$ in the LGW model. Moreover, at the fixed point, both LGW and NLσ models display the same critical exponents. For example, the exponent ν is given to lowest non trivial order in $1/N$ by:

$$\nu_{1/N}(D) = \frac{1}{D-2}\left(1 - \frac{1}{ND}12(D-1)S_D\right), \tag{113}$$

$$S_D = \frac{\sin\left(\pi(D-2)/2\right)\Gamma(D-1)}{2\pi\Gamma(D/2)^2}. \tag{114}$$

By expanding Eq.(113) to lowest order in ϵ, we find that $\nu_{1/N}$ coincides with $\nu_{4-\epsilon}(N)$ (Eq.(50)) and $\nu_{2+\epsilon}(N)$ (Eq.(111)) to lowest order in $1/N$. The same type of expansion can be done on the other exponents with the same results.

We may thus conclude as in the ferromagnetic case that, when the fixed point exists near $D = 2$ and near $D = 4$, we can follow it smoothly from $D = 4 - \epsilon$ down to $D = 2 + \epsilon$.

5. Interpolating between $D = 2 + \epsilon$ and $D = 4 - \epsilon$

As we have seen, there is clearly a mismatch between the RG results obtained in $D = 4 - \epsilon$ and in $D = 2 + \epsilon$ so that, so far, we are not able to conclude for the physically interesting case $D = 3$, $N = 3$. As we shall see, as soon as we take into account the different results for all values of N and D we are able, under a minimal number of assumptions, to draw a consistent picture for the physics as both N and D vary.

Let us first summarize the situation. In the preceding sections we have found that:

- when $N = 3$, both ϵ expansions disagree. While a first order transition near $D = 4$ was predicted by the LGW analysis of both two and three vector models, the results obtained from the NLσ approach strongly suggests that the transition should be of second order and in the $SO(4)$ universality class near $D = 2$.

- when N is sufficiently large, i.e. when $N > N_{c,1,2}(D)$, one finds a stable infrared fixed point in both LGW models suggesting that the mismatch between both ϵ expansions does not persist when N is large. This is indeed what happens in the two-vector LGW model and the $SO(N) \otimes SO(2)$ NLσ model. Although results for the $SO(N) \otimes SO(3)$ NLσ model are still lacking, there is no doubt that this should be also the case for the three vector LGW model.

We shall now focus on the $SO(N) \otimes SO(2)$ NLσ and two vector LGW models. In this case, we found in the preceding section that the critical exponents obtained either from the $\epsilon = 4 - D$ expansion of the two-vector model, the $\epsilon = D - 2$ expansion of the NLσ and the large N expansion agree. We can therefore conclude that, as it is the case in colinear systems, there exists a unique fixed point $C(N, D)$ that can be followed smoothly, when N is sufficiently large, between $D = 2$ and $D = 4$. This fixed point coincides with the one found in perturbation theory of both $SO(N) \otimes SO(2)$ NLσ and two-vector LGW models when N is sufficiently large near respectively $D = 2$ and $D = 4$. This fact naturally leads us to draw the following picture[24,12].

In the whole space $E = \{(M) = \text{coupling constants}, D, N\}$ there should exist a domain Z where the transition is of second order and which is governed by the unique fixed point $C(N, D)$. In the complementary of Z the transition is expected to be of first order. On the boundary Γ between these two domains the transition should be tricritical in the simplest hypothesis. This is the scenario we propose to account for all the different perturbative results. Note that we do not require additional fixed points which are not seen in perturbation theory, in contrast with an other proposal[3]. We only require that the physical quantities are continuous as a function of D, N and M. Note that we cannot specify what is the relevant coupling space M on the basis of perturbation theory. The point is that we expect the boundary Γ to be a non universal domain which depends on the different microscopic Hamiltonians. The reason for this is that no universality is expected in tricritical phenomena. It is clearly the goal of future investigations to determine the nature of Γ, M and Z and in particular the relevant directions that flow away from the tricritical fixed point. Needless to say, such an investigation goes far beyond standard pertubation theory, and some new physical insight is clearly needed.

One can summarize the situation in the plane (N, D) of the number of components and dimension. The $4 - \epsilon$ findings in section 3.1 have shown that there is a curve $N_c(D)$ separating a second order region from a first order one. If one believes that the $2 + \epsilon$ results survive perturbation theory then the neigborhood of $D = 2$ belongs to the second order region for all $N \geq 3$. As a consequence, the line $N_c(D)$ should intersect the $N = 3$ axis somewhere between $D = 2$ and $D = 4$. This defines a critical dimension D_c that we do not expect to be universal. In this case, the transition should be of the $SO(4)$ universality class for all $D < D_c$. At this point one may object that, as we have repeatedly stated, the two vector model cannot be $SO(4)$ symmetric. This is true in pertubation theory where one takes into account only renormalizable terms. If a continous $SO(4)$ transition is likely to occur for $D < D_c$, we expect naively irrelevant terms to become important and nothing prevents the $SO(4)$ symmetry to be realized. We have already discussed such a possibility when studying the three vector model. Making the hypothesis that perturbation theory has captured all the relevant fixed points we can distinguish now, for a given microscopic model and when $N = 3$, $D = 3$, between three different possibilities:

• The critical value D_c is between $D = 3$ and $D = 4$. This implies that the physical case $N = 3$ and $D = 3$ always undergoes an $SO(4)$ transition.

• The critical value D_c is between $D = 2$ and $D = 3$. Then the physical case undergoes a first order transition induced by fluctuations. It cannot be excluded that actually $D_c = 2$ in which case, the perturbation theory of the NLσ model would be irrelevant.

• The critical value $D_c = 3$. In this case, a tricritical mean-field like behavior could be seen for $N = 3$.

We stress that if $D_c \sim 3$, it should be difficult experimentally or numerically to distinguish between a tricritical transition and either an $SO(4)$ or first order behavior. In particular, if $D_c \sim 3^-$, the transition could be extremely weakly first order and a minimum in the renormalization group flow could fake a true fixed point[31]. The same behavior is expected if $D_c \sim 3^+$: even though the model lies in the basin of attraction of the $SO(4)$ fixed point, $SO(4)$ behavior may be extremely difficult to observe. Finally, let us add for completeness, that one may observe all different behaviors by tuning appropriately some parameter on a given lattice model since we expect D_c to be non universal.

We are now in a position to discuss Monte Carlo results on STA and BCT lattices with Heisenberg spins. As mentioned above, both models are found to undergo a continuous transition with critical exponents[37,38,39] $\nu \sim 0.59$ and $\gamma \sim 1.17$[†]that are clearly different from those of the $N = 4$ universality class[40] ($\nu = 0.74$, $\gamma = 1.47$) and from tricritical mean field exponents ($\nu = 0.5$, $\gamma = 1.$). Within our scenario we are led to conclude that both STA and BCT models lie near the boundary Γ separating the first order region from the basin of attraction of the $SO(4)$ fixed point. In this case it should be difficult to observe either $SO(4)$ scaling or a weakly first order transition within accessible lattice sizes and Monte Carlo steps. It should be noted that a different interpretation of these results has been made by Kawamura[3]. He conjectured a new universality class for frustrated magnets which is defined by the critical exponents found in the MC simulations. From the theoretical point of

---

[†]For more details, see the chapter of Plumer et al. in this book.

view, this would imply the existence of a new fixed point that cannot be seen in perturbation theory. Although one cannot exclude such a possibility, we find that it is highly speculative and does not rely, as seen in the preceding sections, on any theoretical calculation. In this respect, the results of Antonenko and Sokolov[30] and Zumbach[31] make very hard to believe that the two-vector LGW model belongs to the hypothetisized new universality class.

To our knowledge, no evidence of $SO(4)$ behavior has been observed up to now. However, we stress that one has to be careful in concluding on the only basis of numerical results until one can rely on a well established finite size scaling theory. Advances in this direction will be more than welcome. Let us finally stress that our scenario is consistent with all the perturbative results and does not require additional fixed points not seen in any perturbative analysis. In this sense, it is minimal and is the only reasonable scenario that can be drawn from the theoretical analysis. The whole picture relies of course on the hypothesis that the perturbation theories can be trustred at least in the vicinity of the critical dimensions $D = 2$ and $D = 4$. In any cases, there should be some values of the couplings where either the NLσ or LGW approaches are irrelevant and where pertubation theory breaks down at some finite distance from either $D = 2$ or $D = 4$. The question that thus remains is to inquire what can be responsible for one or the other perturbative approach to fail. As discussed precedently, we make important assumptions with both LGW and NLσ models that one has now to discuss further.

Concerning the NLσ model, we have neglected so far terms of order $\exp(-1/T)$ which take into account of the global structure of the manifold $SO(3)$. These are precisely these terms that take into account of the nature of the massive modes and of the representation r of $SO(3)$ spanned by the order parameter. Such terms may be relevant for the physics. It has been argued recently[41,42] that models with continuous symmetry cannot disorder within the $\epsilon = D - 2$ expansion. The reason advocated is that at a given order in the ϵ expansion, an infinite number of high gradient irrelevant operators become relevant at sufficiently high temperature. The increase of such an infinite number of terms would not allow the spins to deviate from a fixed direction. In addition, Halperin[43] and next Cardy and Hamber[44] argued that topological excitations are necessary to disorder $O(N)$ models when $N \leq D$. The extrapolation of these arguments at a finite distance from $D = 2$ is unclear to us. However, in the frustrated case, where one may expect that the ϵ expansion may fail for sufficiently small values of N, irrelevant operators may play a crucial role. This may be even worse in the $N = 3$ case where, as already said in the preceding sections, there exist stable topological defects in both $D = 2$ and $D = 3$. We shall return latter on this problem.

In the LGW approach, one takes into account explicitly of the representation r of the order parameter but one makes the assumption that the collective excitations responsible for the phase transition belong to the same representation r. From simple power counting, one finds that the couplings of the interaction terms in either the two or three vector models scale in the infrared as $s^{D-4}, s \sim 0$, so that we have a strongly interacting theory in the low energy limit. Therefore, there is no guarantee that the relevant long distance degrees of freedom are still those that enter in the bare Lagrangian. In colinear systems, the situation is safe because one finds a fixed point in the RG recursion relation. This is not the case in frustrated models, so that one may not exclude the possibility that we are not using the correct fields to describe the low energy physics. If the transition is likely to be of second order, one may thus expect that a different LGW theory for these modes can be written. It has been recently suggested, that if topological excitations were irrelevant, the relevant massive collective excitations could span the $S = 1/2$ representation of $SO(3)$ rather that the vector one[45,46]. These are the spinons excitations $z = (z_1, z_2)$ that span the fundamental representation of $SU(2)$. The effective action for these modes writes:

$$S = \int d^d x \left[\frac{\Lambda^{d-2}}{2f_1} \partial_\mu z^\dagger \partial_\mu z + \frac{\Lambda^{d-2}}{2f_2} \left(z^\dagger \overset{\leftrightarrow}{\partial}_\mu z \right)^2 \right], \tag{115}$$

where the z field is subject to the constraint: $z^\dagger z = 1$. This defines the NLσ model $SU(2) \otimes U(1)/U(1)$ which only differs from the $SO(3) \otimes O(2)/O(2)$ model by its topological properties. As a consequence, as exposed in section 4., it shares the same renormalization properties in the $\epsilon = D - 2$ expansion. The corresponding LGW model is obtained as ususal by replacing the constraint by a potential $V(z) \sim (z^\dagger z - 1)^2$. Now, simple power counting arguments indicate that again the current term $\left(z^\dagger \overset{\leftrightarrow}{\partial}_\mu z \right)^2$ is irrelevant near $D = 4$. It can then be easily shown that this LGW model reduces to that of the standard $N = 4$ vector model. Therefore, provided that the relevant critical excitations are spinons, we have a completely consistent picture of the critical physics between $D = 2$ and $D = 4$ as it is the case for colinear systems and thus we expect an $SO(4)$ transition in $D = 3$. This hypothesis is far from being trivial and in particular it supposes that topological excitations are irrelevant since we did not distinguish between $SO(3)$ and $SU(2)$.

As seen, both LGW and NLσ effective low energy theories suffer from basic approximations that one cannot control safely. In the absence of more elaborate theoretical tools, very few can be said on the reasons of the mismatch between the different approaches. However, in the Heisenberg case, when $N = 3$, we found that topological excitations are natural candidates for the breakdown of perturbation theory. The question of the relevance of topological excitations in frustrated Heisenberg models becomes thus of central importance. While it is now difficult to state about their relevance in three dimensions, they may be relevant in $D = 2$ since a phase transition of the Berezinsky-Kosterlitz-Thouless (BKT) type, resulting from the unbinding of the Z_2 vortices cannot be excluded[28]. One may thus seriously question the relevance of the NLσ approach at least in $D = 2$. It is the object of the next section to inquire the low temperature physics of two dimensional frustrated systems.

6.   Frustrated models in two dimensions

As discussed in the preceding section, one of the possible reason for the failure of perturbation theory is the presence of stable topological excitations. This is what happens in the XY model where the renormalized spin wave theory alone is unable to predict a massive phase. In $D = 2$, it is sufficient to have recourse to the vortex unbinding process to account for mass generation. This is the celebrated Berezinsky-Kosterlitz-Thouless (BKT) phase transition. Let us emphasize that the $O(2)$ NLσ model correctly describes the physics at sufficiently low temperature (in the vortex condensate phase) in predicting an infinite correlation length and the correct temperature dependence of the anomalous dimension η.

In non abelian NLσ models the correlation length ξ is predicted to be finite and there is no need to have recourse to topological excitations to account for mass generation as it is the case for example of the $O(3)/O(2)$ model. However in the $SO(3)$ model, since the first homotopy group of $SO(3)$ is non trivial: $\Pi_1(SO(3)) = Z_2$, these systems have topological stable point defects (vortices) in $D = 2$ which may affect the physics in a non trivial way. This seems to happen in the Antiferromagnetic Heisenberg model on the triangular lattice[28] where a two phase structure was observed in Monte-Carlo simulations. Associated to the liberation of these Z_2 vortices at some temperature T_o, a BKT type phase transition cannot be excluded. If a genuine BKT phase transition occurs, as in the $O(2)$ case, one expects that the correlation length diverges in the condensate phase which would invalidate perturbation theory even at very low temperature. Although this might be possible, it is more reasonable to think that the vortex unbinding process is a smooth crossover rather than a phase transition and that the low temperature phase, $T \ll T_o$, is correctly described by the $SO(3)$ NLσ model. Therefore it is of crucial interest to test the validity of the NLσ model approach. This is usually done by measuring the correlation length ξ in Monte-Carlo simulations. However, since the transition may take place at very low temperature it may be rather difficult to measure ξ while the spin stiffness should remain an measurable quantity at low temperature.

In the following, we compute both the two loop correlation length[18] and the finite size dependence of the spin stiffness[47] of $O(3) \otimes O(2)/O(2)$ NLσ model which are relevant for AFT model and Helimagnets.

In this case, there are only two different couplings p_1 and p_3 in the matrix $P = (p_1, p_1, p_3)$ so that the effective action is $O(3) \otimes O(2)$ invariant. The corresponding metric tensor η_{ab} has also only two different components: $\eta_{ab} = \text{diag}(\eta_1, \eta_1, \eta_3)$, where $\eta_1 = p_1 + p_3$ and $\eta_3 = 2p_1$. The two loop recursion equations are:

$$\begin{cases} \dfrac{\partial \eta_1}{\partial \log \lambda} = \dfrac{1}{2\pi}\beta_1^{(1)} + \left(\dfrac{1}{2\pi}\right)^2 \beta_1^{(2)} \\[4mm] \dfrac{\partial \eta_2}{\partial \log \lambda} = \dfrac{1}{2\pi}\beta_2^{(1)} + \left(\dfrac{1}{2\pi}\right)^2 \beta_2^{(2)} \end{cases} \qquad (116)$$

where the $\beta_i^{(j)}$ are the β functions for the i component η_i at the $j^{th} = 1, 2$ order in the loop expansion which can be obtained from Eq.(109):

$$\begin{cases} \beta_1^{(1)} = 1 - \dfrac{1}{2}\dfrac{\eta_3}{\eta_1} \\[2ex] \beta_3^{(1)} = \dfrac{1}{2}\left(\dfrac{\eta_3}{\eta_1}\right)^2 \\[2ex] \beta_1^{(2)} = \dfrac{5}{8}\dfrac{\eta_3^2}{\eta_1^3} - \dfrac{3}{2}\dfrac{\eta_3}{\eta_1^2} + \dfrac{1}{\eta_1} \\[2ex] \beta_3^{(2)} = \dfrac{1}{8}\dfrac{\eta_3^3}{\eta_1^4} \end{cases} \tag{117}$$

Note that we have reestablished the factor 2π since it is relevant in the following. It is convenient to introduce a new set of variables: $\alpha = \eta_3/\eta_1 - 1$ and η_3. At one loop, the recursion equations read, when expressed in terms of our new variables:

$$\begin{cases} \dfrac{\partial \eta_3}{\partial \log \lambda} = -\dfrac{1}{4\pi}(1+\alpha)^2 \\[2ex] \dfrac{\partial \alpha}{\partial \log \lambda} = -\dfrac{1}{2\pi\eta_3}\alpha(1+\alpha)^2 \end{cases} \tag{118}$$

6.1.   The correlation length

The recursion relations (118) are easy to solve at one loop order. The invariant curves are solutions of:

$$\frac{\partial \eta_3}{\partial \alpha} = \frac{1}{2\alpha}\eta_3 \tag{119}$$

which implies:

$$2\eta_3^{-1}(\lambda)\sqrt{|\alpha(\lambda)|} = K = \text{constant} \tag{120}$$

Plugging this result into the recursion relation on α, we get the two sets of solutions depending on the sign of α:

$$\begin{cases} \alpha > 0, \quad a_+(\alpha(\log \lambda)) - a_+(\alpha_0) = -\frac{1}{4\pi}K\log\lambda \\[2ex] a_+(\alpha) = \dfrac{\sqrt{\alpha}}{1+\alpha} + \arctan\sqrt{\alpha} \end{cases} \tag{121}$$

and:

$$\begin{cases} \alpha < 0, \quad a_-(\alpha(\log \lambda)) - a_-(\alpha_0) = -\frac{1}{4\pi}K\log\lambda \\[2ex] a_-(\alpha) = \dfrac{\sqrt{-\alpha}}{1+\alpha} + \arg\tanh\sqrt{-\alpha} \end{cases} \tag{122}$$

The corresponding flow diagram is shown in Fig.(5).

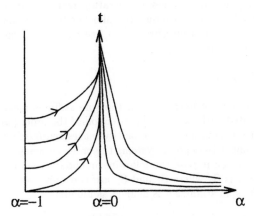

Fig. 5.   The flow diagram of the NLσ model in $D = 2$

In the long distance limit, $\lambda \to 0$. Therefore, irrespective to the sign of α_0, $|\alpha(\lambda)|$ decreases toward the asymptotic value $\alpha = 0$ while $\eta_3^{-1}(\lambda)$ increases until the perturbative regime is left. This is consistent with the fact that at sufficiently large distance, all the models with parameter α become equivalent to the one with $\alpha = 0$ which is $O(3) \otimes O(3)/O(3) = SO(4)/O(3)$ symmetric. This is what happens in $D = 2 + \epsilon$ at the critical point. Finally, as readily seen, the Eqs.(122) cease to be satisfied when λ is of order ξ^{-1}. This defines the correlation length at one loop order as:

$$\xi \sim \Lambda^{-1} e^{4\pi a_{\pm}(\alpha_0(\Lambda))/K}, \tag{123}$$

which is of course solution of Eq.(90).

To calculate the correlation length at two loop order, we have solved Eq.(90) using the two loop β functions of Eq.(117) to obtain[18]:

$$\xi = \Lambda^{-1} C_\xi \sqrt{\frac{2}{2\pi\eta_3}} (1 + \alpha) \exp\left(-\frac{1}{4}\left(\frac{(1+\alpha)^2}{\sqrt{\alpha}} a_+(\alpha) - 2\right)\right) \exp\left(2\pi\eta_3 \frac{a_+(\alpha)}{\sqrt{\alpha}}\right) \tag{124}$$

where C_ξ is a constant that depends only on the regularization scheme. If not for the non trivial α dependence, the correlation length of $O(3) \otimes O(2)/O(2)$ models behaves qualitatively like that of the $SO(4)/O(3)$ model since the prefactor of the exponential term in Eq.(124) scales as $1/\sqrt{2\pi\eta_3}$. A result which could not have been obtained in the one loop approximation. The measurement of ξ in a Monte-Carlo simulation should provide a decisive test for the validity of the NLσ model in $D = 2$. In particular, by measuring ξ for different models (i.e. by varying both α and η_3) one may test nonetheless the qualitative $SO(4)$ behavior, $\xi \sim BT^{1/2} \exp(A/T)$, but also the α dependence of the coefficients A and B. Unfortunately, because of the exponential behaviour as a function of $\beta = 1/kT$ and the computationally accessible lattice sizes, studying the very low temperature regime is very demanding, or even impossible. The relevant physical quantity allowing to reach this regime for accessible sizes is the spin-stiffness, a measure of the free energy increment under twisting of the boundary conditions.

6.2.   The spin stiffness

In a finite system, of linear size L, the spin stiffness measures the free energy increment associated with twisting the direction of the order parameter by imposing suitable boundary conditions. This quantity makes sense only for sizes $a \ll L \ll \xi$. In the case of colinear Heisenberg systems the order parameter is a three component vector $\mathbf{S}$. Thus we can impose the following boundary condition in the x direction: $\mathbf{S}(x)=\mathbf{S}_1$, $\mathbf{S}(x = L)=\mathbf{S}_2$ where $\mathbf{S}_1$ and $\mathbf{S}_2$ are two unit vectors making an angle θ, $\mathbf{S}_1.\mathbf{S}_2=\cos\theta$. The spin stiffness γ is defined as follows:

$$\gamma(L) = \frac{\partial^2 F(\theta)}{\partial \theta^2}|_{\theta=0} \tag{125}$$

and satisfies the one loop RG equation[48]:

$$\frac{\partial \gamma}{\partial \log L} = -\frac{1}{2\pi} \tag{126}$$

The important point is that $\gamma(L)$ is scale independent in D=2 with logarithmic corrections which amplitude is determined by the $O(3)/O(2)$ NLσ model. The crucial point in measuring $\gamma(L)$ is that its predicted size dependence is all the more valid since $L \ll \xi$. Therefore, in the very low temperature regime we can hope to test formula (126) by using a large range of relatively small lattice sizes. In contrast, measuring the temperature dependence of ξ requires $\xi \leq L$ and therefore relatively high temperatures for accessible sizes, a regime where the validity of the perturbation theory becomes less controlled. A most important point to notice is that at the very low temperatures considered here the physics of the model is entirely controlled by collective excitations - spin waves- and therefore we must take great care of these large-scale moves in any simulation of the model ("beating" the critical slowing down).

A recent Monte Carlo simulation on the $D = 2$ Heisenberg model is in excellent agreement with the perturbative prediction[49]. We now turn to the frustrated models[47].

In this case, the order parameter is a rotation matrix and the spin stiffness is a tensor. A general twist is now defined as: $R(x = 0)=R_0$, $R(x = L)=R_0 e^\theta$ where $\theta = \theta^a T_a$ and T_a are the generators of the Lie algebra of $SO(3)$. The free energy for this configuration is given by:

$$\Xi(L, \theta) = \int_{R(x=0)=R_0}^{R(x=L)=R_0 e^{\theta}} d\mu(R) e^{-\frac{1}{2} \int_0^L d^D x \mathcal{L}(R)} \tag{127}$$

Here, $d\mu(R)$ is the Haar measure on $SO(3)$. As in the preceding subsection, let us change the variable with $R(x)=R_{cl}(x)e^{\varphi(x)}$ where $R_{cl}(x)$ is some classical solution of the equations of motion which satisfy the suitable boundary conditions (127). When expressed in terms of the $\omega^a_{cl \ \mu}$ fields defined by $R_{cl}^{-1}\partial_\mu R_{cl} = \omega^a_{cl \ \mu}T_a$, $T_a \in \text{Lie}[SO(3)]$ they read:

$$\partial_\mu \omega^a_{cl \ \mu} = \epsilon^{abc} \frac{\eta_b - \eta_c}{2\eta_a} \omega^b_{cl \ \mu}\omega^c_{cl \ \mu} \tag{128}$$

The field $\varphi(x)$ represents the thermal fluctuations around this classical configuration and satisfies the boundary conditions: $\varphi(x = 0)=\varphi(x = L) =0$. The calculation proceeds exactly as in the preceding subsection, and we get at one loop order for the free energy:

$$\Xi(L, \theta) = e^{-\frac{1}{2} \int_0^L d^D x \left(\eta_{ab} - \frac{1}{L^D}R_{ab} \sum' \frac{1}{q^2} \right) \omega^a_{cl \ \mu}\omega^b_{cl \ \mu} + O(\omega^4_{cl}, \partial \omega_{cl})} \tag{129}$$

where R_{ab} is the Ricci tensor of the manifold $O(3) \otimes O(2)/O(2)$ which is the expected counter-term at the one loop order. The primed sum means that some infrared cut-off $1/L$ is present in the x-direction. The spin stiffness tensor is defined at the scale of some ultraviolet cut-off, such as for example the lattice spacing a, as:

$$\eta_{ab}(a) = \frac{1}{2}\frac{\delta^2 S(\omega^a)}{\delta\omega^a_\mu\delta\omega^b_\mu}|_{\omega=0} \tag{130}$$

It follows that the spin stiffness tensor at scale L, $\eta_{ab}(L)$ is given by:

$$\eta_{ab}(L) = \frac{\delta^2 S_{eff}(L, \omega^a)}{\delta\omega^a_{cl \ \mu}\delta\omega^b_{cl \ \mu}}|_{\omega=0} = \left(\eta_{ab}(a) - 1/L^D R_{ab} \sum' \frac{1}{q^2} \right) \tag{131}$$

or,

$$\eta_{ab}(L) = \eta_{ab}(a) - \zeta_{ab}(L/a) - \frac{1}{2\pi}R_{ab}\log(L/a) \tag{132}$$

$-\zeta_{ab}(L/a)$ is the finite part of $\sum '1/q^2$. It is a function of L/a which depends on the lattice and on the particular choice of boundary conditions. When $L \gg a$, ζ_{ab} becomes independent of L/a. As a consequence the asymptotic scaling equation for the spin stiffness is:

$$\frac{\partial \eta_{ab}}{\partial \log L} = -\frac{1}{2\pi}R_{ab} \tag{133}$$

These equations are nothing but the recursion relations (74), as expected. They are valid in the regime $a \ll L \ll \xi$. They are a set of coupled non linear equations of the η_a's. As a consequence, contrary to the $O(3)/O(2)$ case, their scale dependence is not just linear in $\log L$. However, as we shall show for $O(3) \otimes O(2)/O(2)$ models, there exists a model dependent scale beyond which one recovers a universal linear $\log L$ dependence.

6.2.1. The effective $O(3) \otimes O(2)/O(2)$ spin stiffness tensor

We are interested in the case $\alpha = \eta_3/\eta_1 - 1 \geq 0$ which is relevant for the AFT model. From Eq.(121), we obtain:

$$
\begin{cases}
a_+(\alpha(L)) = \dfrac{1}{4\pi} K \log \xi/L \\[2mm]
\eta_1(L) = \dfrac{2}{K}\sqrt{\alpha(L)}/(1 + \alpha(L)) \\[2mm]
\eta_3(L) = \dfrac{2}{K}\sqrt{\alpha(L)}
\end{cases}
\tag{134}
$$

where the flow invariant K and the one loop correlation length ξ are respectively given by Eq.(120) and Eq.(123). As seen, $\eta_1(L)$ and $\eta_3(L)$ are non trivial functions of $\log L$ and depend not only through ξ on the initial conditions. However, in the long distance limit, $a_+(\alpha(L)) = 2\sqrt{\alpha(L)} \ll 1$ so that:

$$
\eta_1 = \eta_3 = \frac{1}{4\pi} \log \frac{\xi}{L}
\tag{135}
$$

Therefore, in this limit we recover a universal simple logarithmic behavior. This behavior is expected since, as discussed in the previous section, in the long distance regime, all $O(3) \otimes O(2)/O(2)$ models become equivalent to the $O(3) \otimes O(3)/O(3) = SO(4)/O(3)$ model. The slope in (135) is $1/4\pi$ and is 4 times lower than that found with the conventional parametrization of the $SO(4)/O(3)$ in term of a 4-component unit vector. This difference can be easily explained by noticing that there is just a redefinition of a factor 4 between the energy scale of both parametrizations in terms of a rotation matrix and a 4-component unit vector. Let us finally remark that the linear $\log L$ behavior is valid only in the large L limit. For smaller values of L/a but still large, such that the RG scaling applies, the behavior of the spin stiffness is governed by the non linear function a_+ and is model (i.e. α) dependent. One recovers universal behavior only in the asymptotic regime.

We shall now consider the particular case of the AFT model. Dombre and Read have obtained the bare values of η_{ab} at the scale of the lattice spacing a[25]. They found $\alpha(a) = 1$. The one loop spin stiffnesses $\eta_1(L)$ and $\eta_3(L)$ are given in Eqs.(134). There is an other quantity of physical interest which is the mean spin stiffness $\gamma = (2\eta_1(L) + \eta_3(L))/3$. The physical meaning of this quantity is the following. Let apply a uniform twist with $R(x = L) = R(x = 0) \exp(\theta n^a T_a)$ where $\vec{n}$ is some unit vector which specifies the axis of the rotation. Since η_{ab} is a tensor, the response to the above twist is just $\gamma = \eta_{\bar{n}} = n_a \eta_{ab} n_b$. Taking $\vec{n} = (1,1,1)/\sqrt{3}$ one sees that γ is the response to a twist in the (1,1,1) direction. These quantities have been recently computed in a Monte Carlo simulation on the AFT Heisenberg model by Southern and Young[50]. We see in Fig.(6) that their results are in good agreement with our RG prediction.

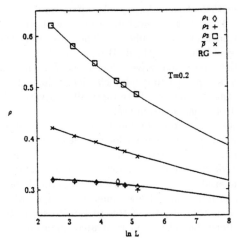

Fig. 6. Monte Carlo data for the spin stiffnesses η_1, η_1, η_3 and the mean spin stiffness $\bar{\eta}$ for the AFT model as a function of $\ln L$ for L between 12 and 180 and $T=0.2$ as obtained by Southern and Young. The solid lines are obtained from our two loop RG results with $\eta_1 = \eta_2$. Note that the stiffnesses are denoted by ρ in the figure instead of η.

We may thus conclude that, at least at sufficiently low temperature, the vortices are irrelevant and therefore the NLσ model approach is relevant in $D = 2$ dimension. As a result one may have some confidence on the $2+\epsilon$ findings. We are of course aware that this result can certainly not constitute a proof that vortices are irrelevant for the critical physics in $D = 3$. In particular, as discussed in section six, one do indeed expect that they may play some role in dimension 3. However, one may imagine that their effect may be not relevant for all values of the couplings and that the $SSO(4)$ behavior may be observable for some microscopic models. In any case, until one can conclude, one needs a consistent theory of the frustration effect that include vortices.

7. Quantum frustrated spin systems in two dimensions.

The preceding sections have been devoted to the study of classical frustrated magnets. In real systems, spins are quantum entities, and there are "a priori" no reasons to neglect the quantum nature of the order parameter particularly at low temperature and/or when the spin quantum number S is small. Of course, at sufficiently high temperature and in high dimensional systems, one naturally expects quantum fluctuations to be irrelevant, in particular at the critical point. This justifies the classical formalism used up to now. The main reason for this stems from a renormalization group argument. If we start, at the scale of the lattice spacing Λ^{-1}, with spins of low magnitude, say $S = 1/2$, we expect that after a sufficiently large number of block spin iterations and if the system is strongly correlated, we end up with an effective long distance Hamiltonian in which the effective spins have a much larger magnitude S_{eff}. Therefore, we expect that at some large scale, apart from some possible irrelevant quantum effects ($\sim 1/S_{eff}$), a classical description with effective parameters renormalized by quantum fluctuations is sufficient. The cross-over scale λ_D between the quantum and the classical behavior is the de Broglie wavelength of the spin wave, $\lambda_D = \hbar c/k_B T$ where c is the typical spin wave velocity and T is the temperature. At scales $\lambda \gg \lambda_D$ the system is expected to display classical behavior, in particular at the transition point T_c, so that the nature of the thermally activated phase transition should be identical to that of a classical spin system.

It is worth stressing that the above argument relies on the assumption that, in the low temperature phase, the system displays long range order, and in particular at zero temperature where quantum fluctuations are important. This is not always the case. In antiferromagnets, where the order parameter is not a constant of motion, one may seriously question the existence of long range order for small values of S. This question becomes even more important at low dimension and/or when one includes the effect of frustration. It is the purpose of this section to study the relevance of quantum fluctuations in dimension $D = 2$ in frustrated quantum antiferromagnets defined by the Hamiltonian:

$$\hat{H} = J \sum_{<i,j>} \hat{\mathbf{S}}_i.\hat{\mathbf{S}}_j \tag{136}$$

where $J > 0$, $\hat{\mathbf{S}}_i^2 = S(S+1)$ and $[\hat{S}_i, \hat{S}_j] = i\epsilon_{ijk}\hat{S}_k$, the sum being performed over all nearest neighbors of the triangular lattice.

Let us first consider the case of a ferromagnet. The quantum ground state corresponds to the classical one and is rotationally degenerate. In addition, the magnetization:

$$\hat{\mathbf{M}} = \sum_i \hat{\mathbf{S}}_i \tag{137}$$

commutes with the Hamiltonian and is therefore a constant of motion. At zero temperature, there is spontaneous symmetry breaking of rotational invariance as the system chooses one of the degenerate ground state and there is long range order for any value of the spin S. At finite T, the symmetry is restored as a result of thermal fluctuations and at a sufficiently large scale, the system is described by a classical NLσ model as in a pure classical system[51]. However, ferromagnets are almost unique in this respect, since in general the ground state of a quantum mechanical system is a non degenerate singlet and the order parameter is *not* a constant of motion. Thus, one expects strong quantum fluctuations around the classical ground state so that, even at zero temperature, the system may disorder.

Let us now consider the case of an antiferromagnet on a bipartite lattice. Classically, the ground state is obtained when the spins of the two sublattices, say A and B, are antiparallel: this is the Néel state. It is rotationally degenerate as in ferromagnets. However, the order parameter, which is the staggered magnetization

$$\hat{\mathbf{M}}_{\mathbf{sta}} = \sum_{i \in A/B} \hat{\mathbf{S}}_i . \tag{138}$$

does not commute with the Hamiltonian and is not a constant of motion. In fact, the quantum analog of the classical Néel state is not an eigenstate of $\hat{H}$. The exact ground state is a singlet and is not degenerate, as we shall show in the following. We thus expect strong fluctuations around the classical Néel state. The standard theory which accounts for these quantum fluctuations is the well known Spin Wave (SW) theory which allows to compute quantum corrections in a $1/S$ expansion. As an example, one gets at leading order for the ground state expectation value of the order parameter per spin[52] $< \hat{\mathbf{M}}_{\mathbf{sta}} > /N$:

$$< \hat{\mathbf{M}}_{\mathbf{sta}} >= S - 0.197 + O(1/S) \tag{139}$$

The effect of quantum fluctuations is a reduction of about 40 percent of $< \hat{\mathbf{M}}_{\mathbf{sta}} > /N$ from its classical value as a result of zero point motion of the order parameter. Eq.(139) suggests that there should exist a critical value of S, S_c, below which quantum fluctuations completely disorder the system. As S is varied, we expect the occurence of a quantum phase transition between the semiclassical order and the quantum disorder. While at leading order, $S_c \sim 0.2$, there is no guarantee that inclusion of higher order terms in the SW series do not modify significantly its value. The question of the magnitude of S_c has been the starting point of a large amount of works in the recent time, particularly because of the RVB theory[13,17] proposed by Anderson for high T_c superconductors. There is now no doubt that the $S = 1/2$ antiferromagnet on the square lattice displays long range order and the search for systems with a quantum disordered ground state has pointed recently in the direction of frustrated magnets.

Let us now consider the effect of frustration. In this section, we shall be mainly concerned with the quantum AFT model while our results should apply to the interesting J_1, J_2 model which is reviewed in the chapter by Schulz et al. in this book. As discussed in the preceding sections, the classical ground state displays the 120 degree structure of Fig.(1). Again it is not an eigenstate of the Heisenberg Hamiltonian and the order parameter is not a constant of motion. If we now compute the zero point contraction per spin at the leading SW approximation we find a reduction of about 60 percent[53]. The net effect of frustration is, as expected, to enhance quantum fluctuations. This is the main reason for thinking that frustrated magnets are natural candidates for exotics quantum ground states. While a lot of works have been done toward this direction there is, up to now, no definite conclusions about the nature of the ground states of frustrated non colinear magnets. It is the purpose of this section to review part of the results known at present. Most of the results discussed in this section are our own and concern mostly the NLσ model approach of frustrated quantum antiferromagnets. We refer the reader interested in other important approaches such as large N Schwinger bosons formalism to recent works[54].

Our aim in the following is threefold:

• first we want to have an effective theory that contrarily to the SW theory, preserves rotational invariance and allows to account systematically for quantum fluctuations of frustrated quantum magnets and therefore to characterize the nature of the quantum phase transition that occurs in these systems,

• the second question we shall address is the spontaneous breaking of the rotational symmetry in systems where the order parameter is not a constant of motion. Apart from evident theoretical importance this problem is of crucial interest for interpreting the data obtained from exact diagonalizations of finite systems[55],

• finally, we shall characterize the low temperature behavior of frustrated antiferromagnets in both Néel and quantum disordered phases.

7.1. Effective theory for quantum frustrated antiferromagnets

As discussed above, we expect a quantum phase transition, as the spin S is varied, between a semi-classical Néel phase and a quantum disordered one. While, the SW theory certainly predicts the occurence of such a transtion, it has two severe drawbacks. First, it breaks rotational invariance and therefore cannot describe the symmetric phase, and second, the $1/S$ series is not controlled contrary to the low T expansion of the NLσ model. What we would like is a relativistic field theory that, as in the case of a temperature driven phase transition, enables us to take full advantage of the standard tools of renormalization theory. In this respect, the main problem with quantum spins is that it is not obvious to build momentum and position operators to write down a conventional relativistic Lagrangian suitable for a $1/S$ semi-classical renormalizable expansion. It turns out, that in *antiferromagnets*, i.e. when the net magnetization is zero, the effective long distance Lagrangian is relativistic (i.e. the spectrum is linear in k). This crucial observation was first made by Haldane for one dimensional antiferromagnets and the extended application to 2D non frustrated antiferromagnets can be found in the beautiful work of Chakravarty, Halperin and Nelson[16]. We now turn to frustrated antiferromagnets.

The relevant effective Lagrangian has been obtained by Dombre and Read[25] for the quantum AFT model, in the large S limit thanks to the coherent state path integral formalism. In the following, we choose to present a derivation based on the Hamiltonian formalism which, in our opinion, enlightens the quantum and symmetry properties of the system. The basic remark is that since we are interested in the low energy physics, the true Hamiltonian is not necessary. There may exist a simpler long distance effective Hamiltonian more suited for the discussion of the low lying spectrum of the theory. As we shall now show this effective Hamiltonian is nothing but that of *quantum* symmetric tops with ferromagnetic-like interactions.

7.2. *The quantum top model*

As we have pointed out above, the difficulty when dealing with quantum spins, is that one cannot define momentum and position operators. The main idea which is behind what follows, is that after a sufficiently large number of block spin iterations, the effective Hamiltonian is expressed in terms of combinations of large clusters of spins and that we should be able to build out of these, an angular momentum density together with the associated position operators. Since, of course, such an exact RG transformation is out of our computational ability, we take the continuum limit, in the large S limit, on an elementary cell. The validity of this procedure relies, as usual, on the presence of a fixed point, the existence of which will be proved in the following.

Let us now introduce on each cell the following fields:

$$\begin{cases} \hat{\mathbf{J}}(\mathbf{r}) = \hat{\mathbf{S}}_1 + \hat{\mathbf{S}}_2 + \hat{\mathbf{S}}_3 \\ \hat{e}_1(\mathbf{r}) = \dfrac{3}{\sqrt{2}S}\left[-\dfrac{\sqrt{3}+1}{2}\hat{\mathbf{S}}_1 + \dfrac{\sqrt{3}-1}{2}\hat{\mathbf{S}}_2 + \hat{\mathbf{S}}_3 \right] \\ \hat{e}_2(\mathbf{r}) = \dfrac{3}{\sqrt{2}S}\left[\dfrac{\sqrt{3}-1}{2}\hat{\mathbf{S}}_1 - \dfrac{\sqrt{3}+1}{2}\hat{\mathbf{S}}_2 + \hat{\mathbf{S}}_3 \right] \\ \hat{e}_3(\mathbf{r}) = \hat{e}_1(\mathbf{r}) \times \hat{e}_2(\mathbf{r}). \end{cases} \tag{140}$$

Quantum mechanically, the components of $\mathbf{J}$ on a fixed orthonormal basis $\mathbf{x}^i$, $J_L^i = \mathbf{J}.\mathbf{x}^i$, $i = 1, 2, 3$, act as generators of the usual $O(3)$ (left) rotations:

$$[J_L^i, J_L^j] = i\epsilon^{ijk}J_L^k \quad , \quad [J_L^i, e_a^j] = i\epsilon^{ijk}e_a^k, \quad a = 1, 2, 3 . \tag{141}$$

The commutators and the products of the e_a^i's are less trivial but they simplify in the large S limit:

$$[e_a^i, e_b^j] = O(J/S) \quad , \quad e_a.e_b = \delta_{ab} + O(J/S) . \tag{142}$$

To go further, we need the components of $\mathbf{J}$ with respect to the e_a's, $J_{aR} = \mathbf{J}.e_a$ which in the large S limit obey the commutation relations:

$$[J_{aR}, J_{bR}] = -i\epsilon_{abc}J_{cR} + O(J/S) \quad , \quad [J_{aR}, e_b^i] = -i\epsilon_{abc}e_c^i + O(J/S) . \tag{143}$$

We see that the J_{aR}'s act as generators of $O(3)$ right rotations that mix the e_a's together. When $J \ll S$, the commutators Eqs. (141) to (143) become exact and we recognize the algebra of a quantum top.

The low energy, long distance effective Hamiltonian is now obtained by taking the continuum limit. Substituting in (136) the spin operators by their expression in terms of $\mathbf{J}$ and (e_1, e_2, e_3), we obtain the following Hamiltonian when $S \gg 1$:

$$H_N = \int_{N^2} d^2x \ \left[\tfrac{1}{2\chi_1}\mathbf{J}^2(x) + \left(\tfrac{1}{2\chi_3} - \tfrac{1}{2\chi_1} \right) J_{3R}^2(x) + p_1 \left((\nabla e_1(x))^2 + (\nabla e_2(x))^2 \right) \right.$$

$$\left. + p_3(\nabla e_3(x))^2 \right] . \tag{144}$$

H_N is the Hamiltonian of N^2 quantum symmetric tops with principal axis $e_1(x), e_2(x)$ and $e_3(x)$ and angular momentum $\mathbf{J}(\mathbf{x})$. In Eq.(144), (χ_1, χ_1, χ_3) are the principal inertia momenta of the tops and $(p_1, p_1, p_3) \propto S^2$ are stiffness constants. At this point, one may wonder how the antiferromagnetic nature of the interaction plays a crucial role in our derivation of Eq.(144) since, after all, Eqs. (141) to (143) do not depend on the nature of the interaction scheme. The point is that the classical minimum of H_N is obtained with $\mathbf{J} = \mathbf{0}$ and is stable since the χ's are positive. If we had been dealing with ferromagnetic interactions, the kinetic term would have had the opposite sign, indicating an unstability toward the formation of a finite magnetization.

Let us define:

$$\hat{\mathbf{J}}_T = \int_{N^2} d^2\mathbf{r}\, \hat{\mathbf{J}}(\mathbf{r})$$

$$\hat{J}_{R3T} = \int_{N^2} d^2\mathbf{r}\, \hat{J}_{R3}(\mathbf{r})$$

(145)

which are respectively the total angular momentum and its projection on the symmetric axis $e_3(x)$. We have:

$$[\hat{\mathbf{J}}_T, \hat{H}] = [\hat{J}_{R3T}, \hat{H}] = 0.$$

(146)

The Hamiltonian (144) is invariant under the action of the left $O(3)$ group and under the action of the right $O(2)$ group generated by J_{3RN}. This $O(2)$ invariance reflects the original C_{3v} symmetry of the triangular lattice.

To the Hamiltonian H_N can be associated a Lagrangian density:

$$L_N = -\frac{1}{2} \int_{N^2} d^2x \left[\mathrm{Tr} \left(P_0 \left(R^{-1}\partial_0 R \right)^2 + P_\perp \left(R^{-1}\partial_i R \right)^2 \right) \right].$$

(147)

where $R(x,\tau) = (e_1, e_2, e_3)$ is a $SO(3)$ matrix depending of space and time coordinates. In Eq.(147) we have used $\partial_\mu = (\partial_0, \partial_i) = (\partial/\partial\tau, \partial/\partial x_i)$; $i = 1, 2$. The matrices $P_\mu = \mathrm{diag}(p_{1\mu}, p_{1\mu}, p_{3\mu})$, $\mu = 0, \perp$ are diagonal matrices which contain the coupling constants. Whereas the $p_\perp$'s are related to the spin siffnesses by $\eta_a = -\mathrm{Tr}(P_\perp T_a^2)$, the p_0's are related to the inertia momenta by $\chi_a = -\mathrm{Tr}(P_0 T_a^2)$ where $T_a \in \mathrm{Lie}[SO(3)]$.

The low temperature as well as the zero temperature properties of H_N can now be investigated with help of the (euclidean) path integral representation of the partition function:

$$Z_N = \int DR(x,\tau) e^{-S_{N,\beta}}$$

(148)

with

$$S_{N,\beta} = \int_0^\beta d\tau L_N,$$

(149)

where β is the inverse temperature. Eqs. (149) and (144) are the basic equations from which will be obtained the relevant low energy physics of frustrated Heisenberg models. They define the *quantum* NLσ model with an order parameter in $SO(3)$. The main difference with the classical case, is that now there is an interaction in the "time" direction which accounts for the quantum fluctuations. Depending on P_μ, it describes all the symmetry breaking patterns compatible with an order parameter belonging to $SO(3)$, i.e. $O(3) \otimes O(p)/O(p)$ with $p = 1, 2, 3$.

In the symmetry broken phase, the Lagrangian Eq.(147) describes the dynamics of three Goldstone modes or spin waves. In general, P_0 is not proportional to $P_\perp$: there exists an anisotropy between space and time and our theory is not "Lorentz" invariant. However, the spectrum of these spin waves is relativistic, i.e. is linear in k as we now show. As in the classical case, we define the currents $\omega_\mu = \omega_\mu{}^a T_a = R^{-1}\partial_\mu R$. The equations of motion write :

$$2\eta_a^\mu \partial_\mu \omega_\mu{}^a = \omega_\mu{}^c \omega_\mu{}^d \epsilon_{adc} \left(\eta_d^\mu - \eta_c^\mu\right). \tag{150}$$

where $\eta_\mu{}^a = \eta^a$ when $\mu = \perp$ and $\eta_\mu{}^a = \chi^a$ when $\mu = 0$. If we choose a coordinate system π, the currrents are expressed as:

$$\omega_\mu{}^a \equiv \omega_i^a \, \partial_\mu \pi^i, \tag{151}$$

and since to first order in π:

$$\omega_\mu{}^a = \left(\delta_i^a + O(\pi^2)\right) \partial_\mu \pi^i \tag{152}$$

we obtain at leading order:

$$\eta_a^\mu \Box_\mu \pi^i = 0 \ . \tag{153}$$

This shows that, in general we have three spin waves with velocities $c_a = \sqrt{\eta_{a,\perp}/\eta_{a,0}}$ For the triangular case we have $\chi_1 = \chi_2 \neq \chi_3$ and $\eta_1 = \eta_2 \neq \eta_3$ so that we have only two differents velocities : $c_1 = c_2 = \sqrt{\eta_1/\chi_1}$ and $c_3 = \sqrt{\eta_3/\chi_3}$.

7.3. The quantum phase transition

We now investigate the zero temperature properties as the spin S is varied. Let us first qualitatively discuss the behavior of this system according to the values of the coupling constants entering in (144). When $p\chi$ is small, we expect that, at large scale the tops are almost decoupled. At these scales, the system consists in independent tops with inertia momenta of order $1+O(1/N)$. There is a gap in the spectrum and no symmetry breaking. For sufficiently large value of $p\chi$, the individual tops are tightly bounded so that, at large scale, the whole system behaves as a single top with an effective inertia momentum $\propto N^2$. Therefore there is no gap in the thermodynamical limit and the system exhibits long-range order. Since $p_a \propto S^2$, we expect, in the original Heisenberg spin system, a phase transition between a semi-classical Néel order and a quantum disorder at a critical value S_c. We now make the above argument more quantitative.

To do so, we need the β functions for the $\eta_\mu{}^a$. The main difficulty compared with the classical case, is that the theory is not Lorentz invariant. We thus expect the β function to depend on space-time indices. While we find that the theory is renormalizable, we cannot use the geometric formalism of Friedan[33], and we must go back to the background field method exposed previously in section 4. To this end we decompose R as:

$$R(\mathbf{r}, \tau) = R_{cl}(\mathbf{r}, \tau) \cdot h(\mathbf{r}, \tau). \tag{154}$$

where R_{cl} is a classical solution and h parametrizes the fluctuations. Integration over h at one loop order leads to the following recursion relations :

$$\lambda \frac{\partial}{\partial \lambda} \eta_\mu^{ab} = -\eta_\mu^{ab} + \frac{1}{2\pi} R_\mu{}^{ab} \tag{155}$$

where R_μ^{ab} is a space-time dependent tensor. Its analytical expression is, in the general case, rather lenghtly. We focus now on the cases where there is an $O(2)$ right invariance as in the AFT model. The recursion relations are, when expressed in terms of inertia and stiffnesses χ and η:

$$\begin{cases} \lambda \dfrac{\partial \chi_1}{\partial \lambda} = \chi_1 - \dfrac{1}{2\pi c_1} \left(1 - \dfrac{\chi_3}{\chi_1} \dfrac{c_3}{c_2 + c_3} \right) \\[3mm] \lambda \dfrac{\partial \chi_3}{\partial \lambda} = \chi_3 - \dfrac{1}{4\pi c_1} \dfrac{\chi_3^2}{\chi_1^2} \\[3mm] \lambda \dfrac{\partial \eta_1}{\partial \lambda} = \eta_1 - \dfrac{c_1}{2\pi} \left(1 - \dfrac{\eta_3}{\eta_1} \dfrac{c_1}{c_2 + c_3} \right) \\[3mm] \lambda \dfrac{\partial \eta_3}{\partial \lambda} = \eta_3 - \dfrac{c_1}{4\pi} \dfrac{\eta_3^2}{\eta_1^2} \end{cases} \tag{156}$$

To study the phase diagram, it is convenient to introduce a new set of variables:
 • $g \equiv 2/\eta_1 \sim 1/S^2$ which is the relevant coupling constant,
 • the spin wave velocities c_1 and c_3 whose difference represents the deviation from Lorentz invariance,
 • $\alpha = \eta_3/\eta_1 - 1$ which represents the deviation from a system of spherical tops.
In terms of these new couplings, Eqs.(156) reads:

$$
\begin{cases}
\lambda \dfrac{\partial c_1}{\partial \lambda} = -\dfrac{g}{8\pi}\dfrac{c_1^3}{c_3}(1+\alpha)\dfrac{c_1 - c_3}{c_1 + c_3} \\[2ex]
\lambda \dfrac{\partial c_3}{\partial \lambda} = \dfrac{g}{16\pi}\dfrac{c_1}{c_3}(1+\alpha)(c_1^2 - c_3^2) \\[2ex]
\lambda \dfrac{\partial g}{\partial \lambda} = -g + \dfrac{g^2}{4\pi}c_1\left(\dfrac{c_3 - \alpha c_1}{c_1 + c_3}\right) \\[2ex]
\lambda \dfrac{\partial \alpha}{\partial \lambda} = \dfrac{g}{8\pi}\dfrac{c_1}{c_1 + c_3}(1+\alpha)\left(c_3(1-\alpha) - (3\alpha + 1)c_1\right)
\end{cases}
\tag{157}
$$

In addition to the trivial infrared fixed point at $g^* = 0$, i.e. at $S^* = \infty$, Eqs.(157) admits a non trivial ultra-violet fixed point at:

- $g_c = 8\pi/c_{3c}$,
- $c_{1c} = c_{3c}$,
- $\alpha = 0 \Leftrightarrow \eta_{1c} = \eta_{3c}$,

that governs a second order phase transition between a semi-classical Néel-like ordered phase at large S and a quantum disordered phase at small S. We show in Fig.(7) the flow diagram corresponding to Eq.(157). At the fixed point we have the following remarkable properties:

- the spin-wave velocities are equal so that Lorentz invariance is recovered,
- the spin-stiffnesses are equal. As it is the case for the classical system, the symmetry is enlarged to $SO(4)$ at the transition.

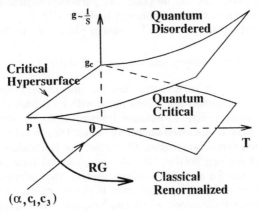

Fig. 7. The flow diagram of the quantum NLσ model in the five dimensional space of coupling constants and temperature

This shows that the nature of the quantum phase transition at $T = 0, D = 2$ is identical to that of the thermally activated transition of a *classical* anisotropic frustrated Heisenberg model at $T = T_c, D = 3$.[‡‡]

What we have just shown, is that frustrated quantum Heisenberg models display a symmetry broken Néel phase for sufficiently large values of S. While our perturbative calculation cannot give the value of S_c as a function of the parameters entering in the original spin Hamiltonian, it is worth stressing that in the spin wave phase, *all models with $S > S_c$ have the same long distance behavior governed by the infrared fixed point at $S = \infty$*. This justifies a posteriori the large S limit used in (144) for *all* systems with $S > S_c$.

In the Néel phase, the Lagrangian Eq.(147) describes three interacting massless modes (spin waves) with bare velocities $c_1 = c_2$, c_3. Their low energy physics is governed by the trivial infrared fixed point at $S = \infty$ so that they are infrared free. In the asymptotic long distance limit, the spin waves do not interact and the effect of quantum fluctuations can be reabsorbed into a (finite) renormalization of the couplings χ and η :

$$\begin{cases} \bar{\chi}_a = \lim_{\lambda \to \infty} \Lambda^d \chi_a(\lambda) \lambda^{-d} \\ \\ \bar{\eta}_a = \lim_{\lambda \to \infty} \Lambda^d \eta_a(\lambda) \lambda^{-d+2} \end{cases} \tag{158}$$

where $\chi_a(\lambda)$ and $\eta_a(\lambda)$ are solutions of the recursion equations Eqs.(156) with initial conditions given by their values at the lattice scale. These quantum renormalized couplings (as the exact value of S_c) cannot be obtained exactly by our perturbative analysis. However, they can be obtained from experiment or exact diagonalization on finite lattices. In any case, they can be taken as "phenomenological input parameters". Our analysis has shown that it is sufficient to define a "finite" number of input parameters to completely characterize the low energy physics in the Néel phase.

7.4. *Spontaneous symmetry breaking and the tower of states*

In the last subsection, we have shown that for sufficiently large values of S, $S > S_c$, there exists a stable semiclassical Néel phase with long range order and spontaneous symmetry breaking (SSB) of rotational invariance. At first sight, this seems to contradict the fact that, as discussed above, we expect the ground state of an antiferromagnet to be a non degenerate singlet. Of course, this last statement holds only in a *finite system* and nothing prevents the thermodynamical limit to be singular. Indeed, in field theory and for an infinite volume system, the well know mechanism of SSB, states that once a classical ground state is chosen among the whole set of degenerate ones, quantum fluctuations, which can be calculated in a systematic $\hbar$ (in our Heisenberg model $1/S$) expansion, are not strong enough to destroy semiclassical

---

[‡‡]This shows, in particular that anisotropies in the classical system like those generated by different couplings in the different space directions are irrelevant at the fixed point as it is the case with the stacked AFT model obtained with $P_0 \sim (1,1,1)$ and $P_\perp \sim (1,1,0)$.

order. The resulting ground state constitutes therefore the vacuum of the theory on which is built the Fock space of the elementary excitations, the Goldstone modes or spin waves. Although the degenerate classical ground states are equivalent, the Fock spaces built above these ground states are all disconected in the thermodynamical limit. This justifies the semi-classical expansion we have done in the preceding subsection.

A well-known example of this phenomenon in condensed matter physics, is the Heisenberg ferromagnet, which is, as discussed above, almost unique, since its quantum ground state nonetheless is degenerate but corresponds to the classical one. This leads to some confusion when discussing SSB in quantum antiferromagnets. The SSB mechanism, we have just described, occurs in the infinite system and gives no insight on the mechanism, which at the (spectral) quantum level allows to understand how the symmetry is broken when we go from a finite volume to the thermodynamical limit. Apart from evident theoretical interest, such a spectral characterization of SSB is of central importance when discussing numerical data obtained from exact diagonalization of finite systems since it .

P.W.Anderson was the first to realize that the SSB mechanism involves a whole tower of excited states that collapse onto the true ground state in the thermodynamical limit[52]. Since these states are in general not rotationally invariant, nothing prevents that some combination of them, which for large systems should persist a very long time, explicitly breaks rotational symmetry. The relevant problem is nonetheless to characterize the nature of this tower of states and to give its size dependence when there is SSB in the infinite volume limit, but also to relate these properties to the RG analysis we made when discussing the field theory.

Therefore, what we want is an effective hamiltonian for the *first* excited states of the quantum frustrated Heisenberg model. The idea is to integrate out in (147) the spatial degrees of freedom, i.e. the $k \neq 0$ modes, to obtain an effective action at the scale N of the lattice size, for the $k = 0$ modes. These modes describe the collective motion of the tops so that a one body effective Hamiltonian follows from this action. To obtain this action we split up the field $R(x, \tau)$ in (147) into $R_0(\tau) \exp e(x, \tau)$ where $R_0(\tau)$ and $e(x, \tau)$ represents the $k = 0$ and $k \neq 0$ modes respectively. By expanding R to order e^2 and by integrating out e in (149), we find, after some algebra, at leading order in N the Hamiltonian of a single symmetric top:

$$H_{eff}(N) = E_{0N} + \frac{1}{N^2}\left[\frac{\mathbf{J}_N^2}{2\chi_1(N)} + \left(\frac{1}{2\chi_3(N)} - \frac{1}{2\chi_1(N)}\right)(J_{3RN})^2\right] \qquad (159)$$

where

$$E_{0N} = \frac{1}{2}(c_3(N) + 2c_1(N))\sum k \qquad (160)$$

is the leading quantum correction to the classical ground state energy. The hamiltonian (159) describes the angular part of the fluctuations of the total order parameter $e_{aN} = \int_{N^2} e_a(x), a = 1, 2, 3$ whose modulus is equal, at leading order in N, to its infinite volume limit. $\mathbf{J}_N$ and J_{3RN} are respectively the total angular momentum and its projection on the symmetric axis e_{3N}. Finally, $c_a(N)$ and $\chi_a(N)$, a = 1,2,3, are the renormalized values of the spin wave velocities and inertia momenta at scale N. We are now in a position to discuss the properties of the tower of states according to the spin S:

- when the system has long range order, i.e. when $S > S_c$, both $c_a(N)$ and $\chi_a(N)$ have finite limits, $\bar{c}_a$ and $\overline{\chi}_a$, when $N \to \infty$. The energy of the first excited states of (144) is:

$$E_{j,m_L,m_R} = E_{0N} + \frac{1}{N^2}\left[\frac{j(j+1)}{2\overline{\chi}_1} + \left(\frac{1}{2\overline{\chi}_3} - \frac{1}{2\overline{\chi}_1}\right)m_R^2\right] \tag{161}$$

where $j(j+1), m_L$ and m_R are the eigenvalues of $\mathbf{J}_N^2$, J_{ZN} and J_{3RN} respectively. For each value of j, there is a total of $(2j+1)^2$ eigenstates each with a degeneracy $2(2j+1)$. Apart from these states, the other low energy modes of (144) have $k \neq 0$ and are the first magnon states which scale as $1/N$. Thus, there are $j_{max} \sim \sqrt{N}$ states in (161) that collapse onto the ground state, when $N \to \infty$, faster than the first magnons and which define the relevant tower of states. With these states one can form the symmetry breaking Néel state in which the order parameter has a non zero mean value with an uncertainty of order $\sim 1/\sqrt{N}$. Of course, this state being not an eigenvector of (161) has a finite life time of order $\sim \sqrt{N}$. Because of the long range order, the system develops a macroscopic collective variable with a "mass" of order N^2 which is localized in the infinite volume limit.

- at the critical point $S = S_c$, the system becomes Lorentz invariant when $N \to \infty$: $c_a(N) \to c^{*}$[19]. In addition we have $\chi_a(N) \to \chi^{*}/N$ as follows from simple dimensional analysis. The predicted scaling for the tower of states changes drastically:

$$E_{j,m_L,m_R} = E_{0N} + \frac{1}{N}\frac{j(j+1)}{2\chi^{*}}. \tag{162}$$

At this point the tower of states collapses onto the ground state *together* with the first magnon states. This is completely consistent with the fact that the theory is critical. This result is independent of the space dimension.

To complete our discussion, we need the leading finite size corrections for the ground state observables: e_N, the total energy per spin and the mean value of the total order parameter (e_{1N}, e_{2N}, e_{3N}). We find for the ground state energy:

$$e_N = e_\infty - \frac{\delta}{2N^3}(\bar{c}_3 + 2\bar{c}_1), \tag{163}$$

where δ is a numerical constant which depends on the lattice. As discussed above, the order parameter is the rotation matrix $R(x,\tau) = (e_1, e_2, e_3)$. As a consequence of the $O(2)$ right symmetry the fields e_1 and e_2 renormalize with the same constant and we have: $\langle(e_{1N})^2\rangle = \langle(e_{2N})^2\rangle = M_N^2$ while $\langle(e_{3N})^2\rangle = \kappa_N^2$. We find for the leading finite size correction of M_N and κ_N :

$$M_N = M_\infty\left(1 - \frac{\gamma}{2N}\left[\frac{1}{\overline{\chi}_1\bar{c}_1} + \frac{1}{\overline{\chi}_3\bar{c}_3}\right]\right), \tag{164}$$

$$\kappa_N = \kappa_\infty \left(1 - \frac{\gamma}{N \overline{\chi}_1 \overline{c}_1} \right) \tag{165}$$

where γ depends on the lattice.

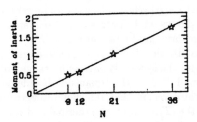

Fig. 8.   The gap of the $S = 1/2$ quantum AFT model as a function of N^2 obtained by exact diagonalization by Bernu et al.[55].

While the scaling given by Eqs. (163), (164) and (165) constitutes the standard test for long range order in the ground state, the observation of the tower of states (161) with the predicted degeneracy and scaling provides a test which involves a *whole set* of low lying excited states. Since it is deeply related to the mechanism of SSB, it constitutes a richer and more effective test for long range order.

Let us now turn to the $S = 1/2$ AFT model. We identify M_N in (164) with the sublattice magnetization and κ_N in (165) with the helicity operator as defined for example by Fujiki and Betts[56]. Numerical results obtained for the ground states quantities e_N, M_N and κ_N for different lattices up to 27 sites have led to contradictory conclusions[57,58,59]. Very recently, extended numerical results have been obtained by exact diagonalization on lattices up to 36 sites[55]. The authors have identified a whole set of low lying excited states that constitute the tower of states (161) but they did not observe the correct degeneracy associated to a symmetric top. This is probably due to the small size of the sample. However they found $(2j+1)^2$ states for each value of the spin j and the predicted scaling $\sim 1/N^2$ for SSB given by (161) as seen Fig.(8). This rules out the possibility that the $S = 1/2$ AFT model is at a critical point as it was suggested[60] and gives strong support for the existence of a Néel ground state in the thermodynamical limit.

7.5. Finite temperature properties.

We have seen in the preceding section that there exists in Heisenberg systems a critical value S_c of the spin S separating at $T = 0$ an ordered phase at large S from a disordered phase at small S[16,19]. We now study in this section the relevance of the nature of the order in the ground state for the thermodynamics of the system. It is actually not obvious that the order, or disorder, in the ground state has any relevance for $T \neq 0$ since anyway from Mermin - Wagner theorem, we know that there is no long range order in $D = 2$ for Heisenberg spins at finite temperature. In fact, we shall see in the following that the "nature of the disorder" at finite temperature is very much different depending on the phase of the system at $T = 0$. At low temperature T, there are three possibilities[16,19], see Fig.(7):

• $S > S_c$, the system is ordered at $T = 0$, and is therefore strongly correlated for $T > 0$ and small. We expect the correlation length ξ to be very large ($\xi(T = 0) = \infty$). We shall show that in fact the system behaves as a *classical* spin system for sufficiently large scales. This is the "renormalized classical region".

• $S < S_c$, the system is disordered at $T = 0$. The quantum fluctuations dominate the thermal fluctuations and the system is only short range ordered. We expect the correlation length to be small. This is the "quantum disordered region".

• There is finally a third region that interpolates between the two preceding regions. In this domain of the (T, S) space, the thermal and the quantum fluctuations are typically of the same order of magnitude. This is the "quantum critical region" which is reached in practice when S is close to S_c and T is not too small.

Let us study in more details the first region. For $S > S_c$, the long distance physics at scale L, $a \ll L \ll \xi$, is obtained from the microscopic physics at scale a by blocking. After a sufficiently large number of RG transformations, the quantum fluctuations become negligible since T, which is the inverse of the size of the time direction, Eq.(149), is a relevant quantity: when blocking, T increases and the size of this direction becomes eventually of the order of the lattice spacing (times the inverse of the spin wave velocity) thus forbidding quantum fluctuations. At that scale, which is typically the thermal de Broglie wavelength $\lambda_D \sim \hbar c/kT$, the time direction does not play any role so that we can trivially average on it and the model becomes effectively bi-dimensional and thus classical. The parameters of this model are those of the microscopic quantum model transformed by the RG flow at the scale λ_D. They are renormalized by thermal *as well as quantum fluctuations*. Then, for scales larger than λ_D (but smaller than ξ), the whole analysis of the two dimensional classical Heisenberg systems developed in the previous section can be used by just taking λ_D as the effective lattice spacing and by taking as input parameters, the renormalized parameters at scale λ_D.

A specific difficulty of the frustrated systems is that there are several different spin wave velocities and the cross over scale λ_D is a non trivial combination of the two de Broglie wavelengths associated with the two different spin wave velocities c_1, c_2. Thus λ_D must be computed explicitly by a one loop calculation. Let us sketch how match the classical and quantum NLσ models in the renormalized classical region and how the cross over scale can be computed. Since we want to integrate out the quantum fluctuations, we use a modified version of the background field method where we write the matrix $R(x, \tau)$ as:

$$R(x, \tau) = R(x)h(x, \tau) \qquad (166)$$

where $R(x)$ contains only the "classical", i.e. $\omega = 0$, mode and $h(x, \tau) = \exp \varphi(x, \tau)$ contains the "quantum", $\omega \neq 0$, modes where ω are the Matsubara frequencies. Since

h does not involve the $\omega = 0$ mode we must have:

$$\int_0^\beta d\tau \varphi(x,\tau) = 0 \qquad (167)$$

This constraint allows to eliminate the linear terms in φ in the action thus leaving at the lowest non trivial order only a gaussian integral on φ. After integration, this produces an effective action on $R(x)$ alone. The effective partition function reads for the quadratic part in $R(x)$:

$$Z = \int DR(x) e^{-\frac{1}{2}\int d^2x \, \eta_{cd}^{eff} \omega_i^c \omega_i^d} \qquad (168)$$

where $\omega_i = R^{-1}(x)\partial_i R(x) = \omega_i^c T^c$. This is nothing but the classical NLσ model with effective stiffnesses η_{cd}^{eff}. The computation of both η_{cd}^{eff} and λ_D in terms of the couplings $\bar{\eta}_a$ and $\bar{\chi}_a$ renormalized by quantum fluctuations at zero temperature, as defined in Eq.(158), requires a careful analysis of the finite parts coming from the intergration over the quantum fluctuations. The calculation is rather lengthy and cumbersome and we shall only give here the results and refer to previous works for the details[18,61]. We find:

$$\lambda_D = \beta \bar{c}_1 \exp\left(-\frac{1+\bar{\alpha}}{16\bar{\alpha}}[(1+\bar{\alpha})G(\bar{\alpha}) - G(0)]\left(1 + \frac{\log\frac{\bar{c}_3^2}{\bar{c}_1^2}}{1 - \frac{\bar{c}_3^2}{\bar{c}_1^2}}\right)\right) \qquad (169)$$

As seen, it is a non trivial function of the renormalized spin wave velocities $\bar{c}_1$ and $\bar{c}_3$. At this scale η_{cd}^{eff} is just given, at leading order in β, by its asymptotic value $\beta\bar{\eta}_{cd}$ as given by Eq.(158) with $\lambda = \beta$. Once the matching scale λ_D and the corresponding value of the effective coupling η_{cd}^{eff} are known it is not difficult to obtain the correlation length of the quantum model thanks to that of the classical one of Eq.(124). The result is:

$$\xi = \frac{C_\xi}{\sqrt{8k_BT}} (\lambda_D T) \sqrt{(1+\bar{\alpha})\bar{t}} \exp\left(-\frac{1}{8}[(1+\bar{\alpha})G(\bar{\alpha}) - G(0)]\right) \exp\left(\frac{2\pi G(\bar{\alpha})}{k_BT\bar{t}}\right)) \qquad (170)$$

where $\bar{t} = 2/\bar{\eta}_1$. In equations (169) and (170), we have introduced a function $G(\bar{\alpha})$ analogous to the function a_+ introduced in the preceding section:

$$G(\bar{\alpha}) = 2 + 2(1+\bar{\alpha})\frac{\arctan(\sqrt{\bar{\alpha}})}{\sqrt{\bar{\alpha}}} \qquad (171)$$

We conclude on these expressions that in the quantum ordered phase, the quantum correlation length diverges exponentially with $1/T$ as the classical one but differs in the preexponential factor by a power of T as readily seen on Eqs.(124,170). The difference comes, of course, from the fact that the lattice spacing of the classical model is replaced by λ_D that varies with the temperature as $1/T$. In our expression of ξ, the preexponential factor behaves as $1/\sqrt{T}$ and differs from that of both Heisenberg ferromagnet where it scales as $\sqrt{T}$ and of Heisenberg colinear antiferromagnets[14] where it turns out to be temperature independent.

In the quantum critical region, the correlation length is predicted to behave as[45]:

$$\xi \sim \frac{1}{T} \tag{172}$$

while in the quantum disordered case, the correlation length tends to a constant as $T \to 0$. It follows from the above discussion that the behavior of the correlation length as a function of T is also a criterion of the nature of the order at $T = 0$ which is different of the existence of the tower of states.

The correlation length of the $S = 1/2$ quantum AFT model has been calculated by Elstner and Young[62] in a high temperature expansion. These authors have fited their results with the following form:

$$\xi = \frac{A}{\sqrt{T}} e^{B/T} \tag{173}$$

They estimate that $B = 0.1$ while the naive calculation of B from the bare parameters gives $B = 1.7$. This result supports the existence of long range order in the $S = 1/2$ AFT model at $T = 0$ in agreement with the results of exact diagonalizations[55]. We note however the strong effect of quantum fluctuations since B is found to be considerably reduced from its classical value. This may suggest that we are close to the quantum critical region[45].

Let us finally conclude by quickly reviewing the results obtained in the other phases. In the non colinear case, most of these results have been obtained by Chubukhov, Read, Sachdev and their collaborators[63,54,45].

The quantum critical region is a high as well as a low temperature region: high with respect to any energy scale measuring the deviation from the critical point at $T = 0$ (the spin stiffness η for instance in the ordered region) and low with respect to the microscopic scales such as the exchange coupling J. If this region is not small then it is possible to reach at sufficiently high temperature, both in the ordered and in the disordered phases, a regime where the quantum fluctuations are comparable with the thermal ones and where the system does not know if it is ordered or disordered in its ground state. In the disordered region the relevant energy scale is a spin-gap Δ above which are supposed to propagate low-lying quasi-particles excitations with non-zero spin. Both η and Δ are supposed to vanish around the critical point at $T = 0$ as[54,45]:

$$\eta, \Delta \sim |g - g_c|^\nu \tag{174}$$

The important assumption behind Eq.(174) is that the quantum phase transition is continuous and that there exists a well defined effective field theory controlled by a fixed point. Since we expect that the nature of this transition is analogous to that of the three dimensional classical thermally activated one, such an effective theory may well not exist. We saw that, within perturbation theory, if the quasi-particle excitations are taken to be in the vector representation of the $SO(3)$ group, like in the two- or the three- LGW models of section 2, no fixed point exists. In section 5., we have seen that if instead the quasi-particles are hypothetisized to be $S = 1/2$ spinons[63,45,46] then an effective theory for the critical region exists and is controlled by the $SO(4)$ fixed point.

At the quantum level, these spinons are obtained through the Schwinger boson representation of the spin operators:

$$S_i = \frac{1}{2} b_\alpha^\dagger \sigma_{\alpha\beta} b_\beta \qquad (175)$$

Chubukov and coworkers[45] have shown within the $1/N$ expansion that in the quantum disordered region the quanta corresponding to the operators b are deconfined spin $1/2$ bosonic spinons and are the quasi-particles that leave above the gap Δ. The main result of their analysis is that all quantities obey universal scaling relations in the quantum critical-region, that are reminiscent of the $SO(4)$ fixed point, with only five independent parameters in the non colinear case: the two stiffnesses, the two susceptibilities and the order parameter condensate. All these results are still waiting for numerical and experimental support.

7.6. Concluding remarks

We have shown that there are different kinds of test for the order at $T = 0$ for an antiferromagnet: and we have reviewed in this chapter the analysis of the spectrum and the behavior as a function of T of different thermodynamical quantities as the correlation length. One of the most effective signature at finite size of the symmetry breaking is the existence of a tower of states with the correct scaling with the size and the correct degeneracy. On the other hand, the exponential behavior of the correlation length as a function of the inverse temperature provides also a decisive test for long range order at $T = 0$. Moreover, when $S > S_c$ and T not too large, we showed that the finite temperature properties can be obtained thanks to the classical two dimensional NLσ model with suitable effective parameters renormalized by quantum fluctuations.

More problematic is the analysis of the quantum critical region which relies on the hypothesis that the quantum phase transition is continuous. As discussed extensively in this chapter, this may be not the case. At this point, it is worth stressing that all the properties of the symmetry broken phase, at sufficiently low temperature, do not rely on any assumption on the nature of the transition. They rely only on the existence of an infrared fixed at $S = \infty$ that can be hardly questioned. It is important to notice that even when $S > S_c$, and at sufficiently high temperature, the system may enter in the quantum critical region as seen in Fig.(7). In this case, one may observe a crossover between exponential behavior and inverse power law of the correlation length as a function of T that may also provide a test for the existence of an $SO(4)$ fixed point in frustrated magnets.

As discussed above there are now strong evidences that the $S = 1/2$ quantum AFT model is ordered at $T = 0$. Finite temperature properties should be accessible experimentally since the AFT model modelizes adsorbed ^{3}He on graphite substrate[64,65]. Therefore, one may hope to observe $SO(4)$ behavior in a real system. It will also be very interesting to study systems that are disordered at $T = 0$. One promising candidate is the $S = 1/2$ Heisenberg antiferromagnet on the Kagome lattice[66,67,68,69].

8. Conclusion

In this chapter, we have addressed the question of the existence of an effective theory of both three dimensional classical and two dimensional quantum Heisenberg frustrated systems. We have shown, that in the symmetry broken phase, i.e. when $S > S_c$ or $T < T_c$ such a theory exists and is the Non Linear Sigma model with an order parameter in $SO(3)$. It describes three Goldstone modes which low energy behavior is well controlled by the infrared fixed point at $S = \infty$ or $T = 0$. Thanks to this theory, one is able to compute relevant physical quantities such as for example the correlation length. In addition, it allows to give a consistent picture of the symmetry breaking mechanism in the quantum case. We have established the existence of a tower of states that collapses onto the singlet ground state in the thermodynamical limit. The non trivial aspect of frustration manifests itself when one tries to describe the critical region. We have seen, that in this region while a stable $SO(4)$ fixed point was predicted by the NLσ model no fixed point can be found in the naive LGW strategy. We have proposed a scenario that reconciles both perturbative approaches in which the phase transition may be either first order or continuous with $SO(4)$ or tricritical mean-field like behavior depending on the microscopic parameters of the model. In particular, from the theoretical analysis we find unlikely that a new fixed point, not seen in any perturbation theory, can describe a new universality class for frustrated Heisenberg models. The very problem is to understand what is physically responsible for the mismatch of both approaches.

In fact, both theories suffer from basic approximations in perturbation theory. While the NLσ approach is insensitive to topological properties of the order parameter space, one has to make an *ansatz* on the nature of the relevant critical fluctuations as soon as one writes a LGW effective action. In contrast with colinear systems, there may exist different such actions in the frustrated case, as for example the two- and three-vector models of section 2.. Both models differ by the number of massive modes and do not describe the same excitations and none of them possesses a fixed point. An other intriguing possibility is that the relevant excitations are bosonic spin 1/2 spinons in which case the corresponding LGW model possesses a stable fixed point which agrees with that found in the NLσ model. Up to now, there is no numerical evidence that this should be the case since no evidence for $SO(4)$ behavior is seen in Monte Carlo simulations of $D = 3$ classical frustrated models. The problem may be due to the effect of topological excitations. While in $D = 2$ and at sufficiently low temperature the NLσ model is likely to correctly describe the physics there is no evidence that this should be the case in $D = 3$. In this respect, it has been recently suggested that Z_2 vortices lines could favor a first order transition in nematic system which possesses the same topological defects than frustrated Heisenberg models[29].

Let us finally conclude that this mismatch between different perturbative approaches is by no means particular to frustrated spin systems. It occurs also in the abelian gauge theory of the superconducting transition[70,71], and in non abelian gauge theories relevant for the phase transition of the primordial universe[72]. It could well be that what happens with frustrated systems is generic and that colinear systems are in this respect quite particular.

9. Acknowledgements

We want to thank all our collaborators, F. Delduc, T. Jolicoeur and D. Mouhanna who contributed to the different works presented in this chapter. We want particularly to acknowledge D. Mouhanna who has contributed directly to the section 7. on quantum frustrated spin systems and with whom we have discussed all the topics developped in this chapter. Interesting discussions are also acknowledged with N. Andrei, M. Caffarel, A. Chubukov, F. Delduc, H.T. Diep, A. Dobry, H. Godfrin, T. Jolicoeur, P. Lecheminant, C. Lhuillier, M. Plumer, S. Sachdev and T. Ziman.

10. References

1. T. Garel and P. Pfeuty, *J. Phys. C* , L9 (1976) 245.
2. D. Bailin, A. Love, and M.A. Moore, *J. Phys.C*, 10 (1977) 1159.
3. H. Kawamura, *Phys. Rev. B*, 38 (1988) 4916.
4. H.T. Diep, *Phys. Rev. B*, 39 (1989) 3973.
5. D.A. Tindall, M.O. Steinitz, and M.L. Plumer, *J. Phys. F*, 7 (1977) 263.
6. D.A. Tindall, M.O. Steinitz, and M.L. Plumer, *J. Phys. F*, 4 (1992) 9927.
7. J. Eckert and G. Shirane, *Solid State Commun.*, 19 (1976) 911.
8. E. Loh, C.L. Chien, and J.C. Walker, *Phys. Letters A*, 49 (1974) 357.
9. B.D. Gaulin, M. Hagen, and H.R. Child, *J. de Physique C*, 8 (1988) 327.
10. S.W. Zochowski, D.A. Tindall, J. Genossar M. Kahrizi, and M.O.Steinitz, *J. Magn. Mater.*, 54 (1986) 707.
11. P. Azaria, B. Delamotte, and T. Jolicoeur, *Phys. Rev. Lett.*, 64 (1990) 3175.
12. P. Azaria, B. Delamotte, F. Delduc, and T. Jolicoeur, *Nucl. Phys. B*, 408 (1993) 485.
13. P.W. Anderson, *Mat. Res. Bull.*, 8 (1973) 153.

14. S. Chakravarty, B.I. Halperin, and D.R. Nelson, *Phys. Rev. Lett.*, **60** (1988) 1057.
15. P.W. Anderson. *Science*, **235** (1987) 1196.
16. S. Chakravarty, B.I. Halperin, and D.R. Nelson, *Phys. Rev. B*, **39** (1989) 2344.
17. P. Fazecas and P.W. Anderson, *Phil. Mag.*, **30** (1974) 423.
18. P. Azaria, B. Delamotte, and D. Mouhanna, *Phys. Rev. Lett.*, **68** (1992) 1762.
19. P. Azaria, B. Delamotte, and D. Mouhanna, *Phys. Rev. Lett*, **70** (1993) 2483.
20. K.G. Wilson and J. Kogut, *Physics Report C*, **12** (1974) 75.
21. J. Zinn-Justin, *Quantum Field Theory and Critical Phenomena*, Oxford University Press, New York (1989).
22. A.M. Polyakov, *Phys. Lett. B*, **59** (1975) 79.
23. E. Brézin and J. Zinn-Justin, *Phys. Rev. Lett*, **36** (1976) 691.
24. P. Azaria, B. Delamotte, and T. Jolicoeur, *J. Appl. Phys.*, **69** (1991) 6170.
25. T. Dombre and N. Read, *Phys. Rev. B*, **39** (1989) 6797.
26. N.D. Mermin, *Rev. Modern Phys.*, **51** (1979) 591.
27. P. Azaria, P. Lecheminant, and D. Mouhanna, *In preparation*.
28. H. Kawamura and S. Miyashita, *J. Phys. Soc. Jpn.*, **53** (1984) 4138.
29. P.E. Lammert, D.S Rokhsar, and J. Toner, *Phys. Rev. Lett.*, **70** (1993) 1650.
30. S.A. Antonenko and A.I. Sokolov, *Renormalisation Group 91 - Second International Conference*, Edited by D.V. Shirkov and V.B. Priezzhev, World Scientific (1991).
31. G. Zumbach, *Phys. Rev. Lett.*, **71** (1994) 2421.
32. Landau and Lifchitz, *Physique Statistique, Editions Mir Moscou (1980)*.
33. D.H. Friedan, *Ann. Phys.*, **163** (1985) 318.
34. C. Becchi, A. Blasi, G. Bonneau, R. Collina, and F. Delduc, *Commun. Math. Phys.*, **120** (1988) 121.
35. A.M. Polyakov, *Gauge fields and strings*, Harwood Academic Publishers, (1987).
36. D.R.T. Jones, A. Love, and M.A. Moore, *J. Phys. C*, **9** (1976) 743.
37. H. Kawamura. *J. Phys. Soc. Jpn.*, **61** (1992) 1299.
38. D. Loison and H.T. Diep, *Preprint*.
39. A.Mailhot, M.L.Plumer, and A. Caillé, *Phys.Rev.B*, submitted.
40. J.C. Le Guillou, *Private communication*.
41. F. Wegner, *Z. Phys. B*, **78** (1990) 33.
42. G. E. Castilla and S. Chakravarty, *Phys. Rev. Lett.*, **71** (1993) 384.
43. B.I. Halperin, *in Physics of defects, Les Houches Summer School edited by R. Balian et al. (North Holland)*.
44. J.L.Cardy and H.W.Hamber, *Phys. Rev. Lett.*, **45** (1980) 499.
45. A. Chubukov, S. Sachdev, and T. Senthil, *"Quantum phase transitions in frustrated quantum antiferromagnets"*, *Preprint 94*.
46. P. Azaria, P. Lecheminant, and D. Mouhanna, *In preparation*.
47. P. Azaria, B. Delamotte, T. Jolicoeur, and D. Mouhanna, *Phys. Rev. B*, **45** (1992) 12612.

48. S. Chakravarty, *Phys. Rev. Lett.*, **66** (1991) 481.
49. M. Caffarel, P. Azaria, B. Delamotte, and D. Mouhanna, *To be published in Europhysics Letters.*
50. B.W. Southern and A.P. Young. *Phys. Rev. B*, **48** (1993) 13170.
51. P. Kopietz, P. Scharf, M.S. Skaf, and S. Chakravarty, *Europhys. Lett.*, **9** (1989) 465.
52. P.W. Anderson, *Phys. Rev.*, **86** (1952) 694.
53. T. Jolicoeur and J.C. Leguillou, *Phys. Rev. B*, **40** (1989) 2727.
54. S. Sachdev, *Lectures presented at the summer course on "Low Dimensional Quantum Field Theories for Condensed Matter Physicists 24 Aug. to 4 Sep. 1992, Trieste, Italy.*
55. B. Bernu, C. Lhuillier, and L. Pierre, *Phys. Rev. Lett.*, **69** (1992) 2590.
56. S. Fujiki and D.D. Betts, *Can. J. Phys.*, **65** (1987) 76.
57. H. Nishimori and H. Nakanishi, *J. Phys. Soc. Jpn.*, **73** (1985) 18.
58. S. Fujiki, *Can. J. Phys.*, **65** (1987) 489.
59. M. Imada, *J. Phys. Soc. Jpn.*, **56** (1987) 311.
60. R.R.P. Singh and D.A. Huse, *Phys. Rev. Lett*, **68** (1992) 1766.
61. D.Mouhanna, *Doctorate Thesis, L'université Pierre et Marie Curie (1994).*
62. N. Elstner, R.R.P. Singh, and A.P. Young, *Phys. Rev. Lett.*, **71** (1993) 1629.
63. N. Read and S. Sachdev, *Phys. Rev. Lett.*, **62** (1989) 1694.
64. R.E. Rapp and H. Godfrin, *Phys. Rev. B*, **47** (1993) 12004.
65. H. Godfrin, *Preprint.*
66. C. Broholm, G. Aeppli, G.P. Espinoza, and A.S. Cooper, *Phys. Rev. Lett.*, **65** (1990) 3173.
67. V. Elser, *Phys. Rev. Lett.*, **62** (1989) 2405.
68. B. Bernu, P. Lecheminant, C. Lhuillier, and L. Pierre, *Physica Scripta*, **49** (1993) 192.
69. S. Sachdev, *Phys. Rev. B*, **45** (1992) 12377.
70. B.I. Halperin, T.C. Lubensky, and S.H. Ma, *Phys. Rev. Lett.*, **32** (1974) 292.
71. I.D. Lawrie and C. Athorne, *J. Phys. A*, **16** (1983) 587.
72. P. Ginsparg, *Nucl. Phys. B*, **170** (1980) 388.

Magnetic System with Competing Interaction
Ed. H. T. Diep
©1994 World Scientific Publishing Co.

CHAPTER III

FRUSTRATION IN QUANTUM ANTIFERROMAGNETS

H. J. Schulz

Laboratoire de Physique des Solides,* Université Paris-Sud, 91405 Orsay, France.

and

T. A. L. Ziman and D. Poilblanc

Laboratoire de Physique Quantique,* Université Paul Sabatier, 31602 Toulouse Cedex, France.

1. Introduction

We discuss in this article the physics of frustrated quantum Heisenberg anti-ferromagnets at zero temperature and in low dimension, specifically one and two spatial dimensions. For simplicity we shall restrict discussion to models of Heisenberg type, that is a set of vector spins coupled with pairwise exchange interactions which are bilinear in the spin operators and isotropic under simultaneous rotation of all spins. We shall concentrate on the so–called $J_1 - J_2$ model in one and two dimensions where J_1 refers to a nearest neighbor exchange on the lattice and J_2 to a second nearest neighbor in one or two dimensions. We shall be particularly interested in the case of spin 1/2 for which quantum fluctuations should be most noticeable, but we shall also consider higher spin when results are known. The Hamiltonian of the model is

$$H = J_1 \sum_{\langle i,j \rangle} \boldsymbol{S}_i \cdot \boldsymbol{S}_j + J_2 \sum_{\langle i,j' \rangle} \boldsymbol{S}_i \cdot \boldsymbol{S}_{j'} \ . \tag{1}$$

Here $\langle i,j \rangle$ and $\langle i,j' \rangle$ represents summation over nearest– and next–nearest neighbor pairs on a one– or two–dimensional lattice (see fig.1), the spin operators obey $\boldsymbol{S}_i \cdot \boldsymbol{S}_i = S(S+1)$, and $J_1 = 1$ throughout this review.

The history of frustrated quantum antiferromagnets goes back at least to a paper by Fazekas and Anderson[1] who suggested, based on what by present day standards would be considered rather primitive numerical calculations, that the square antiferromagnet might be disordered for spin 1/2. Later there was a great deal of attention paid to classical frustrated antiferromagnets motivated in large part by experiments on spin glasses. While the main part of the activity was focused on the classical thermodynamics of such materials and the delicate question of whether there is a phase transition at a well–defined freezing temperature or not, there was some discussion of quantum effects. In this context, the investigation of frustrated models, both classical and quantum, was motivated by the idea that one might distinguish

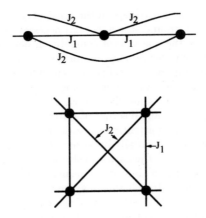

Fig. 1. The frustrated one– and two–dimensional lattices considered in this paper

two issues. One issue is the fact that in a disordered system like a spin glass there is a breaking of translational invariance of the crystal, and the other is that local degeneracies in the ground state might enhance fluctuations. By studying models on regular lattices without disorder, for example based on triangular geometries, it might be possible to give precise answers to some questions without the additional burden of disorder. J. Villain and co–workers[2,3] introduced the concept of "order from disorder" the idea that fluctuations might lift the degeneracy of different possible orderings. Of course to study the effect of quantum fluctuations, it must be remembered that quantum magnets differ from their classical counterparts in that in quantum mechanics local, as distinct from global, degeneracies rarely survive: if two initially degenerate quantum levels have an overlap the degeneracy will be immediately lifted. In addition, the models relevant to quantum magnetism are often, as those discussed here, vector models rather than the anisotropic Ising type variables that figure in most classically frustrated cases. Even classically three vector spins forming a triangle with competing antiferromagnetic couplings may not saturate each bond energy but they may nonetheless find a configuration minimizing the total bond energies with the three spins at angles of $2\pi/3$. This is unique up to a global rotation. Thus many of the interesting features of calculations of classical model systems which relied on a local degeneracy, giving for example a non–zero entropy in the limit of zero temperature, might be removed by quantum effects. Of course global degeneracies associated with macroscopic states with zero overlap in the thermodynamic limit can and do survive. What is the interest of frustration, then, if this is so? The question is interesting because while the above arguments are correct, they do not fully answer the question because of the fact that the mixing of local degeneracies invoked usually does not preserve a long range antiferromagnetic order. Thus while frustration in the sense of the production of large numbers of exactly degenerate ground states will not necessarily survive the introduction of quantum mechanics, the question of competition between local removal of degeneracy and long range order remains important. Frustration can be thought of as increasing the quantum mechanical zero–point fluctuations for an antiferromagnet beyond the most quantum–mechanical value of spin one–half to produce a system as far as possible from the semi–classical case of large spin.

The question of order and disorder of states with spontaneously broken symmetries has gained increasing attention since the earlier epoch inspired by the general question of spin–glasses essentially because of the increased orientation of the solid state physics community to the nature of strong quantum fluctuations in low dimensional solids. This is in part, but by no means entirely, due to the suggestions that strong correlations in electron liquids may be a key to understanding high temperature superconductors.[4] The resolution of the competition between different forms of order is frequently the central issue. In the fractional quantum Hall effect the competition between the dynamics induced by the external magnetic field and the Coulomb interactions between the electrons means one again needs reliable means of deciding between different symmetry states. The relevance of the frustrated quantum spin systems is that they may provide concrete, well–controlled theoretical and experimental systems where we may reliably test general ideas about quantum many body systems, applicable in the rather richer world that includes fermions and charge fluctuations as well as the spin fluctuations of pre–formed magnetic moments. As an example of this we may consider recent works testing hypotheses of describing interacting many body systems by means of random matrix theories.[5] For the case of the two dimensional Heisenberg antiferromagnet we shall produce an estimate for at least one critical Hamiltonian which may be used in the future to test ideas of critical scaling at a quantum critical point.

Returning to more specific application, one observes that the superconducting oxides are, before doping, rather satisfactory antiferromagnets[6] and the order they display most probably disappears as superconductivity appears. The experimental situation here, in echo of the situation of earlier times, is however clouded by the existence of disorder which makes the precise nature of the disappearance unclear, but nevertheless it is clear that an understanding of the question of order is important to fully describe such materials, whether or not it is key to predicting superconducting mechanisms.

Our aim here, then, is to discuss calculations of the last few years which attempt to give correct answers to the question of what happens to a quantum Heisenberg antiferromagnet when we introduce interactions or geometry such that in a classical picture not all antiferromagnetic bonds may be satisfied. In particular we want to answer the question of whether the long range antiferromagnetic order, or rather in the one–dimensional case infinite antiferromagnetic susceptibility associated with power–law quasi–order, is maintained as the frustration is increased, if it disappears in what manner and what is the phase that then occurs? We will see that again quantum mechanical systems are here rather more subtle than classical fluctuations in that while long range correlations may disappear in the original variables in the new phase there may appear a new order in a different order parameter. Thus there may be some confusion amongst readers of the current litterature about what may be called a "spin–liquid". The best definition is probably a state that has neither a non–zero expectation value of spin operators (i.e. spin rotation symmetry is conserved), nor any breaking of translational invariance. We shall below discuss calculations to determine the ground state where there is good evidence that magnetic order is absent.

In the first part of this review we shall discuss the $J_1 - J_2$ model in one dimension, and, following that, the same model in two dimensions on a square lattice. We shall give a shorter review of work concerned more specifically with triangular and Kagomé lattices. The interest of the last is that there is a good experimental realization. From the point of view of calculations there are advantages to the simpler systems based on a square lattice, namely that the techniques used, primarily exact diagonalisation and finite–size scaling, are most probably better adapted to the geometry we have treated, because the unit cells are simpler and the order parameters, at least initially, less complex. The general issues of course remain the same, but we should remember that the problem of acquiring results for sufficiently large systems to reliably give the limit of an infinite solid are much more difficult as the geometry becomes more complex.

A rather more technical question of interest for frustrated magnets is that the quantum mechanics is that of a system in which the node structure of the ground state is unknown. Thus while the unfrustrated bipartite quantum antiferromagnet may be shown to have a nodeless ground state, the same is not true for the frustrated case. Thus one of the reasons for studying such systems is that they are examples in which quantum Monte–Carlo techniques for example are in addition still plagued by what is known in general as the minus–sign problem.

2. One–Dimensional Models

If we take the Hamiltonian written in equation (1), but now restricting the sum over sites to be those on a single chain, and the J_2 coupling to be a second nearest neighbor coupling we have a model whose general properties are relatively well understood, especially for spin one–half. In particular, for vanishing J_2 it is exactly soluble by the Bethe Ansatz[7,8]. The Bethe Ansatz gave first the ground state energy and over the decades a more and more complete description of the excitation spectrum and thermodynamics. One feature that posed difficulties with what one might expect for low dimensional models emerged from this solution. True antiferromagnetic long–range order in the model is prohibited by the Mermin–Wagner theorem,[9] since it would have a spontaneously broken continuous symmetry and this is not possible in one dimension. In intuitive terms any ordered state by Goldstone's theorem implies the existence of gapless excitations and by integrating the number of such excitations over reciprocal space there would be a divergence in the number of such excitations. Nevertheless from the Bethe Ansatz the model has gapless excitations and an infinite antiferromagnetic susceptibility. In fact the spin correlations decay with a power law[10] corrected with a logarithm[11,12] as

$$\langle S_i \cdot S_j \rangle \approx (-1)^{(i-j)} \log(i-j)^{1/2} / |i-j| \ . \tag{2}$$

It was the knowledge of the gapless nature of the spectrum for the integrable model that in fact lead to Haldane's analysis[13] of the quantum antiferromagnets which distinguished between this, at first surprising, fact and the expectation related by mapping onto a the Lagrangian of $1 + 1$ dimensional $SU(2)$ invariant action which one would expect to be massive. In the course of the analysis Haldane advocated that addition of a frustrating J_2 interaction would lead to suppression of the quasi long–range order and appearance via an Kosterlitz–Thouless like transition of a state of spin–Peierls order (or spontaneous dimerization) which would have a gap of magnetic exciton type and spontaneously broken translational order[14]. Each even site is dimerized preferentially with either its left or right neighbor and the preference is preserved throughout the lattice, giving long–range order of Ising type. Finite size studies, first by looking at the behavior of the gap to construct a numerical renormalization group and later including the hypothesis of conformal invariance and the relation to field theories with topological terms, confirmed this picture with increasing degrees of sophistication. For spin $1/2$ the transition to the dimerized state is at a value of J_2 of approximately 0.25.[15] Further confirmation comes from the observation by Majumdar and Ghosh[16] that at $J_2 = 0.5$ while the model is not completely integrable a limited number of exact states can be exactly written down and the ground state has no long range antiferromagnetic correlations.

Later studies confirmed Haldane's original hypothesis that the unfrustrated Heisenberg chain remains massless for any half–integer spin[17,18]. With increasing frustration, there again is a transition to a dimerized state. For spin $3/2$ the critical value is about 0.33.[19] Here finite size scaling was very slowly convergent in essence because of the logarithmic decay of corrections to finite–size scaling mentioned before and it was only by using the full $SU(2) \times SU(2)$ symmetry at the critical point that convergent results could be found for the length of chain that could be diagonalized for spin $3/2$. For large half–integer spin one can compare to order S^2 the energy per bond of the antiferromagnetic state $(J_1 - J_2)S^2$ to the singlet energy from singlet pairing on every second bond $J_1 S(S + 1)/2$. The transition, if it is between a simple antiferromagnetic state and a dimerized state would then be estimated as $J_2 \approx 1/2$. In fact a classical spin would lower its energy further by forming a helical state, as will be discussed below for the integer case which would favor an antiferromagnetic state for larger J_2 so this limit cannot be considered completely understood.

For half–integer spin, then, there is a clear theoretical picture of the effects of frustration on the ground state: a quasi–ordered antiferromagnetic state with gapless excitations whose gaplessness is reminiscent of the Goldstone bosons that may occur in higher dimensions where the symmetry is truly broken. In fact the spin excitations are rather more complex than this in that they are better considered as combinations of pairs of half–spin objects than as elementary excitations.[20,21] Furthermore while one may be tempted to visualize this power law decay as indicating that the state is roughly antiferromagnetic, but with corrections at long distances, one should remember that there are other correlation functions that are also have power law decay: the singlet–singlet correlation function

$$\langle (S_i \cdot S_{i+1})(S_1 \cdot S_0) \rangle_c \sim \frac{(-1)^i \log(i)^{-3/2}}{i^{\eta_s}} \tag{3}$$

decays with the same power law $\eta_s = 1$ but with a different logarithmic amplitude. So while one may consider the state quasi antiferromagnetic one may also say it is "quasi valence–bond ordered".

Finite size studies were based on the detailed theory of finite size scaling predicted by conformal invariance of the fixed point. The full operator algebra can be calculated from the finite–size formula

$$E - E_0 = \frac{2\pi v}{L}(x + n) \tag{4}$$

where E is an energy of a level of a given symmetry in a finite chain length L with periodic boundary conditions, v is a spin–wave velocity defined from levels at finite wave number, n is an integer and x is the scaling dimension of an operator of appropriate symmetry. Finite size studies of the spin chains,[11] both from exact diagonalization of the Hamiltonian (1) and from Bethe Ansatz results for special cases, confirmed that the full operator algebra, i.e. the set of all power correlation functions, was consistent with that following from the conformally invariant model with the SU(2)×SU(2) chiral symmetry. One useful consequence of the knowledge of the operators is that one can examine the stability of the phase to different perturbations. For small J_2 the critical phase is marginally stable against an increase of J_2, as can be seen from the scaling dimension of the corresponding operator. Addition of relevant operators immediately leads to instability of the massless state and to the appearance of a gap in the excitation spectrum. For example a dimerizing term added to the Hamiltonian (1)

$$\delta H = \alpha \sum_{\langle i,i+1 \rangle} (-1)^i S_i \cdot S_{i+1} \tag{5}$$

is relevant and by reading of exponents leads to a gap

$$\Delta \approx \frac{\alpha^{2/3}}{\log(\alpha)^{1/2}} \tag{6}$$

As J_2 increases the massless state remains qualitatively the same until a finite value of J_2, about 0.25 for spin one–half and 0.33 for spin 3/2 at which the marginal operator becomes marginally relevant rather than marginally irrelevant. There then is a continuous transition to the spontaneously dimerized (spin–Peierls) state discussed above. In this state the excitations may be considered magnetic excitons: propagating modes that locally excite the bonds. For the value of $J_2 = 1$ corresponding to a line of back to back triangles the model is thus dimerized at least for spins 1/2 and 3/2. In the limit of infinite J_2 neglecting J_1 would give two disconnected antiferromagnetic chains but reintroduction of the coupling J_1 is a relevant operator, so that the spontaneous spin–Peierls order should persist for any large J_1.

For integer spins the picture is all rather different. First as predicted by Haldane, there is no longer even a quasi–ordered antiferromagnetic state but one with a gap, for which increasingly accurate numerical studies[22,23] have given $0.4105J_1$ for spin one and no frustration. The antiferromagnetic state is, however, in a sense ordered in that there exists an order parameter which is, however, non–local in the spin variables. The explicit form for the string correlation function exhibiting long–range order is[24]

$$\left\langle S_i^z \exp\left[i\pi \sum_{r=i+1}^{j-1} S_r^z \right] S_j^z \right\rangle . \qquad (7)$$

The physical interpretation of this order parameter is rather interesting: for spin 1 the z–component m_z of the spin can be -1, 0, or $+1$. Consider two spins situated at sites i and j, one with $m_z = -1$, the other with $m_z = 1$. Such a configuration gives a negative contribution to (7) if the S^z's on the intermediate sites sum to 0 (mod2), and positive otherwise. More generally, a positive value of (7) indicates that two successive spins, both with $|m_z| = 1$, but separated by an arbitrarily long *string* of $m_z = 0$ are preferentially aligned parallel, whereas a negative sign means preferential antiparallel alignment. It is this second possibility which is realized in the Haldane–gapped state.

Recent finite size scaling studies[25,26] have suggested that there is no phase transition as a function of increasing J_2, but this seems unlikely from the following argument: for large J_2, the system concists essentially of two very weakly coupled chains, each on one sublattice. Each of these subsystems then has an order defined by an order parameter of the type of (7). This state then represents an order of a different type from that present at $J_2 = 0$, and the addition of the nearest–neighbor coupling as a perturbation would be unlikely, because of the Haldane gap and the consequent absence of infinite susceptibilities, to change the order. It may well be that there is a higher order transition between two singlet phases, however question of what happens as the model is frustrated has not been addressed in detail. One question is whether for quantum spins there is any remnant of the classical picture of the $J_1 - J_2$ model with a spiral structure with relative angle $\theta = \arccos(-J_1/4J_2)$: one might expect this to occur for large integer spin, possibly with an increasingly large correlation length but without ever attaining long range order.

In general what do we learn from these one–dimensional models of frustrated magnetism? We see that as stated in the introduction the question is not really order versus disorder but more order of antiferromagnetic type in competition with order of some other type, in particular dimer ordering. Note that dimer formation is a purely quantum effect and has no direct classical analogue. For integer spin, the Haldane phase with no local order parameter may be considered as close to a disordered state but in a sense is highly ordered: the spin excitations for wave vectors close to the zone edge are actually well–defined objects with zero width although they disappear into a continuum for smaller k and higher energy.

3. Two Dimensions and Finite–Size Scaling Analysis

For the two–dimensional case .we take the same Hamiltonian but with nearest neighbors on a two–dimensional lattice and for the most part the J_2 on the diagonals of a square lattice. In the final section we shall discuss results other authors have obtained on the triangular and Kagomé lattices. The model on a square lattice has attracted most attention as a rather crude model[27] of the effects of doping on copper oxide planes. The square sites correspond to the copper sites in the two–dimensional plane of copper and oxygen that is the common structural feature of the family of superconducting copper oxides.

Modelling doped carriers by a moderately frustrating interaction seems nevertheless to be too simplistic, mainly due to the fact that doping introduces a new energy scale into the problem, namely the kinetic energy of the carriers, which is typically at least ten times bigger than the nearest neighbor exchange. However, as in one dimension, the model is of rather more general interest in that a complete understanding would, for a specific and well–defined case, provide a clear answer to several general questions of key interest in the area of quantum phase transitions. We note that in two spatial dimensions one can have long–range order of a vector order parameter at zero temperature. At finite temperatures, of course, any long range order would be eliminated by excitations but with an exponentially diverging correlation length which would amplify the stabilizing effects of relatively weak interplanar coupling.

The first question is that even in a ground state with true long range order of rather a classical looking symmetry, as is the case in an antiferromagnet, how do we show unequivocally that the order really is of long range and not simply local with a long correlation length? In order to compare to experiment how do we calculate physical correlations without resorting to low order perturbation theory? In fact, the renewed interest in recent years in the model was because of doubts that the unfrustrated case would display long–range order in the thermodynamic limit. While such doubts are now relatively rare thanks to extensive numerical calculations and tighter rigorous limits for higher spin and lower symmetry,[28] there is as yet no rigorous proof for the isotropic spin one–half model in two spatial dimensions. Thus the first element to establish convincingly is that the order is indeed present. One then may test the quantitative success of ideas of finite size scaling as applied to numerical diagonalisations limited to small samples. In two dimensions this is much more severe a constraint than in one dimension for two separate reasons. The first is that for a given linear dimension there are obviously many more sites: hence in practice the largest lattices that can be treated are six by six. The second is that in the search for critical phases the conformal invariance of correlations that in one dimension gave extremely detailed descriptions of finite size scaling (and corrections to it) is much less informative. It is true that critical phases may be less common than in one dimension but they are nonetheless important in order to delimit phases of different symmetry. The conformal group in three (two spatial plus one time) dimensions being smaller than in one plus one dimension where it is infinite dimensional we are in effect back to ideas of scaling alone. The history of finite size

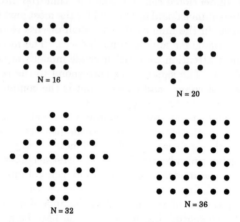

Fig. 2.   The clusters used in finite size studies with sizes ranging from 16 to 36 sites

effects goes back to Anderson in the early fifties,[29] who first invoked the fact that the infinite degeneracy of the ground–state with spontaneously broken continuous symmetry must be manifest in a large number of nearly degenerate states in a large but finite system. This idea of a "tower" of states whose degeneracy corresponds to the symmetry and whose energy scales determine the long distance parameters of the spontaneously broken model of the infinite system has since been made more precise and less dependent on perturbative concepts in the language of non–linear sigma models and chiral perturbation theory.[30,31] The model we consider here has the advantage over, for example, the triangular or Kagomé antiferromagnets[32] that the classical limit has a simpler unit cell and thus the structure of the towers rather simpler. In practice the difficulties of applying finite size studies are still considerable: there are subleading as well as leading corrections which make the ultimate goal of reliable quantitative calculations difficult even here. It is helpful that we may easily stabilize the ordered state to study the disappearance of order in a controlled fashion by applying negative J_2.

A second question is whether the finite size methods developed can be applied all the way to a critical point at which the order may disappear. The first step is to identify critical parameters unequivocally, so far this is still a matter for contention, indeed some Schwinger boson expansions have been interpreted as indicating a first order transition[33] at least for large spin.

A third question, separate from the study of ordered antiferromagnetism, is the question of what happens when this order disappears. In the mapping of quantum interacting ground states to thermodynamics of classical models in higher dimensions there is at first sight a difference in that quantum phase transitions tend to show order–order rather than order–disorder transitions. Of course what one means by "order" is crucial to such a distinction, here an ordered state would be understood to have long range order in a different local order parameter, for example a spin–Peierls dimerization variable or chirality parameter. In section 3.3 below we shall discuss the various forms of order proposed and their likely occurrence.

As we said in the introduction, part of the interest in the model is as providing example of a quantum phase transition. One would then like to disorder the initially ordered state. While it is true that the initial antiferromagnetically order is destroyed by increasing J_2 this does not guarantee a disordered state. What happens for large J_2 is that there is simply a different form of magnetic order. This is not at all surprising in a classical picture where the magnetic ordering vector is given by the minimum of the Fourier transform of the couplings:

$$J(Q) = 2J_1(\cos(q_x) + \cos(q_y)) + 2J_2(\cos(q_x + q_y) + \cos(q_x - q_y))$$

For $J_2 = 0$ the minimum is at (π, π) where it remains for small J_2 and finally there is a transition when J_2 exceeds 0.5 where the minima simply moves from the corner (π, π) of the Brillouin zone to one of two equivalent, but separate values $(\pi, 0)$ or $(0, \pi)$ when J_2 exceeds 0.5. At the classically critical value of 0.5 there are degenerate lines of minima around the edge of the first Brillouin zone, ie for $q_x = \pi, q_y$ arbitrary, or $q_y = \pi, q_x$ arbitrary. This is clearly seen from the factorization valid for this special value:

$$J(Q) = 2J_1\left[(\cos(q_x) + 1)(\cos(q_y) + 1) - 1\right]$$

If we reason more microscopically, for large J_2 we first form a simple antiferromagnet on each the two sublattices formed by connecting diagonals alone. If J_1 is set to zero we have two independent antiferromagnets. With the reintroduction of a relatively small J_1 coupling, classically there would still be an extra degeneracy involving the relative angle of the two sublattices. This is because a given spin has four nearest neighbors of the other sublattice and the average spin of the four will be zero, irrespective of the axis of ordering. This degeneracy is removed when one includes quantum fluctuations, however, as can be seen from a simple argument first made by Shender[34] in three dimensions. The argument is simply that in any given ordered state a contribution to the quantum mechanical ground state energy comes from zero point motion of the spins, from the effects of the transverse couplings. Spin waves in general have two polarisations transverse to the ordering vector. Now if we take the two sublattice vectors of the two antiferromagnetic structures in a general directions the spin waves transverse to one sublattice vector will not be transverse to the other. Only if the two ordering directions are in the same direction will the full energy be gained. This means two neighboring spins are either parallel or anti–parallel.

The present model is rather interesting as an example of order from disorder. As to be expected from the above argument, for sufficiently large J_2 we find a collinearly ordered state, with the magnetization directions of the two sublattices fixed with respect to each other, e.g. the quantum fluctuations have lowered the symmetry of the classical grouns state. In the intermediate case, at exactly the classical transition point $J_2 = J_1/2$, the order, most probably of the dimer type, is another example "order from disorder" in that quantum mechanical transitions select a well defined ground state from amongst the very large manifold of classically degenerate states.

The $J_1 - J_2$ model we have studied here has been investigated previously by a number of techniques. Previous finite–size studies[35,36] found some indication of an intermediate phase without magnetic order, however due to the limitation to $N = 16$ and 20 only, it was impossible to make extrapolations to the thermodynamic limit and to arrive at quantitative statements. Our own previous study,[37] using $N = 16$ and 36, produced results very similar to our current best estimates. However, due to the larger number of clusters we now use (and due to the possibility to ignore the anomalous $N = 16$ cluster), we feel that our conclusions are considerably more reliable.

Lowest order spin–wave theory[38] produces a phase diagram very similar to ours (see fig.7). On the other hand, higher order (in $1/S$) calculations do not seem to be very useful, due to increasingly strong singularities at $J_2 = J_1/2$. It has been attempted to include higher order corrections using a selfconsistently modified spin wave theory.[39,40] These calculations as well as the closely related Schwinger boson approach[33] produce a first order transition between Néel and collinear state. However, their applicability to an $S = 1/2$ system is hard to judge.

Quantum Monte Carlo methods are plagued with the sign problem for frustrated spin systems. Nevertheless, conclusions very similar to the modified spin wave calculations have been reached recently using a quantum Monte Carlo method.[41] However, these results have rather large error bars and in some cases, in particular in the region of intermediate J_2, are in disagreement with our present exact results. The validity of these results thus appears doubtful to us.

Another approach has been via series expansion methods around a lattice covered by isolated dimers.[42] Expanding around a columnar arrangement of dimers, these authors find a phase diagram very similar to ours, at least as far as magnetic order is concerned. However, these results are not without ambiguity: expanding around a staggered dimer arrangement, there appears to be a first order transition between Néel and collinear states. The results of this method may be biased by the starting point of the expansion.

We shall discuss results gained from finite–size diagonalization on clusters of $N = 16, 20, 32, 36$ sites (fig.2) with periodic boundary conditions. These are all the clusters accessible by our present calculational means that both respect the square symmetry of the lattice and do not frustrate the collinear magnetic state expected at large J_2. The obvious way to look for long range order from a finite size calculation is to calculate the magnetic susceptibility, or squared order parameter

$$M_N^2(\boldsymbol{Q}) = \frac{1}{N(N+2)} \sum_{i,j} \langle \boldsymbol{S}_i \cdot \boldsymbol{S}_j \rangle e^{i\boldsymbol{Q} \cdot (R_i - R_j)} \ . \tag{8}$$

The normalization, in particular the factor $N+2$, is chosen so that this quantity is size–independent in a perfect Néel state. The values we use are always expectation values in the *true ground state*, e.g. for large J_2 states of symmetry B ($N = 20$) or B_1 ($N = 36$) are used. The dominant type of magnetic order changes from $\boldsymbol{Q} = (\pi, \pi)$ at relatively small J_2 (Néel state) to $\boldsymbol{Q} = (\pi, 0)$ at larger J_2 (the collinear state). How exactly this change occurs will be clarified in the following section.

3.1. Finite–Size Scaling Analysis of the Small–J_2 Antiferromagnetic Phase

As long as frustration is not too strong, the ground state of the model (1) is expected to have long–range antiferromagnetic order, and then the low–energy long–wavelength excitations are expected to be described by the quantum nonlinear sigma model,[43] with action

$$S = \frac{\rho_s}{2} \int d^2r \int_0^\beta d\tau \left[(\nabla \boldsymbol{n})^2 + \frac{1}{c^2} \left(\frac{\partial \boldsymbol{n}}{\partial \tau} \right)^2 \right] \ . \tag{9}$$

Here $\boldsymbol{n} \equiv \boldsymbol{n}(\boldsymbol{r}, t)$ is the local orientation of the staggered magnetization, with $|\boldsymbol{n}| = 1$, c and ρ_s are the spin wave velocity and stiffness, and the inverse temperature β has to be taken to infinity here as we are interested in ground state properties. Thus there are three important independent physical parameters to describe the physics at low frequencies and temperature, two of which we can take to be c and ρ_s. The third is the relation between the physical staggered magnetization and the normalized vector $\boldsymbol{n}$, i.e. the absolute intensity of magnetic Bragg peaks. Lowest order spin wave theory gives[43,44] $c_0 = \sqrt{2(1 - 2J_2/J_1)}J_1$, $\rho_{s0} = (J_1 - 2J_2)/4$, but there are of course important quantum fluctuation corrections to these quantities. One way to extract these corrections from finite size data will be discussed below. Other methods to obtain these quantities in the unfrustrated case $J_2 = 0$ are described in recent review articles.[45,46] We note that the magnetic susceptibility at $\boldsymbol{q} = 0$ is given by $\chi = \rho_s/c^2$, which in lowest–order spin wave theory equals $1/(8J_1)$.

From the classical limit we expect a transition between a Néel ordered region for $J_2 \lesssim 0.5$ to a state with so–called collinear order (i.e. ordering wavevector $\boldsymbol{Q} = (\pi, 0)$ at $J_2 \gtrsim 0.6$. To analyze the way this transition occurs in more detail, we used finite–size scaling arguments.[47,48] These arguments are based on the fact that once there is a continuous broken symmetry, Goldstone's theorem guarantees certain analytic properties of the correlation functions. The leading effects of finite size are to replace an integral over k space of a given observable by a discrete sum. In other words the finite size acts to eliminate zero point motion of modes that are not present because of the finite boundary conditions. The asymptotic difference gives leading terms in with coefficients that depend on the values of various physical quantities in the thermodynamic limit. Thus one can estimate the long–wavelength constants from the magnitude of the finite–size correction. From the nonlinear sigma model describing the low–energy excitations in a state with Néel order one can then derive the finite–size properties of various physical quantities. The quantity of primary interest here is the staggered magnetization $m_0(\boldsymbol{Q}_0)$ defined by

$$m_0(\boldsymbol{Q}_0) = 2 \lim_{N \to \infty} M_N(\boldsymbol{Q}_0) \ , \tag{10}$$

where $\boldsymbol{Q}_0 = (\pi, \pi)$. The normalization is chosen so that $m_0(\boldsymbol{Q}_0) = 1$ in a perfect Néel state. The leading finite size corrections to m_0 are given by

$$\begin{aligned}
M_N^2(\boldsymbol{Q}_0) &= \frac{1}{4} m_0(\boldsymbol{Q}_0)^2 + 1.2416 \frac{\kappa_1^2}{\sqrt{N}} + ... \\
&= \frac{1}{4} m_0(\boldsymbol{Q}_0)^2 (1 + \frac{0.6208c}{\rho_s \sqrt{N}} + ...) \ ,
\end{aligned} \tag{11}$$

where for the infinite system κ_1 gives the amplitude of the diverging matrix element of the spin operator between the ground state and single magnon states at $\boldsymbol{Q} \approx \boldsymbol{Q}_0$. Least square fits of our finite–size results to eq.(11) are shown in fig.3. For small values of J_2 the scaling law is quite well satisfied: e.g. for $J_2 = 0$ the four data points in fig.3 very nearly lie on the ideal straight line, and the extrapolated value of the staggered magnetization, $m_0(\boldsymbol{Q}_0) = 0.648$, is quite close to the best current estimates, $m_0(\boldsymbol{Q}_0) = 0.615$. Using the same type of finite size extrapolations for other values of J_2, we obtain the results indicated by a dashed line in fig.4. For $J_2 = 0$, a check on the reliability of our method can be obtained by comparing the numerical results with what one would expect from eq.(11), using the rather reliable results for m_0, c, and ρ_s obtained by series expansion techniques.[49-51] The curve expected from eq.(11) is shown as a dash–dotted line in fig.3. It appears that there are sizable but not prohibitively large next–to–leading corrections.

Another measure of the reliability of the finite–size extrapolation can be obtained comparing results obtained by the use of different groups of clusters. For negative J_2, i.e. *nonfrustrating interaction*, the values of $m_0(\boldsymbol{Q}_0)$ are nearly independent of the clusters sizes used, and the results in fig.4 therefore are expected to be quite accurate. In this region the next nearest neighbor interaction stabilizes the antiferromagnetic order and therefore the staggered magnetization tends to its saturation value unity for large negative J_2. On the other hand, for positive J_2 the interaction is frustrating. In this case, the agreement between different extrapolations is never as good. We see, that in all but two cases the staggered magnetization tends to zero as in a second order phase transition, with a critical value of J_2 between 0.3 and 0.5. The question than arises as to which extrapolation to trust most. In fact, none of the clusters considered here is free of some peculiarity: for $N = 16$, there is an extra symmetry when $J_2 = 0$, because this cluster

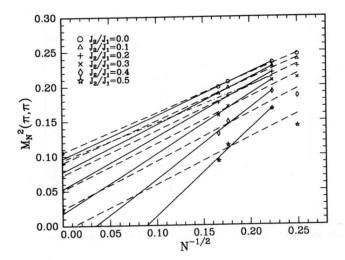

Fig. 3. Finite size results for $M_N^2(\boldsymbol{Q}_0)$ for different values of J_2. The full lines are least squares fits to the data according to eq.(11), using all available clusters. The dashed lines are fits using only $N = 20, 32, 36$. The dash–dotted line is the leading finite size behaviour expected at $J_2 = 0$ (see eq.(11)).

is in fact equivalent to a $2 \times 2 \times 2 \times 2$ cluster on a four–dimensional hypercubic lattice; the $N = 20$ cluster has a lower symmetry than all the others (C_4 instead of C_{4v}); for $N = 20$ and $N = 36$ the ground state changes symmetry with increasing J_2. In addition the 20 and 32 site clusters are unusual in that they are rotated, by different angles, with respect to the lattice directions. The least biased choice

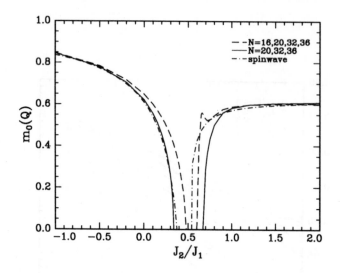

Fig. 4.   Comparison of our finite size fits for the antiferromagnetic and collinear order parameters (left and right curves, respectively) with linear spin wave theory.

then is probably to use all available clusters, as indicated by the full lines in fig.4, leading to a critical value of J_2 for the disappearance of antiferromagnetic order of $J_{2c} \approx 0.48$, rather close to those found earlier using only the $N = 16, 36$ clusters[37] where it was of course difficult to give a systematic estimate of errors. However, examining the scaling results from different clusters we see, see figure 3, that the $N = 16$ results are slightly anomalous, in particular for $J_2 \geq 0.3$. Using the larger ($N = 20, 32, 36$) clusters only gives more consistent results and indicates a critical value of $J_{2c} \approx 0.34$. These more recent results are then more trustworthy and the conclusion that there is a non–magnetic state beyond a certain critical value of the frustration is reinforced, with a moderate shift of the best estimate: from $J_{2c} \approx 0.4$ to $J_{2c} \approx 0.34$.

The ground state energy per site in the thermodynamic limit can be obtained from the finite–size formula for an antiferromagnet[47]

$$E_0(N)/N = e_0 - 1.4372 \frac{c}{N^{3/2}} + \dots ,$$ (12)

where c is the spin wave velocity. This provides a way of obtaining the constant c in the non–linear sigma model. In the collinear state (see the next section), an analogous formula holds, but with c replaced by some anisotropy–averaged value. Away from the "critical" intermediate region, i.e. for $J_2 \leq 0.2$ and $J_2 \geq 0.8$, eq.(12) provides a rather satisfying desription of the results, in particular if the $N = 16$ cluster is disregarded. The fit is even better than that for the order parameters This is certainly in large part due to the much weaker finite size correction to the ground state energy, as compared to those for the order parameters. On the other hand, in the intermediate region $0.4 \leq J_2 \leq 0.7$, the fits are not very good. In this region the ground state energy per site is rather irregular, for example there is generally a *decrease* from $N = 32$ to $N = 36$ contrary to what eq.(12) suggests. The failure of eq.(12) in the intermediate region is of course not surprising, as the analysis of the previous section showed the absence of magnetic order, which implies the non–existence of an effective nonlinear sigma model and therefore the invalidity of the extrapolation formula (12). The result of our extrapolations is shown in fig.5. Over

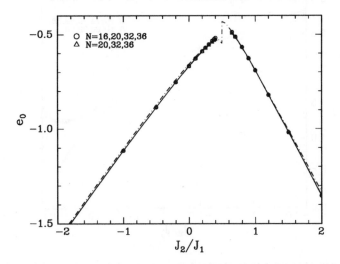

Fig. 5. Ground state energy per site as obtained from finite size extrapolation using eq.(12). In the intermediate region $0.4 < J_2 < 0.65$ the extrapolation can not be used reliably, and no results are shown. Results obtained using different clusters are indistiguishable on the scale of this figure. The dash–dotted line is the spin–wave result.

most of the region shown, results from extrapolations using different clusters are indistinguishable on the scale of the figure. Only close to the critical region is there a spread of about 2 percent in the results. In particular, at $J_2 = 0$ we find values between $e_0 = -0.668$ and $e_0 = -0.670$, very close to the probably best currently available estimate, obtained from large–scale quantum Monte Carlo calculations, of $e_0 = -0.66934$.[52]

The amplitude of the leading correction term in eq.(12) allows for a determination of the spin wave velocity c. Results are shown in fig.6. In this case, there is a wider spread in results. This is certainly not surprising, given that this quantity is derived from the correction term in eq.(12). Nevertheless, the agreement between different extrapolations is reasonable for $J_2 \leq 0$. At $J_2 = 0$ and using all clusters we find $c = 1.44 J_1$, close to but somewhat lower than the spin wave result $c_{SW} = 1.65 J_1$. For positive J_2 the extrapolations give different answers, according to whether the $N = 16$ cluster is included or not. This of course is due to the anomalous behavior of this cluster in the energy extrapolations. An important point should however be noticed: independently of the inclusion of the $N = 16$ cluster, at the critical value J_{2c} for the disappearance of the antiferromagnetic order the spin wave velocity remains finite. The final parameter in the nonlinear sigma model is the spin stiffness

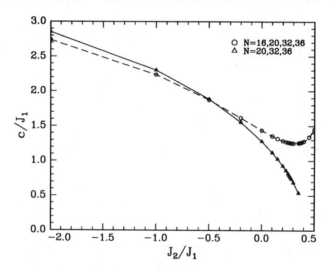

Fig. 6. The spin wave velocity in the antiferromagnetic state as obtained from finite size extrapolation using eq.(12). No results are shown in the region where according to the previous analysis there is no antiferromagnetic order $J_2 > 0.5$ or $J_2 > 0.34$ according to whether the $N = 16$ cluster is included or not.

constant ρ_s. It can be found from our finite size results using[47]

$$\rho_s = \frac{m_0(\mathbf{Q}_0)^2 c}{8\kappa_1^2} , \qquad (13)$$

with κ_1 determined from eq.(11). Results are shown in fig.7. Again, for the same

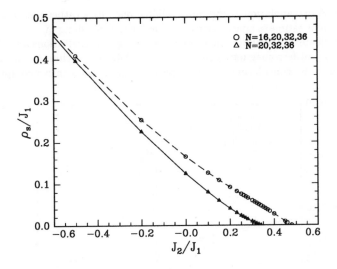

Fig. 7. The spin stiffness in the antiferromagnetic state as obtained from finite size extrapolation using eq.(13). Lines are a guide to the eye.

reasons as before, there is some scatter in the results, because of the use of the correction terms in eqs.(11) and (12). The result at $J_2 = 0$ is about a factor two lower than other estimates.[43,50] The fact that $\rho_s \to 0$ as $J_2 \to J_{2c}$ is again in agreement with expectations from the nonlinear sigma model analysis, but is of course a trivial consequence of eq.(13).

An independent test of the reliability of the results was obtained by calculating the susceptibility χ: even in an antiferromagnetically ordered state, the ferromagnetic susceptibility is finite, whereas for unconventional states (e.g. dimer or chiral), one has a spin gap and therefore a vanishing susceptibility. The vanishing of the susceptibility can thus be associated with the vanishing of the magnetic order parameter. Moreover, in an antiferromagnetic state one has $\chi = \rho_s/c^2$, and we thus have a consistency check on our calculated values for c and ρ_s. At fixed cluster size one has $\chi(N) = 1/(N\Delta_T)$, where Δ_T is the excitation energy of the lowest triplet state (which has momentum $Q = (\pi, \pi)$ in an antiferromagnetic state). An extrapolation of $\chi(N)$ to the thermodynamic limit can be performed using the finite–size formula[48,52] $\chi = \chi(N) - const./\sqrt{N}$, and results are shown in fig.8. Again, the $N = 16$ cluster behaves anomalously in that $\chi(N)$ increases going from $N = 16$ to $N = 20$, whereas for bigger clusters there is the expected decrease. In the present case, this anomaly occurs for nearly the whole range $J_2 > 0$. Also, our result for $J_2 = 0$ and using $N = 20, 32, 36$ is $\chi = 0.671$, very close to both Monte Carlo estimates[52,53] and series expansion results.[49–51] We therefore think that the

$N = 20, 32, 36$ extrapolation is the most reliable one. It is rather pleasing to that this independent estimate gives a critical value for the vanishing of the susceptibility (which indicates the disappearance of gapless magnetic excitations and therefore of long–range antiferromagnetic order) of $J_{2c} \approx 0.42$, quite close to the estimate we found above by considering the order parameter. A quantitative comparison of

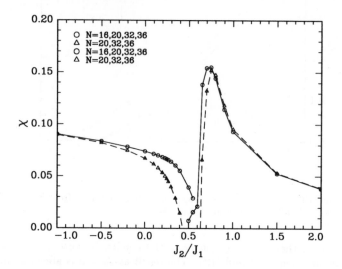

Fig. 8. The susceptibility as obtained from $\chi = 1/(N\Delta_T)$ (Néel region) and $\chi = 2/(N\Delta_T)$ (collinear region) using different extrapolations. As discussed in the text, the $N = 20, 32, 36$ extrapolation is expected to be the most reliable one.

results for the susceptibility obtained either from the excitation gap or from the previously calculated values of c and ρ_s and using $\chi = \rho_s/c^2$ reveals discrepancies of the order of 20 or 30 percent (see fig.8), even well away from the "critical region" $J_2 \approx 0.4$. The most likely explanation for this is that our calculation of c and ρ_s is based on *corrections* to the leading finite–size behavior, whereas χ is obtained directly from the gap. The direct estimate of χ is thus expected to be more precise.

Another way to assess the consistency of the finite–size extrapolations we using is to verify the underlying scaling hypothesis via a "scaling plot". The fundamental constants c and ρ_s of the nonlinear sigma model define a length scale c/ρ_s, and if finite size scaling is verified one therefore expects all finite size corrections to be universal functions of the variable $x = c/(\rho_s\sqrt{N})$. In particular, for the order parameter susceptibility we expect

$$M_N^2(\mathbf{Q}_0) = m_0(\mathbf{Q}_0)^2 \Phi(x) \ . \tag{14}$$

Combining eqs.(11) and (13) the small–x expansion of the scaling function is $\Phi(x) = (1 + 0.6208x)/4$. Plots of our results for $M_N^2(\mathbf{Q}_0)$ as a function of the scaling variable

x are shown in fig.9. One sees that for he $N = 20, 32, 36$ extrapolation the plot is nearly perfect in that nearly all data points are collapsed onto a single curve. The only points that show a significant deviation are those obtained for $N = 16$ close to the phase transition to the nonmagnetic state. This of course is nothing but a manifestation of the anomalous behavior of this cluster already found previously. The behavior for the $N = 16, 20, 32, 36$ extrapolation is clearly less satisfying. A similar scaling plot for the ground state energy produces even better results, due to the better convergence of the corresponding finite–size formula (12). A scaling

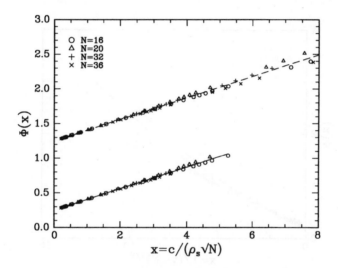

Fig. 9. Scaling plot of $\Phi(x) = M_N^2(\boldsymbol{Q}_0)/m_0(\boldsymbol{Q}_0)^2$ as a function of the variable $x = c/(\rho_s\sqrt{N})$, using the $N = 20, 32, 36$ (lower curve) and $N = 16, 20, 32, 36$ (upper curve) extrapolations for c and ρ_s. For clarity, data for the $N = 16, 20, 32, 36$ extrapolation are shifted upward by 3 units. The straight lines represent the spin wave result $\Phi(x) = (1+0.6208x)/4$

plot like fig.9 permits to assess the consistency of data obtained for clusters of different sizes, however, the form of the scaling function itself is obviously less significant as the coefficients c and ρ_s entering the definition of the scaling variable x are calculated assuming finite–size formulae like (11) and (12) to be valid, i.e. implicitly *assuming* the form $\Phi(x) = (1 + 0.6208x)/4$. An *independent* estimate of Φ can in principle be obtained using independent estimates for c and ρ_s. We do not have currently such an estimate for ρ_s, however we can use our independent results for the susceptibility (fig.8) to rewrite the scaling variable as $x = 1/\sqrt{\chi\rho_s N}$. The plot obtained using estimates for ρ_s and χ from $N = 20, 32, 36$ is shown in fig.10. The collapse of data obtained for different sizes is not as satisfactory as in the previous case, however, this is certainly related to the fact that here we use a

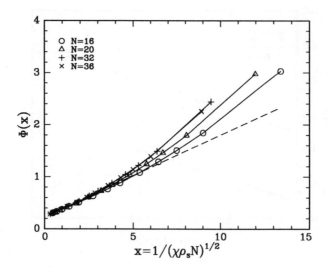

Fig. 10. Scaling plot of $\Phi(x) = M_N^2(\mathbf{Q}_0)/m_0(\mathbf{Q}_0)^2$ as a function of the variable $x = 1/(\chi\rho_s N)^{1/2}$, using the $N = 20, 32, 36$ results for χ (see fig.8) and the $N = 20, 32, 36$ extrapolation and ρ_s. The dashed line represents the spin wave result $\Phi(x) = (1+0.6208x)/4$

second independently estimated quantity, namely χ. Still, for $x \lesssim 5$, the collapse is rather good, showing the consistency of our analysis in this region. For the larger clusters, this region corresponds to $J_2/J_1 \leq 0.2$, i.e. it extends rather close to the transition which occurs at $J_2/J_1 \approx 0.28$. For small x the calculated scaling function essentially agrees with the spin wave results shown by the dashed line in fig.9. For $x \gtrsim 5$, there are discrepancies between results obtained from $M_N^2(\boldsymbol{Q}_0)$ for different N. This probably indicates that at least for the smaller clusters, finite size effects become so important that it is no more sufficient to include the lowest order finite size corrections only. The fact that the numerically found scaling function is larger than the spin wave approximation is not entirely unexpected: in fact, for large x, i.e. in the critical region, one would expect $\Phi(x) \propto x^{1+\eta}$, where η is the correlation exponent of the three–dimensional Heisenberg model. However, we doubt that what we observe in fig.10 is actually a critical effect. First, the numerical value of η is very small: $\eta \approx 0.028$, and one thus expects an extremely smooth crossover. Moreover, in fig.10 we have used the independently calculated susceptibility (see fig(8) which goes to zero only at $J_2/J_1 \approx 0.4$, rather than at $J_2/J_1 \approx 0.28$ where our estimated staggered magnetization vanishes. Consequently, the abscissae of the data points in fig.10 are underestimated, i.e. the data in fig.10 overestimate the true $\Phi(x)$.

3.2. Finite–Size Scaling Analysis of the Large–J_2 Antiferromagnetic Phase

We now follow the same logic to analyze the behavior for larger J_2, where we expect the existence of magnetic order with ordering wavevector $\boldsymbol{Q}_1 = (\pi, 0)$. Of course, this state again breaks the continuous spin rotation invariance, and therefore the low energy excitations are described by a nonlinear sigma model. There is an additional breaking of the discrete lattice rotation symmetry (ordering wavevector $(0, \pi)$ is equally possible), however, this does not change the character of the low–lying excitations, though it may make the non–linear sigma model anisotropic. The corresponding effective parameters therefore cannot been obtained quite as easily from the lowest finite size correction terms. The finite size behavior is again determined by the low energy properties, and therefore we expect a finite size formula analogous to eq.(11):

$$M_N^2(\boldsymbol{Q}_1) = \frac{1}{8}m_0(\boldsymbol{Q}_1)^2 + \frac{const.}{\sqrt{N}} + \cdots \quad . \tag{15}$$

Here the factor $1/8$ (instead of $1/4$ in (11)) is due to the extra discrete symmetry breaking which implies that finite–size ground states are linear combinations of a larger number of basis states. Moreover, the linear sigma model is anisotropic, because of the spontaneous discrete symmetry breaking of the ordering vector, and consequently a precise determination of the coefficient of the $\sqrt{N}$–term is not straightforward. The important point here is however the N–dependence of the correction term in eq.(15). Least square fits of our numerical results to eq.(15) are shown in fig.11, and the extrapolated collinear magnetization $m_0(\boldsymbol{Q}_1)$ is shown in fig.4. For $J_2 \geq 0.9$ eq.(15) provides a satisfactory fit to our data, even though not

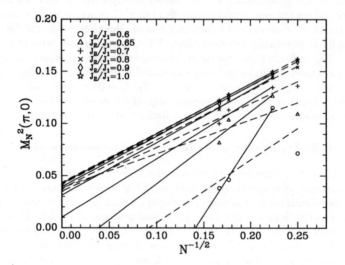

Fig. 11.   Finite size results for $M_N^2(\boldsymbol{Q}_1)$ for different values of J_2. The lines are least squares fits to the data according to eq.(15), using all available clusters. The dashed lines are fits using only $N = 20, 32, 36$.

quite as good as in the region $J_2 \leq 0$ in the Néel ordered case. For smaller J_2 there is a wide spread in the extrapolated results, depending on the clusters used. We notice however that for the majority of clusters used, there is a common feature: $m_0(\boldsymbol{Q}_1)$ remains finite down to $J_2 = 0.65$, and then suddenly drops to zero at $J_2 = 0.6$. This would indicate a *first order transition* to the collinear state somewhere in the interval $0.6 < J_{2c} < 0.65$. ¿From $N = 16, 20, 32, 36$ extrapolation one then obtains a collinear magnetization which is roughly constant above J_{2c} at $m_0(\boldsymbol{Q}_1) \approx 0.6$. Notice that the first–order character of the transition is *not* due to the level crossings occurring in the $N = 20$ and $N = 36$ clusters: if these clusters are omitted from the extrapolation, the first order character is in fact strongest. As in the small J_2 regime scaling indicates that the 16 site cluster is slightly anomalous and neglecting it gave what we consider a best estimate of the critical coupling coupling $J_{2c} \approx 0.68$. More importantly, the phase transition from the nonmagnetic to the collinear state now appears to be of second order.

A calculation of the susceptibility analogous to that of the small J_2 phase can be performed in the region of larger J_2, where the lowest excited triplet state is at $\boldsymbol{Q} = (\pi, 0)$. In this case, because of the double degeneracy of this state, the susceptibility is given by $\chi = 2/(N\Delta_T)$. Using the same finite–size extrapolation as before, and results obtained for different combinations of cluster sizes are shown in fig.8. The 16 site cluster again shows rather anomalous behavior and therefore we did not take it into account in these extrapolations. The results then indicate a transition into a nonmagnetic ($\chi = 0$) state at $J_2/J_1 \gtrsim 0.6$, in approximate agreement with what we obtained from estimates of the order parameter above. The decrease of χ with increasing J_2 is not surprising, as for large J_2 the model consists of two nearly decoupled unfrustrated but interpenetrating Heisenberg models, each with exchange constant J_2, and consequently one has $\chi \propto 1/J_2$. It is more surprising to see the sharpness of the maximum of χ around $J_2/J_1 = 0.7$.

In the finite size study then, the most important finding seems to that there is convincing evidence for the existence of a region of intermediate second nearest neighbor coupling J_2 where no magnetic order, antiferromagnetic, collinear or otherwise, exists (by magnetic order, we mean a state with a non–zero expectation value of a local spin operator). The location of the boundaries of this nonmagnetic region depended on the cluster size involved in the estimate. With $N = 20, 32, 36$ this interval is larger than previous estimates: $0.34 < J_2/J_1 < 0.68$. Given the irregular behavior of the $N = 16$ cluster we often found above, in particular in the region of intermediate J_2, this estimate would appear to be the more reliable than the one based on all availale clusters. In any case, independently of which extrapolation one prefers, there is a nonmagnetic interval.

Beyond the existence of a nonmagnetic region, we have also obtained quantitative estimates for a number of fundamental physical parameters in the magnetically ordered states, antiferromagnetic for small or negative J_2, collinear for large positive J_2. The accuracy of these estimates can best be assessed by comparing with the unfrustrated case $J_2 = 0$, for which case there are currently rather precise results available, mainly from large–scale Monte Carlo calculations and series expansions.

Table I. Comparison of our results at $J_2 = 0$ obtained from the $N = 16, 20, 32, 36$ and $N = 20, 32, 36$ extrapolations with previous estimates from series expansions and quantum Monte Carlo calculations. A more complete compilation of previous results can be found in review articles.[45,46]

	e_0	$m_0(\boldsymbol{Q}_0)$	χ
$N = 16, 20, 32, 36$	-0.6688	0.649	0.0740
$N = 20, 32, 36$	-0.6702	0.622	0.0671
series expansions [a]	-0.6696	0.614	0.0659
quantum Monte Carlo [b]	-0.6693	0.615	0.0669

[a]See refs. 49 and 51.

[b]See refs. 52 and 53.

A summary of our results, together with other recent data, is given in table I.

The results for the ground state energy, the staggered magnetization, and the susceptibility agree to within a tenth of a percent or a percent, respectively, with the best currently available numbers. On the other hand, our estimates for the spin-wave velocity and the spin stiffness are rather imprecise. This is certainly mainly due to the fact that these quantities are obtained from the amplitudes of the leading correction to the asymptotic large–size behavior of the ground state energy and the order parameter susceptibiliy, and these correction are almost certainly estimated less precisely than the leading terms.

We found it instructive to also investigate regions where magnetic order is well-established, i.e. $J_2 \leq 0$ for the antiferromagnetic case and $J_2 \geq J_1$ for the collinear case. In these regions we find that the finite–size formulae like (11) and (12) provide an excellent fit to our numerical results. The progressive worsening of the quality of the fits as the intermediate region is approached certainly is consistent with the existence of a qualitatively different ground state in that region. If to the contrary the transition between antiferromagnetic and collinear order occurred via a strong first order transition, as had suggested by some approximate theories, no such progressive worsening is expected. We also notice in this context that the $N = 16$ cluster is systematically the one exhibiting the largest deviations from the expected behavior, probably due to its unusually high symmetry. This is why we feel that estimates ignoring this cluster may be more reliable.

3.3. The Nature of the Intermediate Phase

We can conclude from the previous two sections then, that there is indeed a nonmagnetic interval with best estimates as to its extent $0.34 < J_2 < 0.68$. We now discuss the interesting, but difficult, question of the exact nature of the order, or disorder in this region. Even at this stage, the magnetic structure factor $S(\boldsymbol{Q}) = (N + 2)M_N^2(\boldsymbol{Q})$ already gives some valuable information: in fact, as shown in fig.12, with increasing J_2 the collinear peak at the X point grows and the Néel peak at the M point shrinks, however there never is a maximum at other points. There is thus no evidence for incommensurate magnetic order. This is in accord

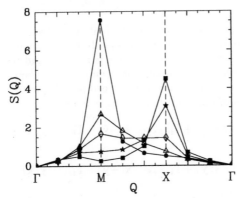

Fig. 12.   Magnetic structure factor, as obtained from the $N = 36$ cluster, in the Brillouin zone for $J_2/J_1 = 0$ (•), 0.55 ($\triangle$), 0.6 ($\diamond$), 0.65 ($\star$), 1 (solid squares). The points Γ, M, X are $\boldsymbol{Q} = 0, \boldsymbol{Q}_0, \boldsymbol{Q}_1$, respectively. Note that nowhere there is a maximum at point different from M or X.

with the classical argument presented earlier: in the absence of a J_3 term, a second nearest neighbor along the axis, the function $J(\boldsymbol{Q})$ has a maxima only at symmetry points, but in conflict with earlier speculations[38,54] that the $J_1 - J_2$ model might have a twisted or spiral state. Finite size studies with a J_3 term promise to be rather difficult because of the incommensurability of the classical turn angle with the boundary conditions.

A state which would be of great interest, if its existence could be demonstrated in such a simple microscopic model, is a chiral state spontaneously breaking parity and time reversal invariance. This is because such symmetry breaking would allow for the existence of anyonic excitations.[55] Several groups looked for evidence of important chiral correlations.[56-59] Early studies on small clusters calculated the following equal-time correlation function,

$$\chi_{ch} = \left\langle \left(\frac{4}{N} \sum_{\mathbf{i}} \Gamma_{\mathbf{i}}(\hat{\mathbf{x}}, \hat{\mathbf{y}}) \right)^2 \right\rangle \tag{16}$$

with

$$\Gamma_{\mathbf{i}}(\vec{\delta}, \vec{\delta'}) = \mathbf{S}_{\mathbf{i}} \cdot (\mathbf{S}_{\mathbf{i}+\vec{\delta}} \times \mathbf{S}_{\mathbf{i}+\vec{\delta}+\vec{\delta'}}) \tag{17}$$

This expression involves a non-local triple product of three spins[2] on triangles which can be expressed in the following way:

$$\mathbf{S_i} \cdot (\mathbf{S_j} \times \mathbf{S_l}) = \frac{1}{2i} S_i^z (S_j^- S_l^+ - S_j^+ S_l^-)$$

$$+ \frac{1}{2i} S_j^z (S_l^- S_i^+ - S_l^+ S_i^-)$$

$$+ \frac{1}{2i} S_l^z (S_i^- S_j^+ - S_i^+ S_j^-) \tag{18}$$

The operator Γ_i defines a direction of circulation (clockwise or anticlockwise) around one of the four triangles of each plaquette of the square lattice. Under the time reversal symmetry T the spin direction is reversed and the operator changes sign.

Working with small systems often requires particular care with symmetries. The four triangle operators that can be defined per plaquette may be combined to form three operators of different symmetries as follows,

$$\mathcal{O}_i^{A_2} = \Gamma_i(\hat{\mathbf{x}}, \hat{\mathbf{y}}) + \Gamma_{i+\hat{\mathbf{x}}+\hat{\mathbf{y}}}(-\hat{\mathbf{x}}, -\hat{\mathbf{y}}) + \Gamma_{i+\hat{\mathbf{y}}}(-\hat{\mathbf{y}}, \hat{\mathbf{x}}) + \Gamma_{i+\hat{\mathbf{x}}}(\hat{\mathbf{y}}, -\hat{\mathbf{x}}), \tag{19}$$

$$\mathcal{O}_i^{B_1} = \Gamma_i(\hat{\mathbf{x}}, \hat{\mathbf{y}}) + \Gamma_{i+\hat{\mathbf{x}}+\hat{\mathbf{y}}}(-\hat{\mathbf{x}}, -\hat{\mathbf{y}}) - \Gamma_{i+\hat{\mathbf{y}}}(-\hat{\mathbf{y}}, \hat{\mathbf{x}}) - \Gamma_{i+\hat{\mathbf{x}}}(\hat{\mathbf{y}}, -\hat{\mathbf{x}}), \tag{20}$$

$$\mathcal{O}_i^E = 2(\Gamma_i(\hat{\mathbf{y}}, \hat{\mathbf{x}}) - \Gamma_{i+\hat{\mathbf{x}}+\hat{\mathbf{y}}}(-\hat{\mathbf{x}}, -\hat{\mathbf{y}})). \tag{21}$$

These three different combinations are illustrated in fig.13 and are all odd under time reversal. The operator with A_2 symmetry is the one of greatest interest since

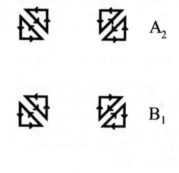

Fig. 13. The different combinations of plaquette operators and their symmetry classes

it is also *odd under reflections on the lattice axes*. It is then appropriate to study parity (P) symmetry breaking, in addition to T. Indeed, a non–zero expectation value of $\mathcal{O}_i^{A_2}$ in the ground state in the thermodynamic limit would be the signature of both P and T symmetry breaking. On the other hand, $\mathcal{O}_i^{B_1}$ is odd under $\pi/2$ rotations but even under axis reflections. $\mathcal{O}_i^E$ is odd under π rotations and is to be associated to the other equivalent component of the doublet obtained by applying a $\pi/2$ rotation (E is a doubly degenerate representation). Moreover, $\mathcal{O}_i^{B_1}$ and $\mathcal{O}_i^E$ exhibit alternating chiralities within each plaquette as can be seen in fig.13 from the orientations of the arrows.

The corresponding equal–time correlation functions is

$$\chi_{pl}^\alpha = \left\langle \left(\frac{1}{N} \sum_i \mathcal{O}_i^\alpha \right)^2 \right\rangle . \tag{22}$$

Note that $4\Gamma_i(\hat{\mathbf{x}}, \hat{\mathbf{y}})$ can be written as the sum of the three previous plaquette operators belonging to orthogonal group representations. Then, symmetry considerations leads to the following sum rule:

$$\chi_{ch} = \chi_{pl}^{A_2} + \chi_{pl}^{B_1} + \chi_{pl}^E . \tag{23}$$

How are these quantities expected to behave with increasing system size in the case of a chiral ground state? If chirality is uniform then we expect both $\mathcal{O}_i^{B_1}$ and $\mathcal{O}_i^E$ to vanish in the thermodynamic limit. From the sum rule it is then true that χ_{ch} and $\chi_{pl}^{A_2}$ will eventually lead to the same results for a large enough lattice. However, for small clusters it is crucial to select the operator with the proper symmetry. The relevant operator to work with is $\mathcal{O}_i^{A_2}$ and not the triangle operator. In fact the first results did not disentangle the different contributions and gave rather featureless results. Subsequent calculations[58] were restricted to 16 site lattices but calculated various related static and dynamic correlation functions with correct separation of the different symmetric combinations.

Since the operator in eq.(16) is invariant under a rotation in spin space the spectral representation of the chiral operators involves only singlet states of zero momentum, as in the ground state. To look for chiral order on the 16 site cluster, we can look at the correlation function between two plaquettes at any arbitrary distance allowed by the periodic boundary conditions. This plaquette–plaquette correlation is defined by

$$C_{pl}^\alpha(\mathbf{r}) = \frac{1}{2N} \sum_i \langle \mathcal{O}_{i+r}^\alpha \mathcal{O}_i^\alpha + \mathcal{O}_i^\alpha \mathcal{O}_{i+r}^\alpha \rangle, \tag{24}$$

where α stands for A_2 or B_1. Since the on–site correlations $C_{pl}^\alpha(0)$ can be strongly J_2–dependent, it is useful to actually plot the *ratios* $C_{pl}^\alpha(\mathbf{r})/C_{pl}^\alpha(0)$ as shown in fig.14. The correlations between the A_2 plaquettes seem to be slightly enhanced around $J_2 \sim 0.55$ but the enhancement remains small at large distances. The conclusion was then that if there was long range chiral order it was small in magnitude.

It is useful, both for experimental purposes and for understanding effects of finite size, to calculate dynamic fluctuations on small lattices. Assuming

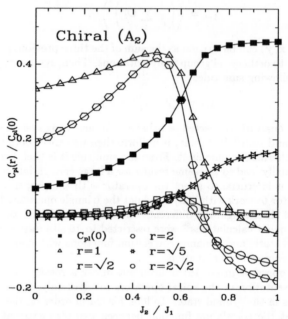

Fig. 14.   The chiral A_2 correlation functions $C_{pl}^{A_2}(\mathbf{0})$ $(\mathbf{r} = \mathbf{0})$ and $C_{pl}^{A_2}(\mathbf{r})/C_{pl}^{A_2}(\mathbf{0})$ for the 4 by 4 cluster for different distances as a function of J_2/J_1

$\mathcal{O}$ is an arbitrary order parameter its fluctuations in real time are given by, $Im \int_{-\infty}^{\infty} \langle \mathcal{O}(t)\mathcal{O}(0)\rangle \exp i\omega t \; dt$. Let us define in this way $I_{ch}(\omega)$ and $I_{pl}^{\alpha}(\omega)$ corresponding to the triangle and plaquette operators,

$$\mathcal{O} = \mathcal{O}_{ch} = \frac{4}{N} \sum_i \Gamma_i(\hat{\mathbf{x}},\hat{\mathbf{y}}), \tag{25}$$

and

$$\mathcal{O} = \mathcal{O}_{pl}^{\alpha} = \frac{1}{N} \sum_i \mathcal{O}_i^{\alpha}, \tag{26}$$

respectively. For the dynamical correlations there is an analogous sum rule:

$$I_{ch}(\omega) = I_{pl}^{A_2}(\omega) + I_{pl}^{B_1}(\omega) + I_{pl}^{E}(\omega) \; . \tag{27}$$

In fig. 15 we show the dynamic correlation function for the chiral operator $I_{pl}^{A_2}(\omega)$. As J_2 is increased from zero the weight of the A_2 symmetry increases to become larger in the intermediate region $0.55 \leq J_2 \leq 0.75$ with a maximum around $J_2 \sim 0.6$, and the frequency becomes smaller.

Can we take the enhancement of the A_2 fluctuations on the 16 site cluster as an indication of chiral long range order in the ground state? Later calculations[60] looked at the scaling properties of the static susceptibility for the 16 and 36 site clusters and did not show the increase with system size than one would expect for a state with long range order (see fig.16; in a long range ordered state there should be an increase of susceptibility proportional to the number of sites). These data thus seem to indicate tha absence of long–range chiral order. This illustrates the difficulty of extracting results from clusters that are too small. An enhancement of the equal time correlation function for a given size is not enough to conclude that there is long range order in the system. It is also crucial to analyze gaps, real space correlation functions, and their size dependence explicitly.

In principle a systematic way of treating the question is by the scaling of the gaps. The basic argument is that at a critical point the correlation length in a finite system is limited only by the linear dimension L of the system, and, given that the gap Δ is inversely proportional to the correlation length, one expects $\Delta(L) \approx 1/L$. Comparing the gaps of clusters of different sizes, at a critical point one thus expects

$$\Delta(L)/\Delta(L') = L'/L \; . \tag{28}$$

In an ordered phase, the gap decreases more quickly with increasing system size, whereas in a disordered state it saturates at a constant value for $L \to \infty$. This analysis was performed by scaling between the 16 and 36 site lattices and the results are shown in figure (17). The relevant gap here is between the ground state and the lowest excited state with A_2 symmetry at $Q = 0$. In the range $0.5 \leq J_2 \leq 0.7$ the gap ratio in fact is slightly smaller than 4/6, indicating the possibility of chiral order. Again that if the order did exist it would be rather small: since the chiral symmetry

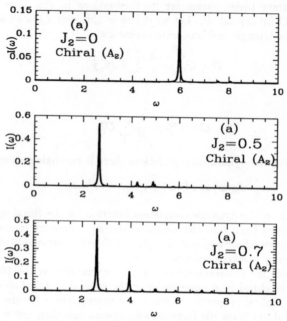

Fig. 15.   The dynamical correlation function for the chiral operator of A_2 symmetry for different values of J_2 on the 16 site cluster.

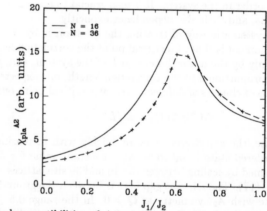

Fig. 16. Chiral susceptibilities of A_2 symmetry calculated on 16 and 36 site clusters

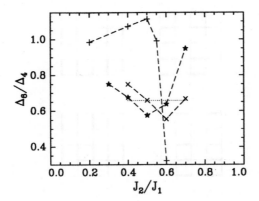

Fig. 17. The gap ratios between the 16 and 36 site clusters. +: gap for the B_1 state at $Q = 0$, relevant for dimerization and collinear magnetic order; ×: gap for the A_2 state at $Q = 0$, relevant for chiral order; $\star$: gap for the A_1 state at $Q = (\pi, 0)$, relevant for dimerization.

is discrete, well developped order should lead to exponentially fast decrease of the gap.

The results for chiral correlations are rather tantalizing: there are enhanced correlations but evidence for true order is slender at most. We now turn to a competing order, namely a spontaneously dimerized phase. The suggested order parameter for the dimer state, in which the spins are coupled in singlet bonds forming columns, is[61]

$$\mathcal{O}_{dim} = \frac{1}{N} \sum_i \mathcal{O}_i^{dim}, \qquad (29)$$

where

$$\mathcal{O}_i^{dim} = \frac{1}{2}((-1)^{i_x} \mathbf{S_i} \cdot \mathbf{S_{i+\hat{x}}} - (-1)^{i_x} \mathbf{S_i} \cdot \mathbf{S_{i-\hat{x}}} \qquad (30)$$
$$+ i(-1)^{i_y} \mathbf{S_i} \cdot \mathbf{S_{i+\hat{y}}} - i(-1)^{i_y} \mathbf{S_i} \cdot \mathbf{S_{i-\hat{y}}}), \qquad (31)$$

where $\mathbf{i} = (i_x, i_y)$. The expectation value of $\mathcal{O}_{dim}$ in the four dimer degenerate states (see fig.18) is $\pm\frac{3}{16}$, $\pm i\frac{3}{16}$ and vanishes for the Néel and the collinear states. This operator contains two independent parts, a real part and an imaginary part each of which acting on a translationally invariant ground state produces vectors with different wave vectors $(\pi, 0)$ and $(0, \pi)$, respectively. It is spin rotationally invariant. The relevant little group in that case is now the C_{2v} group which contains 4 group representations instead of 5 for C_{4v}.[62] Since the dimer operator is odd under a π

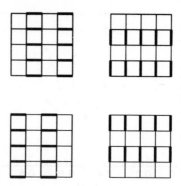

Fig. 18. The four possible dimer configurations

rotation, it then generates transitions between the symmetric ground state and $(\pi, 0)$ (or $(0, \pi)$) singlet excited states which transform according to the B_1 representation of the group, i.e. are odd under π–rotations and even under a reflection along the direction of the momentum. Special care was needed for constructing the spatial correlations of this non–hermitian operator,

$$C_{dim}(\mathbf{r}) = \frac{1}{2N} \sum_{\mathbf{i}} \langle \mathcal{O}_{\mathbf{i+r}}^{dim \, \dagger} \mathcal{O}_{\mathbf{i}}^{dim} + \mathcal{O}_{\mathbf{i}}^{dim \, \dagger} \mathcal{O}_{\mathbf{i+r}}^{dim} \rangle. \tag{32}$$

From fig.19 we see that the correlations are enhanced even at the largest distance available in the 16 site cluster in the parameter range $0.5 \leq J_2 \leq 0.6$. This effect is larger than for the uniform chiral order parameter.

The dynamical correlations of the dimer operator are shown in fig.20. This spectrum is doubly degenerate because of the two independent components of the operator. One difficulty of the study was that for $J_2 = 0$, the spectrum of the dimer operator is identical to an operator characteristic of collinear symmetry because of the extra symmetry specific to the 4×4 cluster. The frustration J_2 breaks this degeneracy and pushes down (up) the $\mathbf{k} = \mathbf{0}$ ($\mathbf{k} = (\pi, 0), (0, \pi)$) B_1 lowest singlet. For J_2 not too large, the lowest energy peak dominates the spectrum while for $J_2 \geq 0.6$, weight is transferred to higher energies.

The later finite–size study[37] using the gaps corresponding to dimer order were rather contradictory. In the thermodynamic limit both the gaps with excited states of B_1 (at $Q = 0$) and A_1 (at $Q = (\pi, 0)$) symmetry should vanish but in practice the ratio for the B_1 gap was considerably larger than the inverse ration of lengths but that for A_1 was less for $0.4 \leq J_2 \leq 0.6$. In addition we calculated the susceptibilities corresponding to dimer order and there was at most a 20% increase, much smaller

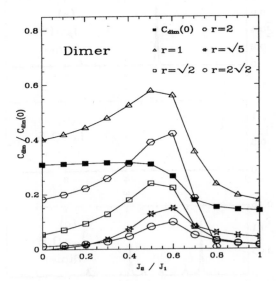

Fig. 19. The dimer correlation functions $C_{dim}(\mathbf{0})$ $(\mathbf{r} = \mathbf{0})$ and $C_{dim}(\mathbf{r})/C_{dim}(\mathbf{0})$ for the 16 site cluster for different distances as a function of J_2/J_1.

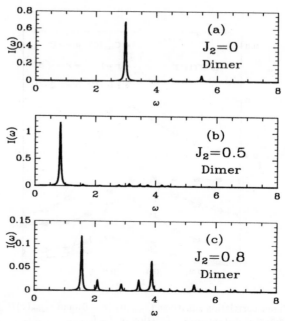

Fig. 20.   The dynamical correlation function for the dimer operator for different values of J_2 on the 16 site cluster.

than the factor 9/4 expected for an ordered state (fig.21) (but cleraly, and contrary to the chiral case, there is at least an increase).

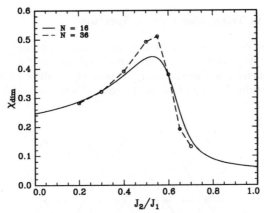

Fig. 21. Susceptibility of the dimer operator for 16 and 36 sites.

What then is the resolution of all this? For now we have to say that the question of which order is present, if any, cannot be considered closed. A promising avenue, as far as finite size studies are concerned, is to complete calculations of static and dynamic correlations of the intermediate clusters, especially that now we have analyzed the results for the magnetic states and seen that indeed the 16 site cluster is rather anomalous. Another approach which looks promising is to perform the equivalent of the effective potential approach to finite size scaling of classical thermodynamics. That is to say, instead of analyzing the scaling behavior of average values of the square of the order parameter in the finite cluster, we should reconstruct the full distribution of weights of its eigenvalues in the ground state. This we have done and preliminary results on the 36 site cluster do indeed show that smooth distributions can be obtained, and furthermore that there is a peak for dimer order, but not for chiral, at a non–zero value. The most probable value of the order parameter is small, much smaller than its average, which may explain previous difficulties to find unambiguous evidence.

In the future it would also be interesting to investigate dynamic correlation functions, in particular in the vicinity of the critical point of the Néel state, $J_2 = 0.34$. One then might gain additional insight into dynamic properties at a quantum critical point.[63,64]

While we have restricted our discussion to the $J_1 - J_2$ model it is clearly interesting to extend the Hamiltonian to include a third–nearest neighbor, i.e. the next neighbor in the lattice direction. This term favors classically helical states with turn angle in general depending on the couplings, and therefore incommensurable.

For the moment finite–size studies looking for such states are clearly problematic but interesting for the future. Another term of physical interest is the four–spin exchange which as well as being possibly essential[65,66] for the interpretation of Raman scattering in the copper oxide systems, is likely to stabilize non–collinear states.

4. Frustration in Triangular Lattice Systems

Lattices based on triangular geometry are of special importance in frustrated systems because they have the appearance of making the frustration look more natural, i.e. geometric, and are probably more likely to allow for experimental realizations by providing actual physical systems. There seem to be two promising

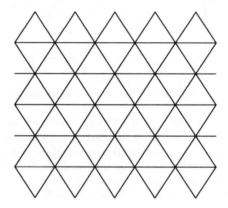

Fig. 22.   The triangular lattice

physical systems that may provide both triangular geometries and relatively isotropic exchange interaction. The first is layers of ^{3}He absorbed onto layers of graphite.[67] This should be a good spin one–half system on the triangular lattice (fig.22) with antiferromagnetic interactions, although one should include multi–spin interactions, in particular three spin exchange, in the Hamiltonian.[68,69] The second is the insulating antiferromagnet $SrCr_{8-x}Ga_{4+x}O_{19}$[72,73] in which the chromium ions have spin 3/2 and form relatively well separated layers in which the sites form a Kagomé lattice, as shown in fig.23. The antiferromagnetic interaction is measured from high temperature to be large and negative (about $-500K$) but at low temperatures only short range antiferromagnetic order is observed. At temperatures below 8K the system appears to be glassy.

Similar methods of finite size scaling have been applied to these systems and the methods are a generalization of those described in detail in the previous section. In conjunction with high order series expansions[70], which are limited in that rotational

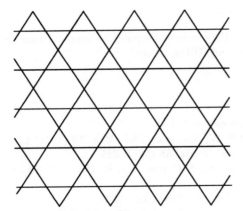

Fig. 23. The Kagomé lattice

invariance is broken by choosing a quantization axis, they lead to the conclusion that the triangular Heisenberg antiferromagnet is probably ordered with a three sublattice order parameter. The Kagomé lattice, which appears more dilute, may be disordered magnetically[71]. The analysis of these cases, as we have said, is considerably more difficult: as the candidate antiferromagnetic state is more complex the structure of the towers of states predicted from field theory is too. This has been analyzed in detail by the authors of ref. 74. ¿From the point of view of comparison to exact diagonalization studies this makes for even greater difficulty. Thus while clusters up to 36 spins[32] have been solved as for the square geometry there is as yet less systematic comparison possible. The advantage is of course the availability of experimental systems and the study of magnetic correlations in those then should be able to at least settle the question of long range magnetic order. The question of non–magnetic states is of course more problematic, as dimer order for example, would be harder to see in terms of magnetic correlations. Furthermore to understand the experiments it is necessary to include finite temperature effects. For the compound $SrCr_{8-x}Ga_{4+x}O_{19}$ the relatively large spin may imply an involved cross–over between the classical and the quantum regimes, neither of which is straightforward in its own right. Furthermore the low temperature phase has glassy characteristics which may be due to fluctuations in the chemical composition.

5. Acknowledgements

We are grateful to the continuing support in computer time and expert help from the CCVR, Palaiseau and IDRIS, Orsay.

6. References

[*] Laboratoires associés au CNRS.
1. P. Fazekas and P. W. Anderson, *Phil. Mag.* **30** (1974) 423.
2. J. Villain, *J. Phys. (Paris)* **38** (1977) 385.
3. J. Villain, R. Bidaux, J. Carton, and R. Conte, *J. Phys. (Paris)* **41** (1980) 1263.
4. P. W. Anderson, *Science* **235** (1987) 1196.
5. D. Poilblanc *et al.*, *Europhys. Lett.* **22** (1993) 537.
6. G. Shirane *et al.*, *Phys. Rev. Lett.* **59** (1987) 1613.
7. H. Bethe, *Z. Phys.* **71** (1931) 205.
8. C. N. Yang and C. P. Yang, *Phys. Rev.* **150** (1966) 321; *ibid.* (1966) 327.
9. D. Mermin and H. Wagner, *Phys. Rev. Lett.* **17** (1966) 1133.
10. A. Luther and I. Peschel, *Phys. Rev.* B **9** (1974) 2911.
11. I. Affleck, D. Gepner, H.J. Schulz, and T. Ziman, *J. Phys.* A **22** (1989) 511.
12. R. R. P. Singh, M. E. Fisher, and R. Shankar, *Phys. Rev.* B **39** (1989) 2562.
13. F. D. M. Haldane, *Phys. Rev. Lett.* **50** (1983) 1153.
14. F. D. M. Haldane, *Phys. Rev.* B **25** (1982) 4925.
15. K. Okamoto and K. Nomura, *Phys. Lett.* A **169** (1992) 433.
16. C. K. Majumdar and D. Ghosh, *J. Math. Phys.* **10** (1991) 1399.
17. H. J. Schulz, *Phys. Rev.* B **34** (1986) 6372.
18. I. Affleck and F. D. M. Haldane, *Phys. Rev.* B **36** (1987) 5291.
19. T. Ziman and H. J. Schulz, *Phys. Rev. Lett.* **59** (1987) 140.
20. L. D. Faddeev and L. A. Takhtajan, *Phys. Lett.* A **85** (1981) 375.
21. For a recent experimental confirmation of this interesting point, see D. A. Tennant, T. G. Perring, R. A. Cowley, and S. E. Nagler, *Phys. Rev. Lett.* **70** (1993) 4003.
22. O. Golinelli, T. Jolicoeur, and R. Lacaze, *Saclay preprint* (1994).
23. S. R. White, *Phys. Rev.* B **48** (1993) 10345.
24. M. den Nijs and K. R. Rommelse, *Phys. Rev.* B **40** (1989) 4709.
25. T. Tonegawa, M. Kaburagai, N. Ichikawa, and I. Harada, *J. Phys. Soc. Jpn.* **61** (1992) 2890.
26. I.Harada, M. Fujikawa, and I. Mannari, *J. Phys. Soc. Jpn.* **62** (1993) 3694.
27. M. Inui, S. Doniach, and M. Gabay, *Phys. Rev.* B **38** (1988) 6631.
28. K. Kubo and T. Kishi, *Phys. Rev. Lett.* **61** (1987) 2585.
29. P. W. Anderson, *Phys. Rev.* **86** (1952) 694.
30. P. Hasenfratz and F. Niedermayer, *Z Phys.* B **92** (1993) 91.

31. H. Leutwyler, *Bern University preprint* (1993).
32. B. Bernu, C. Lhuillier, and L. Pierre, *Phys. Rev. Lett.* **69** (1992) 2590.
33. F. Mila, D. Poilblanc, and C. Bruder, *Phys. Rev. B* **43** (1991) 7891.
34. E. Shender, *Sov. Phys. JETP* **56** (1982) 178.
35. E. Dagotto and A. Moreo, *Phys. Rev. Lett.* **63** (1989) 2148.
36. F. Figueiredo *et al.*, *Phys. Rev. B* **41** (1990) 4619.
37. H. J. Schulz and T. A. L. Ziman, *Europhys. Lett.* **18** (1992) 355.
38. P. Chandra and B. Doucot, *Phys. Rev. B* **38** (1988) 9335.
39. M. Takahashi, *Phys. Rev. B* **40** (1989) 2494.
40. H. Nishimori and Y. Saika, *J. Phys. Soc. Jpn.* **59** (1990) 4454.
41. T. Nakamura and N. Hatano, *J. Phys. Soc. Jpn.* **62** (1993) 3062.
42. M. P. Gelfand, R. R. P. Singh, and D. A. Huse, *Phys. Rev. B* **40** (1989) 10801.
43. S. Chakravarty, B. I. Halperin, and D. R. Nelson, *Phys. Rev. B* **39** (1989) 2344.
44. T. Einarsson and H. Johannesson, *Phys. Rev. B* **43** (1991) 5867.
45. E. Manousakis, *Rev. Mod. Phys.* **63** (1991) 1.
46. T. Barnes, *Int. J. Mod. Phys. C* **2** (1991) 659.
47. H. Neuberger and T. A. L. Ziman, *Phys. Rev. B* **39** (1989) 2608.
48. D. S. Fisher, *Phys. Rev. B* **39** (1989) 11783.
49. R. R. P. Singh, *Phys. Rev. B* **39** (1989) 9760.
50. R. R. P. Singh and D. Huse, *Phys. Rev. B* **40** (1989) 7247.
51. Z. Weihong, J. Oitmaa, and C. J. Hamer, *Phys. Rev. B* **43** (1991) 8321.
52. K. J. Runge, *Phys. Rev. B* **45** (1992) 12292.
53. K. J. Runge, *Phys. Rev. B* **45** (1992) 7229.
54. M. Gabay and P. J. Hirschfeld, *Physica C* **162** (1989) 823.
55. X. G. Wen, F. Wilczeck, and A. Zee, *Phys. Rev. B* **39** (1989) 11413.
56. M. Imada, *J. Phys. Soc. Jpn.* **58** (1989) 2650.
57. H. Shiba and M. Ogata, *J. Phys. Soc. Jpn.* **59** (1992) 2971.
58. D. Poilblanc, E. Gagliano, S. Bacci, and E. Dagotto, *Phys. Rev. B* **43** (1991) 10970.
59. D. Poilblanc and E. Dagotto, *Phys. Rev. B* **45** (1991) 10111.
60. H. J. Schulz, T. A. L. Ziman, and D. Poilblanc, *unpublished results*.
61. S. Sachdev and R. N. Bhatt, *Phys. Rev. B* **41** (1990) 9323.
62. Here we use the center of a plaquette as the origin of the point group elements. Note that one could alternatively define the symmetry operations with respect to a lattice site as e.g. in Ref. 58.
63. S. Sachdev and J. Ye, *Phys. Rev. Lett.* **69** (1992) 2411.
64. A. V. Chubukov, S. Sachdev, and J. Ye, *preprint*.
65. M. Roger and J.M. Delrieu, *Phys. Rev. B* **39** (1989) 2299.
66. S. Sugai *et al.*, *Phys. Rev. B* **42** (1990) 1045.
67. V. Elser, *Phys. Rev. Lett.* **62** (1989) 2405.
68. J.M. Delrieu , M. Roger, and J. H. Hetherington, *J. Low Temp. Phys.* **40** (1980) 71.

69. H. Godfrin, R.R. Ruel, and D.D. Osheroff, *Phys. Rev. Lett.* **60** (1988) 305.
70. R. R. P. Singh and David A. Huse, *Phys. Rev. Lett.* **68** (1992) 1766.
71. C. Zeng and V. Elser, *Phys. Rev. B* **42** (1990) 8436.
72. A. Ramirez, G. Espinosa, and A. Cooper, *Phys. Rev. Lett.* **64** (1990) 2070.
73. C. Broholm, G. Aeppli, G. P. Espinosa, and A. S. Cooper, *Phys. Rev. Lett.* **65** (1990) 3173.
74. P. Azaria, B. Delamotte, and D. Mouhanna, *Phys. Rev. Lett.* **68** (1992) 1762.

Magnetic System with Competing Interaction
Ed. H. T. Diep
©1994 World Scientific Publishing Co.

CHAPTER IV

EXACTLY SOLVED FRUSTRATED MODELS:

REENTRANCE AND PHASE DIAGRAM

H. T. Diep
Groupe de Physique Statistique, Université de Cergy-Pontoise
49, Avenue des Genottes, B.P. 8428, 95806 Cergy-Pontoise Cedex, France.

and

H. Giacomini
Laboratoire de Modèles de Physique Mathématique, Faculté des Sciences et Techniques
Université de Tours, Parc de Grandmont, 37200 Tours, France.

1. Introduction

The study of order-disorder phenomena is a fundamental task of equilibrium statistical mechanics. Great efforts have been made to understand the basic mechanisms responsible for spontaneous ordering as well as the nature of the phase transition in many kinds of systems. In particular, during the last decade, much attention has been paid to frustrated models.[1] The word "frustration" has been introduced [2,3] to describe the situation where a spin (or a number of spins) in the system cannot find an orientation to satisfy all the interactions with its neighboring spins (see definition given in the chapter by Nagai et al, this book). Though this definition was initially applied to Ising spins, the concept of frustration has been extended to vector spins (see the chapter by Plumer et al, this book). In general, the frustration is caused either by competing interactions (such as the Villain model[3]) or by lattice structure as in the triangular, face-centered cubic (fcc) and hexagonal-close-packed (hcp) lattices, with antiferromagnetic nearest-neighbor interaction. The effects of frustration are rich and often unexpected. Many of them are not understood yet at present (see the other chapters of this book).

In addition to the fact that real magnetic materials are often frustrated due to several kinds of interactions (see the chapter by Gaulin, this book), frustrated spin systems have their own interest in statistical mechanics. Recent studies show that many established statistical methods and theories have never before encountered so many difficulties like they do now in dealing with frustrated systems. In some sense, frustrated systems are excellent candidates to test approximations and improve theories. Since the mechanisms of many phenomena are not understood in real systems (disordered systems, systems with long-range interaction, three-dimensional systems, etc), it is worth to search for the origins of those phenomena in exactly solved systems. These exact results will help to understand qualitatively the behavior of real systems which are in general much more complicated.

We are interested here in frustrated Ising spin systems without disorder. A review of early works (up to about 1985) on frustrated Ising systems with periodic interactions, i.e. no bond disorder, has been given by Liebmann.[1] These systems have their own interest in statistical mechanics because they are periodically defined and thus subject to exact treatment. To date, very few systems are exactly solvable. They are limited to one and two dimensions (2D).[4] A few well-known systems showing remarkable properties include the centered square lattice[5] and its generalized versions,[6,7] the Kagomé lattice,[8,9,10] an anisotropic centered honeycomb lattice,[11] and several periodically dilute centered square lattices.[12] Complicated cluster models,[13] and a particular three-dimensional case have also been solved.[14] The phase diagrams in frustrated models show a rich behavior. Let us mention a few remarkable consequences of the frustration which are in connection with what will be shown in this chapter. The degeneracy of the ground state is very high, often infinite. At finite temperatures, in some systems the degeneracy is reduced by thermal fluctuations which select a number of states with largest entropy. This has been called "Order by Disorder",[15] in the Ising case. Quantum fluctuations and/or thermal fluctuations can also select particular spin configurations in the case of vector spins.[16,17] Another striking phenomenon is the coexistence of Order and Disorder at equilibrium: a number of spins in the system are disordered at all temperatures even in an ordered phase.[9] The frustration is also at the origin of the reentrance phenomenon. A reentrant phase can be defined as a phase with no long-range order, or no order at all, occurring in a region below an ordered phase on the temperature scale. In addition, the frustration can also give rise to disorder lines in the phase diagram of many systems as will be shown below.

In this chapter, we confine ourselves to exactly solved Ising spin systems that show remarkable features in the phase diagram such as the reentrance, successive transitions, disorder lines and partial disorder. Other Ising systems are treated in the chapter by Nagai et al. Also, the reentrance in disordered systems such as spin glasses is discussed in the chapter by Gingras.

The systems we consider in this chapter are periodically defined (without bond disorder). The frustration due to competing interactions will itself induce disorder in the spin orientations. The results obtained can be applied to physical systems that can be mapped into a spin language. The chapter is organized as follows. In the next section, we outline the method which allows to calculate the partition function and the critical varieties of 2D Ising models without crossing interactions. In particular, we show in detail the mapping of these models onto the 16- and 32-vertex models. We also explain a decimation method for finding disorder solutions. The purpose of this section is to give the reader enough mathematical details so that, if he wishes, he can apply these techniques to 2D Ising models with non-crossing interactions. In section 3, we shall apply the results of section 2 in some systems which present remarkable physical properties. The systems studied in section 3 contain most of interesting features of the frustration: high ground state degeneracy, reentrance, partial disorder, disorder lines, successive phase transitions, and some aspects of the random-field Ising model. Most of the materials shown in this section are from works that we have recently done. A discussion on the origin of the reentrance phenomenon and concluding remarks are given in section 4.

2. Mapping between Ising models and vertex models

The 2D Ising model with non-crossing interactions is exactly soluble. The problem of finding the partition function can be transformed in a free-fermion model.If the lattice is a complicated one, the mathematical problem to solve is very cumbersome.

For numerous two-dimensional Ising models with non-crossing interactions, there exists another method, by far easier, to find the exact partition function. This method consists in mapping the model on a 16-vertex model or a 32-vertex model. If the Ising model does not have crossing interactions, the resulting vertex model will be exactly soluble. We will apply this method for finding the exact solution of several Ising models in two-dimensional lattices with non-crossing interactions.

Let us at first introduce the 16-vertex model and the 32- vertex model, and the cases for which these models satisfy the free-fermion condition.

2.1. The 16-vertex model

The 16-vertex model which we will consider is a square lattice of N points, connected by edges between neighbouring sites. These edges can assume two states, symbolized by right- and left- or up-and down-pointing arrows, respectively. The allowed configurations of the system are characterized by specifyng the arrangement of arrows around each lattice point. In characterizing these so-called vertex configurations, we follow the enumeration of Baxter[4](see Fig.1).

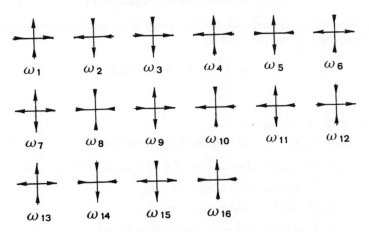

Fig. 1. Arrow configurations and vertex weights of the 16-vertex model.

To each vertex we assign an energy $\epsilon_k (k = 1, 2, ..., 16)$ and a corresponding vertex weight (Boltzmann factor) $\omega_k = e^{\beta \epsilon_k}$, where $\beta = (1)/(k_B T)$, T being the temperature and k_B the Boltzmann constant. Then the partition function is

$$Z = \sum_C e^{-\beta(n_1 \epsilon_1 + ... + n_{16} \epsilon_{16})} \tag{1}$$

where the sum is over all allowed configurations C of arrows on the lattice, n_j is the number of vertex arrangements of type j in configuration C. It is clear from Eq.(1) that Z is a function of the eight Boltzmann weights $\omega_k (k = 1, 2, ..., 16)$:

$$Z = Z(\omega_1, ..., \omega_{16}) \tag{2}$$

So far, exact results have only been obtained for three subclasses of the general 16-vertex model, i.e. the 6-vertex (or ferroelectric) model, the symmetric eight-vertex model and the free-fermion model.[4,18] Here we will consider only the case where the free-fermion condition is satisfied, because in these cases the 16-vertex model can be related to 2D Ising models without crossing interactions. Generally, a vertex model is soluble if the vertex weights satisfy certain conditions so that the partition function is reducible to the S matrix of a many-fermion system.[18] In the present problem these constraints are the following :

$$\begin{aligned}
\omega_1 &= \omega_2 , \; \omega_3 = \omega_4 \\
\omega_5 &= \omega_6 , \; \omega_7 = \omega_8 \\
\omega_9 &= \omega_{10} = \omega_{11} = \omega_{12} \\
\omega_{13} &= \omega_{14} = \omega_{15} = \omega_{16} \\
\omega_1 \omega_3 &+ \omega_5 \omega_7 - \omega_9 \omega_{11} - \omega_{13} \omega_{15} = 0
\end{aligned} \tag{3}$$

If these conditions are satisfied, the free energy of the model can be expressed, in the thermodynamical limit, as follows :

$$f = -\frac{1}{4\pi\beta} \int_0^{2\pi} d\phi \log\{A(\phi) + [Q(\phi)]^{1/2}\} \tag{4}$$

where

$$\begin{aligned}
A(\phi) &= a + c \cos(\phi) \\
Q(\phi) &= y^2 + z^2 - x^2 - 2yz \cos(\phi) + x^2 \cos^2(\phi) \\
a &= \frac{1}{2}(\omega_1^2 + \omega_3^2 + 2\omega_1\omega_3 + \omega_5^2 + \omega_7^2 + 2\omega_5\omega_7) + 2(\omega_9^2 + \omega_{13}^2) \\
c &= 2[\omega_9(\omega_1 + \omega_3) - \omega_{13}(\omega_5 + \omega_7)] \\
y &= 2[\omega_9(\omega_1 + \omega_3) + \omega_{13}(\omega_5 + \omega_7)] \\
z &= \frac{1}{2}[(\omega_1 + \omega_3)^2 - (\omega_5 + \omega_7)^2] + 2(\omega_9^2 - \omega_{13}^2) \\
x^2 &= z^2 - \frac{1}{4}[(\omega_1 - \omega_3)^2 - (\omega_5 - \omega_7)^2]^2
\end{aligned} \tag{5}$$

Phase transitions occur when one or more pairs of zeros of the expression $Q(\phi)$ close in on the real ϕ axis and "pinch" the path of integration in the expression on the right-hand side of Eq.(4). This happens when $y^2 = z^2$, i.e. when

$$\omega_1 + \omega_3 + \omega_5 + \omega_7 + 2\omega_9 + 2\omega_{13} = 2\max\{\omega_1 + \omega_3, \omega_5 + \omega_7, 2\omega_9, 2\omega_{13}\} \qquad (6)$$

The type of singularity in the specific heat depends on whether

$$(\omega_1 - \omega_3)^2 - (\omega_5 - \omega_7)^2 \neq 0 \qquad \text{(logarithmic singularity)}$$

or

$$(\omega_1 - \omega_3)^2 - (\omega_5 - \omega_7)^2 = 0 \qquad \text{(inverse square-root singularity)} \qquad (7)$$

2.2. The 32-vertex model

The 32-vertex model is defined by a triangular lattice of N points, connected by edges between neighboring sites. These edges can assume two states, symbolized by an arrow pointing in or pointing out of a site. In the general case, there are 64 allowed vertex configurations. If only an odd number of arrows pointing into a site are allowed, we have 32 possible vertex configurations. This is the constraint that characterizes the 32-vertex model. To each allowed vertex configuration we assign an energy $\epsilon_k (k = 1, 2, ..., 32))$ and a corresponding vertex weight, defined as it is shown in Fig. 2, where $\omega = e^{-\beta\epsilon_1}$, $\overline{\omega} = e^{-\beta\epsilon_2}$, $\omega_{56} = e^{-\beta\epsilon_3}$, $\overline{\omega}_{56} = e^{-\beta\epsilon_4}$, etc.

This notation for the Boltzmann vertex weights has been introduced by Sacco and Wu,[19] and is used also by Baxter.[4] This model is not exactly soluble in the general case, but there are several particular cases that are soluble.[19] Here we will consider one of such cases, when the model satisfy the free-fermion condition :

$$\begin{aligned}
\omega\overline{\omega} &= \omega_{12}\overline{\omega}_{12} - \omega_{13}\overline{\omega}_{13} + \omega_{14}\overline{\omega}_{14} - \omega_{15}\overline{\omega}_{15} + \omega_{16}\overline{\omega}_{16} \\
\omega\omega_{mn} &= \omega_{ij}\omega_{kl} - \omega_{ik}\omega_{jl} + \omega_{il}\omega_{jk}
\end{aligned} \qquad (8)$$

for all permutations i, j, k, l, m, n of 1, 2, ..., 6 such that $m < n$ and $i < j < k < l$. There are 15 such permutations (corresponding to the 15 choices of m and n), and hence a total of 16 conditions.

The rather complicated notation for the Boltzmann weights is justified by the condensed form of the free-fermion conditions (8).

When these conditions are satisfied, the free energy in the thermodynamical limit can be expressed as

$$f = -\frac{1}{8\pi^2\beta} \int_0^{2\pi} d\theta \int_0^{2\pi} d\phi \log[\omega^2 D(\theta, \phi)] \qquad (9)$$

where

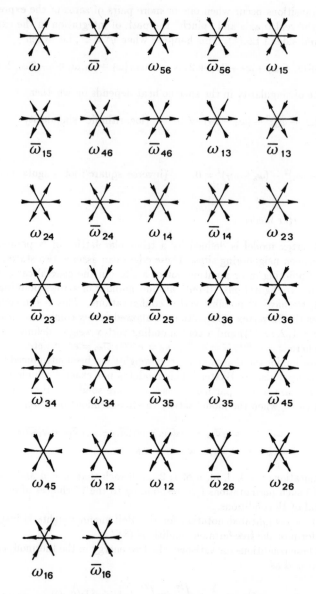

Fig. 2. Arrow configurations and vertex weights of the 32-vertex model.

$$\omega^2 D(\theta, \phi) = \Omega_1^2 + \Omega_2^2 + \Omega_3^2 + \Omega_4^2 - 2(\Omega_1\Omega_3 - \Omega_2\Omega_4)\cos(\theta)$$
$$-2(\Omega_1\Omega_4 - \Omega_2\Omega_3)\cos(\phi) + 2(\Omega_3\Omega_4 - \Omega_5\Omega_6)\cos(\theta + \phi)$$
$$+2(\Omega_5\Omega_6 - \Omega_1\Omega_2)\cos(\theta - \phi) - 4a\sin(\phi)\sin(\theta + \phi)$$
$$-4b\sin(\theta)\sin(\theta + \phi) - 2c\sin^2(\theta + \phi) - 2d\sin^2(\theta) - 2e\sin^2(\phi) \tag{10}$$

with

$$\Omega_1 = \omega + \overline{\omega}, \qquad \Omega_2 = \omega_{25} + \overline{\omega}_{25}$$
$$\Omega_3 = \omega_{14} + \overline{\omega}_{14}, \qquad \Omega_4 = \omega_{36} + \overline{\omega}_{36}$$
$$\Omega_5\Omega_6 = \omega_{15}\omega_{24} + \overline{\omega}_{15}\overline{\omega}_{24} + \omega_{14}\overline{\omega}_{25} + \omega_{25}\overline{\omega}_{14}$$
$$a = \omega_{12}\omega_{45} + \overline{\omega}_{12}\overline{\omega}_{45} - \omega\overline{\omega}_{36} - \overline{\omega}\omega_{36}$$
$$b = \omega_{23}\omega_{56} + \overline{\omega}_{23}\overline{\omega}_{56} - \omega\overline{\omega}_{14} - \overline{\omega}\omega_{14}$$
$$c = \omega\overline{\omega} + \omega_{13}\overline{\omega}_{13} - \omega_{12}\overline{\omega}_{12} - \omega_{23}\overline{\omega}_{23}$$
$$d = \omega\overline{\omega} + \omega_{26}\overline{\omega}_{26} - \omega_{16}\overline{\omega}_{16} - \omega_{12}\overline{\omega}_{12}$$
$$e = \omega\overline{\omega} + \omega_{15}\overline{\omega}_{15} - \omega_{56}\overline{\omega}_{56} - \omega_{16}\overline{\omega}_{16} \tag{11}$$

The critical temperature is determined from the equation

$$\Omega_1 + \Omega_2 + \Omega_3 + \Omega_4 = 2\max(\Omega_1, \Omega_2, \Omega_3, \Omega_4) \tag{12}$$

We will show now how different 2D Ising models without crossing interactions can be mapped onto the 16-vertex model or the 32-vertex model, with the free-fermion condition automatically satisfied in such cases.

Let us consider at first an Ising model defined on a Kagomé lattice, with two-spin interactions between nearest neighbors (nn) and between next-nearest neighbors (nnn), J_1 and J_2, respectively, as shown in Fig. 3.

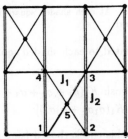

Fig. 3. Kagomé lattice. Interactions between nearest neighbors and between next-nearest neighbors, J_1 and J_2, are shown by single and double bonds, respectively. The lattice sites in a cell are numbered for decimation demonstration.

The Hamiltonian is written as

$$H = -J_1 \sum_{(ij)} \sigma_i \sigma_j - J_2 \sum_{(ij)} \sigma_i \sigma_j \tag{13}$$

where and the first and second sums run over the spin pairs connected by single and double bonds, respectively.

The partition function is written as

$$Z = \sum_\sigma \prod_c \exp[K_1(\sigma_1\sigma_5 + \sigma_2\sigma_5 + \sigma_3\sigma_5 + \sigma_4\sigma_5 + \sigma_1\sigma_2 + \sigma_3\sigma_4) + K_2(\sigma_1\sigma_4 + \sigma_3\sigma_2)] \tag{14}$$

where $K_{1,2} = J_{1,2}/k_B T$ and where the sum is performed over all spin configurations and the product is taken over all elementary cells of the lattice.

Since there are no crossing bond interactions, the system can be transformed into an exactly solvable free-fermion model. We decimate the central spin of each elementary cell of the lattice. In doing so, we obtain a checkerboard Ising model with multispin interactions (see Fig. 4).

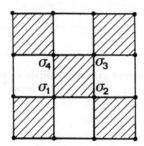

Fig. 4. The checkerboard lattice. At each shaded square is associated the Boltzmann weight $W(\sigma_1, \sigma_2, \sigma_3, \sigma_4)$, given in the text.

The Boltzmann weight associated to each shaded square is given by

$$\begin{aligned}
W(\sigma_1, \sigma_2, \sigma_3, \sigma_4) &= 2\cosh(K_1(\sigma_1 + \sigma_2 + \sigma_3 + \sigma_4))\exp[K_2(\sigma_1\sigma_4 + \sigma_2\sigma_3) \\
&\quad + K_1(\sigma_1\sigma_2 + \sigma_3\sigma_4)]
\end{aligned} \tag{15}$$

The partition function of this checkerboard Ising model is given by

$$Z = \sum_\sigma \prod W(\sigma_1, \sigma_2, \sigma_3, \sigma_4) \tag{16}$$

where the sum is performed over all spin configurations and the product is taken over all the shaded squares of the lattice.

In order to map this model onto the 16-vertex model, let us introduce another square lattice where each site is placed at the center of each shaded square of the checkerboard lattice, as shown in Fig. 5.

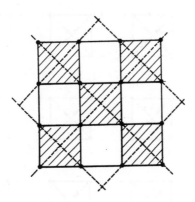

Fig. 5. The checkerboard lattice and the associated square lattice with their bonds indicated by dashed lines.

At each bond of this lattice we associate an arrow pointing out of the site if the Ising spin that is traversed by this bond is equal to +1, and pointing into the site if the Ising spin is equal to -1, as it is shown in Fig. 6.

In this way, we have a 16-vertex model on the associated square lattice. The Boltzmann weights of this vertex model are expressed in terms of the Boltzmann weights of the checkerboard Ising model, as follows

$$\begin{aligned}
\omega_1 &= W(-,-,+,+) & \omega_5 &= W(-,+,-,+)\\
\omega_2 &= W(+,+,-,-) & \omega_6 &= W(+,-,+,-)\\
\omega_3 &= W(-,+,+,-) & \omega_7 &= W(+,+,+,+)\\
\omega_4 &= W(+,-,-,+) & \omega_8 &= W(-,-,-,-)\\
\omega_9 &= W(-,+,+,+) & \omega_{13} &= W(+,-,+,+)\\
\omega_{10} &= W(+,-,-,-) & \omega_{14} &= W(-,+,-,-)\\
\omega_{11} &= W(+,+,-,+) & \omega_{15} &= W(+,+,+,-)\\
\omega_{12} &= W(-,-,+,-) & \omega_{16} &= W(-,-,-,+)
\end{aligned}$$

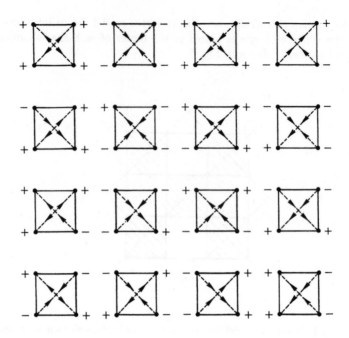

Fig. 6. The relation between spin configurations and arrow configurations of the associated vertex model.

Taking (15) into account, we obtain

$$\omega_1 = \omega_2 = 2e^{-2K_2+2K_1}$$
$$\omega_3 = \omega_4 = 2e^{2K_2-2K_1}$$
$$\omega_5 = \omega_6 = 2e^{-2K_2-2K_1}$$
$$\omega_7 = \omega_8 = 2e^{2K_2+2K_1}\cosh(4K_1)$$
$$\omega_9 = \omega_{10} = \omega_{11} = \omega_{12} = \omega_{13} = \omega_{14} = \omega_{15} = \omega_{16} = 2\cosh(2K_1) \qquad (18)$$

As can be easily verified, the free-fermion conditions (3) are identically satisfied by the Boltzmann weights (18), for arbitrary values of K_1 and K_2. If we replace (18) in (4) and (5), we can obtain the explicit expression of the free energy of the model. Moreover, by replacing (18) in (6) we obtain the critical condition for this system :

$$\frac{1}{2}\left[\exp(2K_1+2K_2)\cosh(4K_1)+\exp(-2K_1-2K_2)\right] \ +$$
$$\cosh(2K_1-2K_2)+2\cosh(2K_1)=2\max\{\frac{1}{2}\left[\exp(2K_1 \ + \ 2K_2)\cosh(4K_1)+\right.$$

$$\exp(-2K_1 - 2K_2)] \; ; \; \cosh(2K_2 - 2K_1) \quad ; \quad \cosh(2K_1)\} \qquad (19)$$

which is decomposed into four critical lines depending on the values of J_1 and J_2.

The singularity of the free energy is everywhere logarithmic.

Now, we will consider another 2D Ising model with two-spin interactions and without crossing bonds. This model is defined on a centered honeycomb lattice, as shown in Fig. 7.

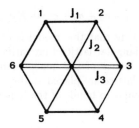

Fig. 7. Unit cell of the centered honeycomb lattice: heavy, light, and double-light bonds denote the interactions J_1, J_2, and J_3, respectively. The sites on the honeycomb are numbered from 1 to 6 for decimation demonstration (see text).

The Hamiltonian of this model is as follows :

$$H = -J_1 \sum_{(ij)} \sigma_i \sigma_j - J_2 \sum_{(ij)} \sigma_i \sigma_j - J_3 \sum_{(ij)} \sigma_i \sigma_j \qquad (20)$$

where $\sigma_i = \pm 1$ is an Ising spin occupying the lattice site i , and the first, second, and third sums run over the spin pairs connected by heavy, light, and doubly light bonds, respectively (see Fig. 7). When $J_2 = J_3 = 0$, one recovers the honeycomb lattice, and when $J_1 = 0 = J_2 = J_3$, one has the triangular lattice.

Let us denote the central spin in a lattice cell, shown in Fig. 7, by σ, and number the other spins from σ_1 to σ_6. The Boltzmann weight associated to the elementary cell is given by

$$W = \exp[K_1(\sigma_1\sigma_2 + \sigma_2\sigma_3 + \sigma_3\sigma_4 + \sigma_4\sigma_5 + \sigma_5\sigma_6 + \sigma_6\sigma_1)+$$

$$K_2\sigma(\sigma_1 + \sigma_2 + \sigma_4 + \sigma_5) + K_3\sigma(\sigma_3 + \sigma_6)] \qquad (21)$$

The partition function of the model is written as

$$Z = \sum_\sigma \prod_c W \qquad (22)$$

where the sum is performed over all spin configurations and the product is taken over all elementary cells of the lattice. Periodic boundary conditions are imposed. Since there is no croossing-bond interaction, the model is exactly soluble. To obtain the exact solution, we decimate the central spin of each elementary cell of the lattice. In doing so, we obtain a honeycomb Ising model with multispin interactions.

After decimation of each central spin, the Boltzmann factor associated to an elementary cell is given by

$$W' = 2\exp[K_1(\sigma_1\sigma_2 + \sigma_2\sigma_3 + \sigma_3\sigma_4 + \sigma_4\sigma_5 + \sigma_5\sigma_6 + \sigma_6\sigma_1)] \times$$

$$\cosh[K_2(\sigma_1 + \sigma_2 + \sigma_4 + \sigma_5) + K_3(\sigma_3 + \sigma_6)] \qquad (23)$$

We will show in the following that this model is equivalent to a special case of the 32-vertex model on the triangular lattice that satify the free-fermion condition.

Let us consider the dual lattice of the honeycomb lattice, i.e. the triangular lattice.[4] The sites of the dual lattice are placed at the center of each elementary cell and their bonds are perpendicular to bonds of the honeycomb lattice, as it is shown in Fig. 8.

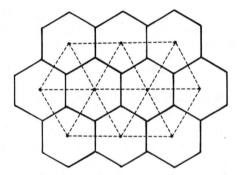

Fig. 8. The honeycomb lattice and the dual triangular lattice, with their bonds indicated by dashed lines.

Each site of the triangular lattice is surrounded by 6 sites of the honeycomb lattice. At each bond of the triangular lattice we associate an arrow. We take the arrow configuration shown in Fig. 9 as the standard one. We can establish a two-to-one correspondence between spin configurations of the honeycomb lattice and arrow configurations in the triangular lattice. This can be done in the following way : if the spins on either side of a bond of the triangular lattice are equal (different), place an arrow on the bond pointing in the same (opposite) way as the standard. If we do this for all bonds, then at each site of the triangular lattice there must be an even number of non-standard arrows on the six incident bonds, and hence an odd number of incoming (and outgoing) arrows. This is the property that characterize the 32 vertex model on the triangular lattice.

Fig. 9. The standard arrow configuration for the triangular lattice.

In Fig. 10 we show two cases of the relation between arrow configurations on the triangular lattice and spin configurations on the honeycomb lattice.

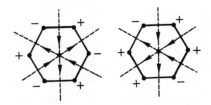

Fig. 10. Two cases of the correspondence between arrow configurations and spin configurations.

In consequence, the Boltzmann weights of the 32-vertex model will be a function of the Boltzmann weights $W'(\sigma_1, \sigma_2, \sigma_3, \sigma_4, \sigma_5, \sigma_6)$, associated to a face of the honeycomb lattice. By using the relation between vertex and spin configurations described above and expression (23), we find

$$
\begin{aligned}
\omega &= W'(+,-,-,-,+,+) = 2e^{2K_1} \\
\overline{\omega} &= W'(+,+,-,+,+,-) = 2e^{-2K_1}\cosh(4K_2 - 2K_3) \\
\omega_{56} &= W'(+,-,+,-,+,+) = 2e^{-2K_1}\cosh(2K_3) \\
\overline{\omega}_{56} &= W'(+,+,+,+,+,-) = 2e^{2K_1}\cosh(4K_2) \\
\omega_{15} &= W'(+,+,+,-,+,+) = 2e^{2K_1}\cosh(2K_2 + 2K_3) \\
\overline{\omega}_{15} &= W'(+,-,+,+,+,-) = 2e^{-2K_1}\cosh(2K_2) \\
\omega_{46} &= W'(+,-,+,+,+,+) = 2e^{2K_1}\cosh(2K_2 + 2K_3) \\
\overline{\omega}_{46} &= W'(+,+,+,-,+,-) = 2e^{-2K_1}\cosh(2K_2) \\
\omega_{13} &= W'(+,+,+,+,-,+) = 2e^{2K_1}\cosh(2K_2 + 2K_3)
\end{aligned}
$$

$$\overline{\omega}_{13} = W'(+,-,+,-,-,-) = 2e^{-2K_1}\cosh(2K_2)$$
$$\omega_{24} = W'(+,-,-,-,-,-) = 2e^{2K_1}\cosh(2K_2+2K_3)$$
$$\overline{\omega}_{24} = W'(+,+,-,+,-,+) = 2e^{-2K_1}\cosh(2K_2)$$
$$\omega_{14} = W'(+,+,+,+,+,+) = 2e^{6K_1}\cosh(4K_2+2K_3)$$
$$\overline{\omega}_{14} = W'(+,-,+,-,+,-) = 2e^{-6K_1}$$
$$\omega_{23} = W'(+,-,-,-,+,-) = 2e^{-2K_1}\cosh(2K_3)$$
$$\overline{\omega}_{23} = W'(+,+,-,+,+,+) = 2e^{2K_1}\cosh(4K_2)$$
$$\omega_{25} = W'(+,-,-,+,-,-) = 2e^{-2K_1}\cosh(2K_3)$$
$$\overline{\omega}_{25} = W'(+,+,-,-,-,+) = 2e^{2K_1}$$
$$\omega_{36} = W'(+,-,+,+,-,+) = 2e^{-2K_1}\cosh(2K_3)$$
$$\overline{\omega}_{36} = W'(+,+,+,-,-,-) = 2e^{2K_1}$$
$$\overline{\omega}_{34} = W'(+,+,-,+,-,-) = 2e^{-2K_1}\cosh(2K_2-2K_3)$$
$$\omega_{34} = W'(+,-,-,-,-,+) = 2e^{2K_1}\cosh(2K_2)$$
$$\overline{\omega}_{35} = W'(+,+,-,-,-,-) = 2e^{2K_1}\cosh(2K_3)$$
$$\omega_{35} = W'(+,-,-,+,-,+) = 2e^{-2K_1}$$
$$\overline{\omega}_{45} = W'(+,+,-,-,+,-) = 2e^{-2K_1}\cosh(2K_2-2K_3)$$
$$\omega_{45} = W'(+,-,-,+,+,+) = 2e^{2K_1}\cosh(2K_2)$$
$$\overline{\omega}_{12} = W'(+,-,+,-,-,+) = 2e^{-2K_1}\cosh(-2K_2+2K_3)$$
$$\omega_{12} = W'(+,+,+,+,-,-) = 2e^{2K_1}\cosh(2K_2)$$
$$\overline{\omega}_{26} = W'(+,+,+,-,-,+) = 2e^{2K_1}\cosh(2K_3)$$
$$\omega_{26} = W'(+,-,+,+,-,-) = 2e^{-2K_1}$$
$$\omega_{16} = W'(+,+,-,-,+,+) = 2e^{2K_1}\cosh(2K_2)$$
$$\overline{\omega}_{16} = W'(+,-,-,+,+,-) = 2e^{-2K_1}\cosh(2K_2-2K_3) \tag{24}$$

Using the above expressions in (9), (10) and (11) we can obtain the expression of the free energy of the centered honeycomb lattice Ising model.

Taking into account (24), (11) and (12), the critical temperature of the model is determined from the equation :

$$e^{2K_1} + e^{-2K_1}\cosh(4K_2-2K_3) + 2e^{-2K_1}\cosh(2K_3) + 2e^{2K_1} +$$
$$e^{6K_1}\cosh(4K_2+2K_3) + e^{-6K_1} = 2\max\{e^{2K_1} + e^{-2K_1}\cosh(4K_2-2K_3) ;$$
$$e^{2K_1} + e^{-2K_1}\cosh(2K_3) \quad ; \quad e^{6K_1}\cosh(4K_1+2K_3) + e^{-6K_1}\} \tag{25}$$

The solutions of this equation are analyzed in the next section.

We think that with the two cases studied above, the reader will be able to apply this procedure to other 2D Ising models without crossing bonds as, for instance, the Ising model on the centered square lattice. After decimation of the central spin in each square, this model can be mapped into a special case of the 16-vertex model, by following the same procedure that we have employed for the honeycomb lattice model.

2.3. Disorder solutions for two-dimensional Ising models

Disorder solutions are very useful for clarifyng the phase diagrams of anisotropic models and also imply constraints on the analytical behaviour of the partition function of these models.

A great variety of anisotropic models (with different coupling constants in the different directions of the lattice) are known to posses remarkable submanifolds in the space of parameters, where the partition function is computable and takes a very simple form. These are the disorder solutions.

All the methods applied for obtaining these solutions rely on the same mechanism : a certain local decoupling of the degrees of freedom of the model, which results in an effective reduction of dimensionality for the lattice system. Such a property is provided by a simple local condition imposed on the Boltzmann weights of the elementary cell generating the lattice.[27]

Some completely integrable models present disorder solutions, e.g. the triangular Ising model and the symmetric 8-vertex model. But very important models that are not integrable, also present this type of solutions, e.g. the triangular Ising model with a field, the triangular q-state Potts model, and the general 8-vertex model. Here we will consider only two dimensional Ising models.

In order to introduce the method, we will annalyse, at the first place, the simplest case, i.e. the anisotropic Ising model on the triangular lattice (see Fig. 11).

Fig. 11. The elementary cell of the triangular lattice, with three interactions K_1, K_2, and K_3.

The Boltzmann weight of the elementary cell is

$$W(\sigma_1, \sigma_2, \sigma_3) = \exp[\frac{1}{2}(K_1\sigma_1\sigma_3 + K_2\sigma_2\sigma_3 + K_3\sigma_1\sigma_2)] \qquad (26)$$

In every case, the local criterion will be defined by the following condition : after summation over some of its spins (to be defined in each case), the Boltzmann weight associated with the elementary cell of the model must not depend on the remaining spins any longer. For instance, for the triangular lattice, we will require

$$\sum_{\sigma_3} W(\sigma_1, \sigma_2, \sigma_3) = \lambda(K_1, K_2, K_3) \tag{27}$$

where λ is a function only of K_1, K_2 and K_3 (it is independent of σ_1 and σ_2). By using (26) we find

$$\sum_{\sigma_3} W(\sigma_1, \sigma_2, \sigma_3) = \exp(\frac{1}{2}K_3\sigma_1\sigma_2)\cosh[\frac{1}{2}(K_1\sigma_1 + K_2\sigma_2)] \tag{28}$$

But, as it is well known, we can write

$$\cosh[\frac{1}{2}(K_1\sigma_1 + K_2\sigma_2)] = A\exp(K\sigma_1\sigma_2) \tag{29}$$

with

$$A = [\cosh(\frac{K_1+K_2}{2})\cosh(\frac{K_1-K_2}{2})]^{\frac{1}{2}} \tag{30}$$

$$K = \frac{1}{2}\log[\frac{\cosh(\frac{K_1+K_2}{2})}{\cosh(\frac{K_1-K_2}{2})}] \tag{31}$$

In order that $\sum_{\sigma_3} W(\sigma_1, \sigma_2, \sigma_3)$ be independent of σ_1 and σ_2 we must impose the condition $K = -\frac{1}{2}K_3$. From this condition we find

$$e^{K_3}\cosh(\frac{K_1+K_2}{2}) = \cosh(\frac{K_1-K_2}{2}) \tag{32}$$

from which we can determine the expression of λ:

$$\lambda(K_1, K_2, K_3) = [\cosh(\frac{K_1+K_2}{2})\cosh(\frac{K_1-K_2}{2})]^{\frac{1}{2}} \tag{33}$$

It is easy to verify that Eq.(32) can be written as

$$\tanh(K_1)\tanh(K_2) + \tanh(K_3) = 0 \tag{34}$$

This 2D subvariety in the space of parameters is called the disorder variety of the model.

Let us now impose particular boundary conditions for the lattice (see Fig. 12) : on the upper layer, all interactions are missing, so that the spins of the upper layer only interact with those of the lower one. It immediately follows that if one sums over all the spins of the upper layer and if one requires the disorder condition (34) , the same boundary conditions reappear for the next layer.

Iterating the procedure leads one to an exact expression for the partition function, when restricted to subvariety (34):

$$Z = \lambda(K_1, K_2, K_3)^N \tag{35}$$

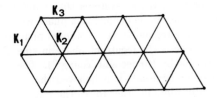

Fig. 12. Two layers of the triangular lattice.

where N is the number of sites of the lattice. The free energy in the thermodynamic limit is given by

$$f = -\frac{1}{2\beta} \log[\cosh(\frac{K_1 + K_2}{2}) \cosh(\frac{K_1 - K_2}{2})] \tag{36}$$

The partition function (35) corresponds to lattices with unusual boundary conditions. In the physical domain, where the coupling constants are real, these do not affect the partition function per site (or the free energy per site) in the thermodynamic limit, and the expression (36) also corresponds to the free energy per site with standard periodic boundary conditions. On the contrary, in the non-physical domain (complex coupling constants), the boundary conditions are known to play an important role, even after taking the thermodynamic limit.

Let us consider now the Kagomé lattice Ising model with two-spin interactions between nn and nnn, studied in the next section. If we apply the same procedure that for the triangular lattice Ising model, we obtain for the disorder variety:

$$e^{4K_2} = \frac{2(e^{4K_1} + 1)}{e^{8K_1} + 3} \tag{37}$$

This disorder variety does not have intersection with the critical variety of the model.

Following the method that we have exposed for the Ising model on the triangular lattice, the reader will be able to find the disorder varieties for other 2D Ising models with anisotropic interactions.

3. Reentrance in exactly solved frustrated Ising spin systems

In this section, we show and discuss the phase diagrams of several selected 2D frustrated Ising systems that have been recently solved. For general exact methods, the reader is referred to the book by Baxter,[4] and to the preceding section. In the following, we consider only frustrated systems that exhibit the **reentrance phenomenon**. A reentrant phase can be defined as a phase with no long-range order, or no order at all, occurring in a region below an ordered phase on the temperature (T) scale. A well-known example is the reentrant phase in spin-glasses (see the chapter by Gingras[21] and the review by Binder and Young[20]). The origin of the reentrance in spin-glasses is not well understood. It is believed that it is due to a combination of frustration and bond disorder. In order to see the role of the frustration alone, we show here the exact results on a number of periodically frustrated Ising systems. The idea behind the works shown in this section is to search for the ingredients responsible for the occurrence of the reentrant phase. Let us review in the following a few models showing a reentrant phase. Discussion on the origin of the reentrance will be given in the conclusion.

3.1. Centered square lattice

Even before the concept of frustration was introduced,[2] systems with competing interactions were found to possess rich critical behavior and non-trivial ordered states. Among these models, the centred square lattice Ising model (see Fig. 13), introduced by Vaks et al,[5] with nn and nnn interactions, J_1 and J_2, respectively, is to our knowledge the first exactly soluble model which exhibits successive phase transitions with a reentrant paramagnetic phase at low T. Exact expression for the free energy, some correlation functions, and the magnetization of one sublattice were given in the original work of Vaks et al.

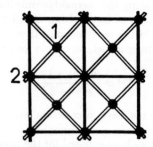

Fig. 13. Centered square lattice. Interactions between nn and nnn, J_1 and J_2, are denoted by white and black bonds, respectively. The two sublattices are numbered 1 and 2.

We distinguish two sublattices 1 and 2. Sublattice 1 contains the spins at the square centers, and sublattice 2 generates a square lattice with interaction J_2 in both horizontal and vertical directions. Spins of sublattice 1 interacts only with spins of sublattice 2 via diagonal interactions J_1. The ground state properties of this model are as follows : for $a = J_2/ \mid J_1 \mid > -1$, spins of sublattice 2 orders ferromagnetically and the spins of sublattice 1 are parallel (antiparallel) to the spins of sublattice 2 if $J_1 > 0$ (< 0); for $a < -1$, spins of sublattice 2 orders antiferromagnetically, leaving the centered spins free to flip.

3.1.1. Phase diagram
The phase diagram of this model is given by Vaks et al.[5] Except for $a = -1$, there is always a finite critical temperature.

When J_2 is antiferromagnetic (> 0) and J_2/J_1 is in a small region near 1, the system is successively in the paramagnetic state, an ordered state, the **reentrant** paramagnetic state, and another ordered state, with decreasing temperature (see Fig. 14).

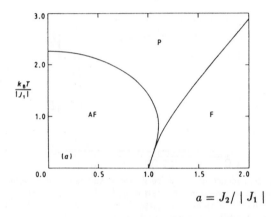

$$a = J_2/\mid J_1 \mid$$

Fig. 14. Phase diagram of centered square lattice.[5]

The centered square Ising lattice has been generalized to include several kinds of interaction.[6,7] For example, when the vertical interaction J_2 is different from the horizontal one, say J_3, the phase diagram becomes more complicated.

3.1.2. Nature of ordering and disorder solutions
For the sake of simplicity, let us consider hereafter the case of nn and nnn interactions only, namely J_1 and J_2 ($J_3 = J_2$). Note that though an exact critical line was obtained,[5] the order parameter was not calculated, though the magnetization of one sublattice were given in the original work of Vaks et al.[5] Very recently Choy and Baxter [22] have obtained the total magnetization for this model in the ferromagnetic region. However, the ordering in the antiferromagnetic (frustrated) region has not been exactly calculated, despite the fact that it may provide an interesting ground for

understanding the reentrance phenomenon. We have studied this aspect by means of Monte Carlo (MC) simulations.[23] The question which naturally arises is whether or not the disorder of sublattice 1 at $T = 0$ remains at finite T. If the spins of sublattice 1 remains disordered at finite T in the antiferromagnetic region we have a remarkable kind of ordered state: namely the coexistence between order and disorder. This behavior has been observed in three-dimensional Ising spin models[24,25] and in an exactly soluble model (the Kagomé lattice).[9] In the latter system, which is similar to the present model (discussed in the next subsection), it was shown that the coexistence of order and disorder at finite T shed some light on the reentrance phenomena. To verify the coexistence between order and disorder in the centered square lattice, we have performed Monte Carlo (MC) simulations. The results for the Edwards-Anderson sublattice order parameters q_i and the staggered susceptibility of sublattice 2 , as functions of T, are shown in Fig. 15 in the case $a = -2$.

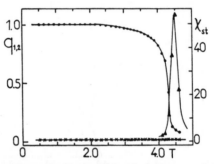

Fig. 15. Temperature dependence of sublattice Edwards-Anderson order parameters, q_1 and q_2 (crosses and black circles, respectively) in the case $a = J_2/ \mid J_1 \mid = -2$, by Monte Carlo simulation. Susceptibility calculated by fluctuations of magnetization of sublattice 2 is also shown. The lattice used contains $N = 2 \times 60 \times 60$ spins with periodic boundary conditions.[23]

As is seen, sublattice 2 is ordered up to the transition at T_c while sublattice 1 stays disordered at all T. This result shows a new example where order and disorder coexists in an equilibrium state. This result supports the conjecture formulated by Azaria et al,[9] namely the coexistence of order and disorder is a necessary condition for the reentrant behavior to occur. The partial disorder just compensates the loss of entropy due to the partial ordering of the high-T phase. In a previous paper,[9] the importance of the disorder line in understanding the reentrance phenomenon has been emphasized. There has been suggested that this type of line may be necessary for the change of ordering from the high-T ordered phase to the low-T one. In the narrow reentrant paramagnetic region, preordering fluctuations with different symmetries exist near each critical line. Therefore the correlation functions change their behavior as the temperature is varied in the reentrant paramagnetic region. As a consequence of the change of symmetries there exist spins for which the two-point correlation function (between nn spins) has different signs, near the two critical lines , in the reentrant paramagnetic region. Hence it is reasonable to expect that it has to vanish at a disorder temperature T_D . This point can be considered as a non-critical transition point which separates two different paramagnetic phases. The two-point correlation function defined above may be thought of as a non-local 'disorder parameter'. This particular point is just the one which has been called a disorder point by Stephenson[26] in analysing the behavior of correlation functions for systems with competing interactions. For the centered square lattice Ising model considered here, the Stephenson disorder line is[23]

$$\cosh(4J_1/k_B T_D) = \exp(-4J_2/k_B T_D) \tag{38}$$

The two-point correlation function at T_D between spins of sublattice 2 separated by a distance r is zero for odd r and decay like $r^{-1/2}[\tanh(J_2/k_B T_D)]^r$ for r even.[26] However,there is *no dimensional reduction* on the Stephenson line given above. Usually, one defines the disorder point as the temperature where there is an effective reduction of dimensionality in such a way that physical quantities become simplified spectacularly.[27] In general, these two types of disorder line are equivalent, as for example , in the case of the Kagomé lattice Ising model (see below). This is not the case here. In order to calculate this disorder line for the centered square lattice, we recall that this model is equivalent to an 8-vertex model that verifies the free-fermion condition.[28] The disorder line corresponding to dimensional reduction, was given for the general 8-vertex model by Giacomini.[29] When this result is applied to the centered square lattice, one finds that the disorder variety is given by

$$\exp(4J_2/k_B T) = (1 - i\sinh(4J_1/k_B T))^{-1} \tag{39}$$

where $i^2 = -1$. This disorder line lies on the unphysical (complex) region of the parameter space of this system. When calculated on the line (39), the magnetization of this model, evaluated recently by Choy and Baxter [22] in the ferromagnetic region, becomes singular, as it is usually the case for the disorder solutions with dimensional reduction.[30] Since in the centered square lattice, the two kinds of disorder line are not equivalent , we conclude, according to the arguments presented above ,that the Stephenson disorder line (38) is the relevant one for the reentrance phenomenon.

Disorder solutions have recently found interesting applications, as for example in the problem of cellular automatas (for a review see Rujan[31]). Moreover, they also serve to built a new kind of series expansion for lattice spin systems.[27]

3.2. Kagomé lattice

3.2.1. Model with nn and nnn interactions

Another model of interest is the Kagomé lattice shown in Fig. 3. The Kagomé Ising lattice with nn interaction J_1 has been solved a long time ago[8] showing no phase transition at finite T when J_1 is antiferromagnetic. Taking into account the nnn interaction J_2 [see Fig.3 and Eq. (13)], we have solved[9] this model by transforming it into a 16-vertex model which satisfies the free-fermion condition.

The critical condition is given by Eq. (19). For the whole phase diagram, the reader is refered to the paper by Azaria et al.[9] We show in Fig. 16 only the small region of J_2/J_1 in the phase diagram which has the reentrant paramagnetic phase and a disorder line.

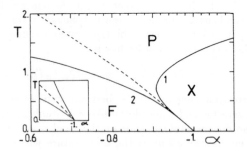

Fig. 16. Phase diagram of the Kagomé lattice with nnn interaction in the region $J_1 > 0$ of the space ($\alpha = J_2/J_1, T$). T is measured in the unit of J_1/k_B. Solid lines are critical lines, dashed line is the disorder line. P, F and X stand for paramagnetic, ferromagnetic and partially disordered phases, respectively. The inset shows schematically enlarged region of the endpoint.

The phase X indicates a partially ordered phase where the central spins are free (the nature of ordering was determined by MC simulations).[9] Here again, the reentrant phase takes place between a low-T ordered phase and a partially disordered phase. This suggests that a partial disorder at the high-T phase is necessary to ensure that the entropy is larger than that of the reentrant phase.

3.2.2. Generalized Kagomé lattice

When all the interactions are different in the model shown in Fig. 3, i.e. the horizontal bonds J_3, the vertical bonds J_2 and the diagonal ones are not equal (see Fig. 17), the phase diagram becomes complicated with new features:[10] in particular, we show that the reentrance can occur in an *infinite region* of phase space. In addition, there may be *several reentrant phases* occurring for a given set of interactions when T varies.

The Hamiltonian is written as

$$H = -J_1 \sum_{(ij)} \sigma_i\sigma_j - J_2 \sum_{(ij)} \sigma_i\sigma_j - J_3 \sum_{(ij)} \sigma_i\sigma_j \qquad (40)$$

where $\sigma_i = \pm 1$ is an Ising spin occupying the lattice site i , and the first, second, and third sums run over the spin pairs connected by diagonal, vertical and horizontal bonds, respectively. When $J_2 = 0$ and $J_1 = J_3$, one recovers the original nn Kagomé lattice.[8] The effect of J_2 in the case $J_1 = J_3$ has been shown above.

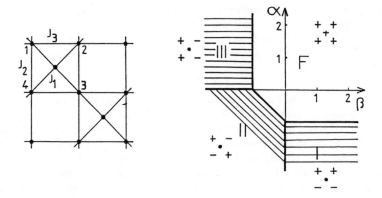

Fig. 17. Left: Generalized Kagomé lattice: diagonal, vertical and horizontal bonds denote the interactions J_1, J_2 and J_3, respectively. Right: Phase diagram of the ground state in the plane ($\alpha = J_2/J_1, \beta = J_3/J_1$). Heavy lines separate different phases and spin configuration of each phase is indicated (up, down and free spins are denoted by +, - and •, respectively). The three kinds of partially disordered phases and the ferromagnetic phase are denoted by I, II , III and F, respectively.

The phase diagram at temperature $T = 0$ is shown in Fig.17 in the space ($\alpha = J_2/J_1$, $\beta = J_3/J_1$) for positive J_1. The ground- state spin configurations are also displayed. The hatched regions indicate the three partially disordered phases (I, II, and III) where the central spins are free. Note that the phase diagram is mirror-symmetric with respect to the change of the sign of J_1. With negative J_1 , it suffices to reverse the central spin in the spin configuration shown in Fig. 17. Furthermore, the interchange of J_2 and J_3 leaves the system invariant, since it is equivalent to a $\pi/2$ rotation of the lattice. Let us consider the effect of the temperature on the phase diagram shown in Fig. 17. Partial disorder in the ground state often gives rise to the reentrance phenomenon as in systems shown above. Therefore, similar effects are to be expected in the present system. As it will be shown below, we find a new and richer behavior of the phase diagram: in particular, the reentrance region is found to be extended to infinity, unlike systems previously studied, and for some given set of interactions, there exist *two disorder lines* which divide the paramagnetic phase into regions of different kinds of fluctuations with a reentrant behavior.

Following the method exposed in section 2, one obtains a checkerboard Ising model with multispin interactions. This resulting model is equivalent to a symmetric 16-vertex model which satisfies the free-fermion condition.[18,32,33] The critical temperature of the model is given by

$$\cosh(4K_1)\exp(2K_2 + 2K_3) + \exp(-2K_2 - 2K_3) = 2\cosh(2K_3 - 2K_2) \pm 4\cosh(2K_1) \tag{41}$$

Note that Eq. (41) is invariant when changing $K_1 \rightarrow -K_1$ and interchanging K_2 and

K_3 as stated earlier. The phase diagram in the three-dimensional space (K_1, K_2, K_3) is rather complicated to show. Instead, we show in the following the phase diagram in the plane $(\beta = J_3/J_1, T)$ for typical values of $\alpha = J_2/J_1$. To describe each case and to follow the evolution of the phase diagram, let us go in the direction of decreasing α :

A. $\alpha > 0$

This case is shown in Fig. 18. Two critical lines are found with a paramagnetic reentrance having a usual shape (Fig. 18a) between the partially disordered (PD) phase of type III (see Fig. 17) and the ferromagnetic (F) phase with an endpoint at $\beta = -1$. The width of the reentrance region $[-1, \beta_1]$ decreases with decreasing α , from $\beta_1 = 0$ for α at infinity to $\beta_1 = -1$ for $\alpha = 0$ (zero width). Note that as α decreases, the PD phase III is depressed and disappears at $\alpha = 0$, leaving only the F phase (one critical line, see Fig. 18b). The absence of order at zero α for β smaller than -1 results from the fact that in the ground state, this region of parameters corresponds to a superdegenerate line separating the two PD phases II and III (see Fig. 17). So, along this line, the disorder contaminates the system for all T. As for disorder solutions, for positive α we find in the reentrant paramagnetic region a disorder line with dimension reduction[26] given by

$$\exp(4K_3) = 2\cosh(2K_2)/[\cosh(4K_1)\exp(2K_2) + \exp(-2K_2)] \qquad (42)$$

This is shown by the dotted lines in Fig. 18.

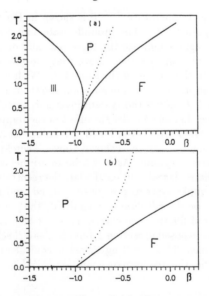

Fig. 18.   Phase diagram in the plane $(\beta = J_3/J_1, T)$ for positive values of $\alpha = J_2/J_1$:(a) $\alpha = 1$, (b) $\alpha = 0$. Solid lines are critical lines which separate different phases: paramagnetic (P), ferromagnetic (F), partially disordrered phase of type III (III). Dotted line shows the disorder line.

B. $0 > \alpha > -1$

In this range of α, there are three critical lines. The critical line separating the F and P phases and the one separating the PD phase I from the P phase have a common horizontal asymptote as β tends to infinity . They form a reentrant paramagnetic phase between the F phase and the PD phase I for positive b between a value β_2 and infinite β (Figs. 19). Infinite region of reentrance like this has never been found before this model. As α decreases, β_2 tends to zero and the F phase is contracted. For $\alpha < -1$, the F phase disappears together with the reentrance.

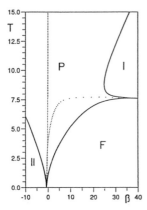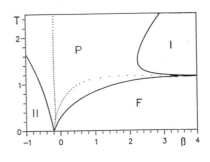

Fig. 19.   Phase diagram in the plane $(\beta = J_3/J_1, T)$ for negative values of $\alpha = J_2/J_1$. Left: $\alpha = -0.25$, Right: $\alpha = -0.8$. Solid lines are critical lines which separate different phases: paramagnetic (P), ferromagnetic (F), partially disordrered phases of type I and III. Dotted lines show the disorder lines.

In the interval $0 > \alpha > -1$, the phase diagram possesses two disorder lines, the first being given by Eq. (42), and the second by

$$\exp(4K_3) = 2\sinh(2K_2)/[-\cosh(4K_1)\exp(2K_2) + \exp(-2K_2)] \qquad (43)$$

These two disorder lines are issued from a point near $\beta = -1$ for small negative α; this point tends to zero as α tends to -1. The disorder line given by Eq. (43) enters the reentrant region which separates the F phase and the PD phase I (Fig. 19, left), and the one given by Eq. (42) tends to infinity with the asymptote $\beta = 0$ as $T \to \infty$. The most striking feature is the behavior of these two disorder lines at low T: they cross each other in the P phase for $0 > \alpha > -0.5$, forming regions of fluctuations of different nature (Fig. 20a). For $-0.5 > \alpha > -1$, the two disorder lines do no longer cross each other (see Fig. 20b). The one given by Eq. (42) has a reentrant aspect: in a small region of negative values of β, one crosses three times this line in the P phase with decreasing T.

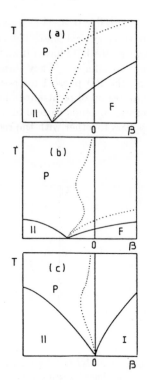

Fig. 20.   The behavior of the disorder lines (dotted) is schematically enlarged in the case (a) $\alpha = -0.25$, (b) $\alpha = -0.8$, (c) $\alpha = -1.5$.

C. $\alpha \leq -1$

For α smaller than -1, there are two critical lines and no reentrance (Fig. 21). Only the disorder line given by Eq. (42) survives with a reentrant aspect: in a small region of negative values of β, one crosses twice this line in the P phase with decreasing T. This behavior, being undistiguishable in the scale of Fig. 21, is schematically enlarged in Fig. 20c. The multicritical point where the P, I and II phases meet is found at $\beta = 0$ and $T = 0$.

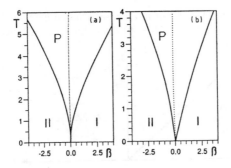

Fig. 21. The same caption as that of Fig. 19 with (a) $\alpha = -1$, (b) $\alpha = -1.5$.

At this stage, it is interesting to note that while reentrance and disorder lines occur along the horizontal axis $\alpha = -1$ and along the vertical axis $\beta = -1$ of Fig. 17 when the temperature is switched on, the most frustrated region ($\alpha < 0$ and $\beta < 0$) of the ground state does not show successive phase transitions (see Fig. 21, for example). Therefore, the existence of a reentrance may require a sufficient frustration, but not overfrustration. Otherwise, the system may have either a PD phase (Fig. 21) or no order at all (Fig. 18b).

The origin of the reentrance phenomenon will be discussed again in the conclusion.

3.3. Centered honeycomb lattice

In order to find common aspects of the reentrance phenomenon, we have constructed a few other models which possess a partially disordered phase next to an ordered phase in the ground state. Let us mention here the anisotropic centered honeycomb lattice shown in Fig. 7.[11] The Hamiltonian is given by (19), with three kinds of interactions J_1, J_2, and J_3 denoting the interactions between the spin pairs connected by heavy, light, and double-light bonds, respectively. We recall that when $J_2 = J_3 = 0$, one recovers the honeycomb lattice, and when $J_1 = J_2 = J_3$ one has the triangular lattice.

Fig. 22 shows the phase diagram at temperature $T = 0$ for three cases ($J_1 \neq J_2 = J_3$), ($J_1 \neq J_3, J_2 = 0$) and ($J_1 \neq J_2, J_3 = 0$). The ground-state spin configurations are also indicated.

The phase diagram is symmetric with respect to the horizontal axis: the transformation $(J_2, \sigma) \to (-J_2, -\sigma)$, or $(J_3, \sigma) \to (-J_3, -\sigma)$, leaves the system invariant. In each case, there is a phase where the central spins are free to flip ("partially disordered phase"). In view of this common feature with other models studied so far, one expects a reentrant phase occuring between the partially disordered phase and its neighboring phase at finite T. As it will be shown below, though a partial disorder exists in the ground state, it does not in all cases studied here yield a reentrant phase at finite T. Only the case $(J_1 \neq J_2, J_3 = 0)$ does show a reentrance.

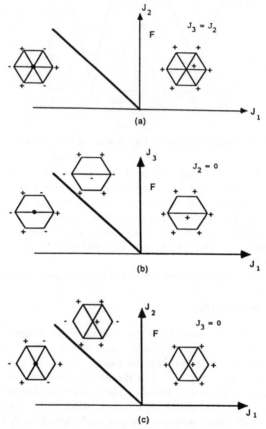

Fig. 22. Phase diagram of the ground state shown in the space: (a) $(J_1, J_2 = J_3)$; (b) (J_1, J_3) with $J_2 = 0$) ; (c) (J_1, J_2) with $J_3 = 0$). Heavy lines separate different phases and spin configuration of each phase is indicated (up, down and free spins are denoted by +, - and •, respectively).

To obtain the exact solution of our model, we decimate the central spin of each elementary cell of the lattice as explicited in the section 2. The resulting model is equivalent to a special case of the 32-vertex model[19] on a triangular lattice that satisfies the free-fermion condition. The explicit expression of the free energy as a function of interaction parameters K_1, K_2, and K_3 is very complicated, as seen by replacing Eq. (23) in Eqs. (9), (10) and (11). The critical temperature is given by Eq. (24).

We have analyzed, in particular, the three cases $(K_1 \neq K_2 = K_3)$, $(K_1 \neq$

$K_3, K_2 = 0$) and ($K_1 \neq K_2, K_3 = 0$).

When $K_2 = K_3$, the critical line obtained from Eq.(24) is

$$\exp(3K_1)\cosh(6K_2) + \exp(-3K_1) = 3[\exp(K_1) + \exp(-K_1)\cosh(2K_2)] \qquad (44)$$

In the case $K_2 = 0$, the critical line is given by

$$\exp(3K_1)\cosh(2K_3) + \exp(-3K_1) = 3[\exp(K_1) + \exp(-K_1)\cosh(2K_3)] \qquad (45)$$

Note that these equations are invariant with respect to the transformation $K_2 \to -K_2$ (see Eq.(44)) and $K_3 \to -K_3$ (see Eq.(45)).

The phase diagrams obtained from Eqs.(44) and (45) are shown in Fig. 23a and Fig. 23b, respectively.

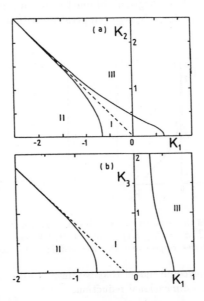

Fig. 23. Phase diagram shown in the space (a) $(K_1, K_2 = K_3)$; (b) $(K_1, K_3, K_2 = 0)$. Solid lines are critical lines which separate paramagnetic (I) , partially disordered (II) and ordered (III) phases. Discontinued lines of slope -1 are the asymptotes.

These two cases do not present the reentrance phenomenon though a partially disordered phase exists next to an ordered phase in the ground state (this is seen by plotting a line from the origin, i.e. from infinite T: this line never crosses twice a critical line whatever its slope, i.e. the ratio $K_{2,3}/K_1$, is). In the ordered phase II, the partial disorder, which exists in the ground state, remains so up to the phase transition. This has been verified by examining the Edwards-Anderson order parameter associated with the central spins in MC simulations.[11] Note that when $K_2 = K_3 = 0$.

one recovers the transition at finite temperature found for the honeycomb lattice.[34] and when $K_2 = K_3 = K_1 = -1$ one recovers the antiferromagnetic triangular lattice which has no phase transition at finite temperature.[35] The case $K_2 = 0$ (Fig. 23b) does not have a phase transition at finite T in the range $-\infty < K_3/K_1 < -1$, and phase II has a partial disorder as that in Fig. 18a.

The case $K_3 = 0$ shows on the other hand a reentrant phase. The critical lines are determined from the equations

$$\cosh(4K_2) = \frac{\exp(4K_1) + 2\exp(2K_1) + 1}{[1 - \exp(4K_1)]\exp(2K_1)} \tag{46}$$

$$\cosh(4K_2) = \frac{3\exp(4K_1) + 2\exp(2K_1) - 1}{[\exp(4K_1) - 1]\exp(2K_1)} \tag{47}$$

Fig. 24 shows the phase diagram obtained from (46) and (47) . The reentrant paramagnetic phase goes down to zero temperature with an end point at $\alpha = -0.5$ (see Fig. 24 right).

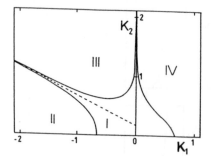 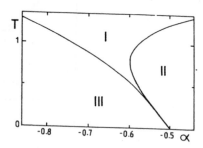

Fig. 24. Phase diagram of the centered honeycomb lattice with reentrance in the space (K_1, K_2) (left) and in the space $(T, \alpha = K_2/K_1)$ (right). I, II, III phases are paramagnetic, partially disordered and ordered phases, respectively. Discontinued line is the asymptote.

Note that the honeycomb model that we have studied here does not present a disorder solution with a dimensional reduction.

3.4. Periodically dilute centered square lattices

In this subsection, we show the exact results on several periodically dilute centered square Ising lattices by transforming them into 8-vertex models of *different vertex statistical weights* that satisfy the free-fermion condition. The dilution is introduced by taking away a number of centered spins in a periodic manner. For a given set of interactions , there may be five transitions with decreasing temperature with two reentrant paramagnetic phases. These two phases extend to infinity in the space of interaction parameters. Moreover, two additional reentrant phases are found, each in a limited region of phase space.[12]

Let us consider several **periodically dilute centered square lattices** defined from the centered square lattice shown in Fig. 25.

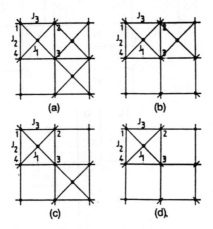

Fig. 25. Elementary cells of periodically dilute centered square lattice: (a) three-center case, (b) two-adjacent-center case, (c) two-diagonal-center case, (d) one-center case. Interactions along diagonal, vertical and horizotal bonds are J_1, J_2, and J_3, respectively.

The Hamiltonian of these models is given by

$$H = -J_1 \sum_{(ij)} \sigma_i \sigma_j - J_2 \sum_{(ij)} \sigma_i \sigma_j - J_3 \sum_{(ij)} \sigma_i \sigma_j \qquad (48)$$

where $\sigma_i = \pm 1$ is an Ising spin occupying the lattice site i , and the first, second and third sums run over the spin pairs connected by diagonal, vertical and horizontal bonds, respectively. All these models have at least one partially disorered phase in the ground state, caused by the competing interactions.

The model shown in Fig. 25c is in fact the generalized Kagomé lattice[10] which is shown above. The other models are less symmetric, require different vertex weights as seen below.

Let us show in Fig. 26 the phase diagrams, at $T = 0$, of the models shown in Figs. 25a, 25b and 25d, in the space (a, b) where $a = J_2/J_1$ and $b = J_3/J_1$. The spin configurations in different phases are also displayed. The three-center case (Fig. 26a), has six phases (numbered from I to VI), five of which (I, II, IV, V and VI) are partially disordered (with, at least, one centered spin being free), while the two-center case (Fig. 26b) has five phases, three of which (I, IV, and V) are partially disordered. Finally, the one-center case has seven phases with three partially disordered ones (I, VI and VII).

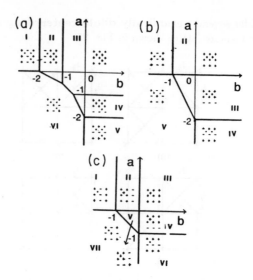

Fig. 26. Phase diagrams in the plane ($a = J_2/J_1$, $b = J_3/J_1$) at $T = 0$ are shown for the three-center case (a), two-adjacent center case (b), and one-center case (c). Critical lines are drawn by heavy lines. Each phase is numbered and the spin configuration is indicated (+, -, and o are up, down, and free spins, respectively). Degenerate configurations are obtained by reversing all spins.

As will be shown later, in each model, the reentrance occurs along most of the critical lines when the temperature is switched on. This is a very special feature of the models shown in Fig. 25 which has not been found in other models.

The partition function is written as

$$Z = \prod_j \sum_\sigma W_j \tag{49}$$

where the sum is performed over all spin configurations and the product over all elementary squares. W_j is the statistical weight of the j-th square. Let us denote the centered spin (when it exists) by σ and the spins at the square corners by σ_1, σ_2, σ_3 and σ_4. If the centered site exists, the statistical weight W_j of the square is

$$W_j = \exp[K_1(\sigma_1\sigma_2 + \sigma_3\sigma_4) + K_2(\sigma_1\sigma_4 + \sigma_2\sigma_3) + K_3\sigma(\sigma_1 + \sigma_2 + \sigma_3 + \sigma_4)] \tag{50}$$

Otherwise, it is given by

$$W_j = \exp[K_1(\sigma_1\sigma_2 + \sigma_3\sigma_4) + K_2(\sigma_1\sigma_4 + \sigma_2\sigma_3)] \tag{51}$$

where $K_i = J_i/k_B T$ ($i = 1, 2, 3$).

In order to obtain the exact solution of these models, we decimate the central spins of the centered squares. The resulting system is equivalent to an eight-vertex model on a square lattice, but with *different vertex weights*. For example, when one center is missing (Fig. 25d), three squares over four have the same weight W_i, and the fourth has a weight $W_i' \neq W_i$. So, we have to define four different sublattices with different statistical weights. The problem has been studied by Hsue, Lin and Wu for two different sublattices[36] and Lin and Wang [37] for four sublattices. They showed that exact solution can be obtained provided that all different statistical weights satisfy the free-fermion condition.[4,18,36,37] This is indeed our case and we get the exact partition function in terms of interaction parameters. The critical surfaces of our models are obtained by

$$\Omega_1 + \Omega_2 + \Omega_3 + \Omega_4 = 2\max(\Omega_1, \Omega_2, \Omega_3, \Omega_4) \qquad (52)$$

where W_i are functions of K_1, K_2 and K_3. We explicit this equation and we obtain a second order equation for X which is a function of K_2 only :

$$A(K_1, K_3)X^2 + B(K_1, K_3)X + C(K_1, K_3) = 0 \qquad (53)$$

with a priori four possible values of A, B and C for each model.

For given values of K_1 and K_3, the critical surface is determined by the value of K_2 which satisfies Eq.(53) through X. X must be real positive. We show in the following the expressions of A, B and C for which this condition is fulfilled for each model. Eq.(52) may have as much as five solutions for the critical temperature,[37] and the system may, for some given values of interaction parameters, exhibit up to five phase transitions. This happens for the model with three centers, when one of the interaction is large positive and the other slightly negative, the diagonal one being taken as unit. In general, we obtain one or three solutions for T_c.

3.4.1. *Model with three centers* (Fig. 25a)

The quantities which satisfy Eq. (53) are given by

$$
\begin{aligned}
X &= \exp(4K_2) \\
A &= \exp(4K_1)\cosh^3(4K_3) + \exp(-4K_1) - \cosh^2(4K_3) - \cosh(4K_3) \\
B &= \pm\{1 + 3\cosh(4K_3) + 8\cosh^3(2K_3) + [\cosh(4K_3) + \cosh^2(4K_3)]\exp(4K_1) \\
&\quad + 2\exp(-4K_1)\} \\
C &= [\exp(2K_1) - \exp(-2K_1)]^2 \\
A &= \exp(4K_1)\cosh^3(4K_3) + \exp(-4K_1) + \cosh^2(4K_3) + \cosh(4K_3) \\
B &= [1 + 3\cosh(4K_3) + 8\cosh^3(2K_3) - (\cosh(4K_3) + \cosh^2(4K_3))\exp(4K_1) \\
&\quad - 2\exp(-4K_1)] \\
C &= [\exp(2K_1) + \exp(-2K_1)]^2
\end{aligned}
\qquad (54)
$$

Let us describe now in detail the phase diagram of the three-center model (Fig. 25a).

For clarity, we show in Fig. 27 the phase diagram in the space ($a = J_2/J_1$, T) for typical values of $b = J_3/J_1$.

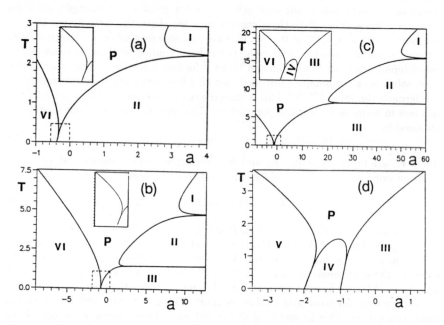

Fig. 27. Phase diagrams in the plane $(T, a = J_2/J_1)$ for several values of $b = J_3/J_1$:(a) $b = -1.25$, (b) $b = -0.75$, (c) $b = -0.25$, (d) $b = 0.75$. Reentrant regions on negative sides of a (limited by discontinued lines) are schematically enlarged in the insets. The nature of ordering in each phase is indicated by a number which is referred to the corresponding spin configuration in Fig. 26. P is paramagnetic phase.

For $b < -1$, there are two reentrances. Fig. 27a shows the case of $b = -1.25$ where the nature of the ordering in each phase is indicated using the same numbers of corresponding ground state configurations (see Fig. 26). Note that all phases (I, II and VI) are partially disordered: the centered spins which are disordered at $T = 0$ (Fig. 26a) remain so at all T. As seen, one paramagnetic reentrance is found in a small region of negative a (schematically enlarged in the inset of Fig. 27a), and the other on the positive a extending to infinity. The two critical lines in this region have a common horizontal asymptote.

For $-1 < b < -0.5$, there are three reentrant paramagnetic regions as shown in Fig. 27b: the reentrant region on the negative a is very narrow (inset), and the two on the positive a become so narrower while a goes to infinity that they cannot be seen on the scale of Fig. 27. Note that the critical lines in these regions have horizontal asymptotes. For a large value of a, one has five transitions with decreasing T: paramagnetic state - partially disordered phase I - reentrant paramagnetic phase

- partially disordered phase II - reentrant paramagnetic phase- ferromagnetic phase (see Fig. 27b). So far, this is the first model that exhibits such successive phase transitions with two reentrances.

For $-0.5 < b < 0$, there is an additional reentrance for $a < -1$: this is shown in the inset of Fig. 27c. As b increases from negative values, the ferromagnetic region (III) in the phase diagram "pushes" the two partially disordered phases (I and II) toward higher T. At $b = 0$, these two phases disappear at infinite T, leaving only the ferromagnetic phase. For positive b, there are thus only two reentrances remaining on a negative region of a, with endpoints at $a = -2$ and $a = -1$, at $T = 0$ (see Fig. 27d).

3.4.2. Model with two adjacent centers (Fig. 25b)

The quantities which satisfy Eq. (53) are given by

$$
\begin{aligned}
X &= \exp(2K_2) \\
A &= \exp(2K_1)\cosh(4K_3) + \exp(-2K_1) \\
B &= 2[\exp(2K_1)\cosh^2(2K_3) - \exp(-2K_1)] \\
C &= \exp(2K_1) + \exp(-2K_1) \\
A &= \exp(2K_1)\cosh(4K_3) - \exp(-2K_1) \\
B &= \pm 2[\exp(2K_1)\cosh^2(2K_3) + \exp(-2K_1)] \\
C &= \exp(2K_1) - \exp(-2K_1)
\end{aligned}
\tag{55}
$$

The phase diagram is shown in Fig. 28.

For $b < -1$, this model shows only one transition at a finite T for a given value of a, except when $a = 0$ where the paramagnetic state goes down to $T = 0$(see Fig. 28a).

However, for $-1 < b < 0$, two reentrances appear, the first one separating phases I and II goes to infinity with increasing a, and the second one exists in a small region of negative a with an endpoint at $(a = -2 - 2b, T = 0)$. The slope of the critical lines at $a = 0$ is vertical (see inset of Fig. 28b).

As b becomes positive, the reentrance on the positive side of a disappears (Fig. 28c), leaving only phase III (ferromagnetic).

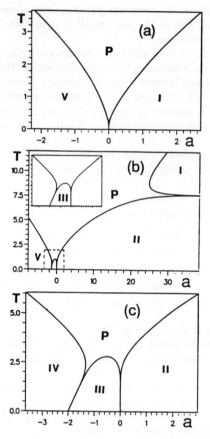

Fig. 28. Two-center model: the same caption as that of Fig. 27 with (a) $b = -1.25$, (b) $b = -0.25$, (c) $b = 2$.

3.4.3. Model with one center (Fig. 25d)

For this case, the quantities which satisfy Eq. (53) are given by

$$
\begin{aligned}
X &= \exp(4K_2) \\
A &= \exp(4K_1)\cosh(4K_3) + \exp(-4K_1) - 2\cosh^2(2K_3) \\
B &= \pm 2\{[\cosh(2K_3) + 1]^2 + [\exp(2K_1)\cosh(2K_3) + \exp(-2K_1)]^2\} \\
C &= [\exp(2K_1) - \exp(-2K_1)]^2 \\
A &= \exp(4K_1)\cosh(4K_3) + \exp(-4K_1) + 2\cosh^2(2K_3) \\
B &= \pm 2\{[\cosh(2K_3) + 1]^2 - [\exp(2K_1)\cosh(2K_3) - \exp(-2K_1)]^2\}
\end{aligned}
$$

$$C = [\exp(2K_1) + \exp(-2K_1)]^2 \qquad (56)$$

The phase diagrams of this model shown in Fig. 29 for $b < -1$, $-1 < b < 0$ and $b > 0$ are very similar to those of the two-center model shown in Fig. 28. This is not unexpected if one examines the ground state phase diagrams of the two cases (Figs. 25b and 25c): their common point is the existence of a partially disordered phase next to an ordered phase. The difference between the one- and two-center cases and the three-center case shown above is that the latter has, in addition, two boundaries, each of which separates two partially disordered phases (see Fig. 26a). It is along these boundaries that the two additional reentrances take place at finite T in the three-center case.

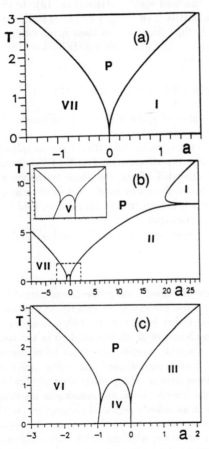

Fig. 29. One-center case: the same caption as that of Fig. 27 with (a) $b = -1.25$, (b) $b = -0.25$, (c) $b = 0.5$.

In conclusion of this subsection, we summarize that in simple models such as those shown in Fig. 25, we have found two reentrant phases occuring on the temperature scale at a given set of interaction parameters. A striking feature is the existence of a reentrant phase between *two partially disordered phases* which has not been found so far in any other model (we recall that in other models, a reentrant phase is found between an ordered phase and a partially disordered phase).

3.5. Random-field aspects of the models

Let us touch upon the random-field aspect of the model. A connection between the Ising models presented above and the random-field problem can be established. Consider for instance the centered square lattice (Fig. 13). In the region of antiferromagnetic ordering of sublattice 2 $(a = J_2/ \mid J_1 \mid < -1)$, the spins on sublattice 1 (centered spins) are free to flip. They act on their neighboring spins (sublattice 2) as an annealed random field h. The probability distribution of this random field at a site of sublattice 2 is given by

$$P(h) = \frac{1}{16}[6\delta(h) + 4\delta(h + 2J_1) + 4\delta(h - 2J_1) + \delta(h + 4J_1) + \delta(h - 4J_1)] \quad (57)$$

The random field at a number of spins is thus zero (diluted). Moreover, this field distribution is somewhat correlated because each spin of sublattice 1 acts on four spins of sublattice 2. Since the spins on sublattice 1 are completely disordered at all T, it is reasonable to consider this effective random-field distribution as quenched. In addition to possible local annealed effects, the phase transition at a finite T of this model may be a consequence of the above mentioned dilution and correlations of the field distribution, because it is known that in 2D random-field Ising model (without dilution) there is no such a transition.[38]

The same argument is applied to other models studied above.

4. Conclusion

In this chapter, we have discussed some properties of periodically frustrated Ising systems. We have limited the discussion to exactly solved models which possess at least a reentrant phase. Other Ising systems which involved approximations are discussed in the chapter by Nagai et al (this book) and in the book by Liebmann.[1].

Let us emphasize that simple models having no bond disorder like those presented in this chapter can possess complicated phase diagrams due to the frustration generated by competing interactions. Many interesting physical phenomena such as successive phase transitions, disorder lines, and reentrance are found. In particular, a reentrant phase can occur in an infinite region of parameters (sections 3.2. and 3.4.). For a given set of interaction parameters in this region, successive phase transitions take place on the temperature scale, with one or two paramagnetic reentrant phases.

The relevance of disorder solutions for the reentrance phenomena has also been pointed out. An interesting finding is the occurrence of two disorder lines which divide the paramagnetic phase into regions of different kinds of fluctuations (section 3.2). Therefore, care should be taken in analyzing experimental data such as correlation functions, susceptibility, etc. in the paramagnetic phase of frustrated systems.

Although the reentrance is found in the models shown above by exact calculations, there is no theoretical explanation why such a phase can occur. In other words, what is the necessary and sufficient condition for the occurrence of a reentrance? We have conjectured[9,12] that the necessary condition for a reentrance to take place is the existence of at least a partially disordered phase next to an ordered phase or another partially disordered phase in the ground state. The partial disorder is due to the competition between different interactions.

The existence of a partial disorder yields the occurrence of a reentrance in most of known cases,[5,6,7,9,10,12] except in some particular regions of interaction parameters in the centered honeycomb lattice (section 3.3.): the partial disorder alone is not sufficient to make a reentrance as shown in Fig. 23a and Fig. 23b, the finite zero-point entropy due to the partial disorder of the ground state is the same for three cases considered in Fig. 22 , i.e. $S_0 = \log(2)/3$ per spin, but only one case yields a reentrance. Therefore, the existence of a partial disorder is a necessary, but not sufficient, condition for the occurrence of a reentrance.

The anisotropic character of the interactions can also favor the occurrence of the reentrance. For example, the reentrant region is enlarged by anisotropic interactions as in the centered square lattice,[7] and becomes infinite in the generalized Kagomé model (section 3.2.). But again, this alone cannot cause a reentrance as seen by comparing the anisotropic cases shown in Fig. 22b and Fig. 22c: only in the latter case a reentrance does occur. The presence of a reentrance may also require a coordination number at a disordered site large enough to influence the neighboring ordered sites. When it is too small such as in the case shown in Fig. 22b (equal to two), it cannot induce a reentrance. However, it may have an upper limit to avoid the disorder contamination of the whole system such as in the case shown in Fig. 23a where the cordination number is equal to six. So far, the 'right' number is four in known reentrant systems shown above. Systematic investigations of all possible ingredients are therefore desirable to obtain a sufficient condition for the existence of a reentrance.

In three dimensions, apart from a particular exactly solved case[14] showing a reentrance, a few Ising systems such as the fully frustrated simple cubic lattice,[24,25] a stacked triangular Ising antiferromagnet[39,40,41,42,43] and a body-centered cubic (bcc) crystal[44] exhibit a partially disordered phase in the ground state. We believe that reentrance should also exist in the phase space of such systems though evidence is found numerically only for the bcc case.[44]

Finally, let us emphasize that when a phase transition occurs between states of different symmetries which have no special group-subgroup relation, it is generally accepted that the transition is of first order . However, the reentrance phenomenon is a symmetry breaking alternative which allows one ordered phase to change into another incompatible ordered phase by going through an intermediate reentrant phase. A question which naturally arises is under which circumstances does a system prefer an intermediate reentrant phase to a first-order transition. In order to analyse this aspect we have generalized the centered square lattice Ising model into three dimensions.[44] This is a special bcc lattice. We have found that at low T the reentrant region observed in the centered square lattice shrinks into a first order transition line which is ended at a multicritical point from which two second order lines emerge forming a narrow reentrant region.[44]

As a final remark, let us mention that although the exactly solved systems shown in this chapter are models in statistical physics, we believe that the results obtained in this work have qualitative bearing on real frustrated magnetic systems. In view of the simplicity of these models, we believe that the results found here will have several applications in various areas of physics.

5. Acknowledgements

The authors wish to thank Patrick Azaria and Marianne Debauche for their close collaborations in the works presented in this chapter.

6. References

1. R. Liebmann, *Statistical Mechanics of Periodic Frustrated Ising Systems*, Lecture Notes in Physics, vol. 251 (Springer-Verlag, Berlin, 1986).
2. G. Toulouse, *Commun. Phys.* **2** (1977) 115.
3. J. Villain, *J. Phys.* C**10** (1977) 1717.
4. R. J. Baxter, *Exactly solved Models in Statistical Mechanics* (Academic, New York, 1982).
5. V. Vaks , A. Larkin and Y. Ovchinnikov, *Sov. Phys. JEPT* **22** (1966) 820.
6. T. Morita, *J. Phys. A* **19** (1987) 1701.
7. T. Chikyu and M. Suzuki, *Prog. Theor. Phys.* **78** (1987) 1242.
8. K. Kano and S. Naya, *Prog. Theor. Phys.* **10** (1953) 158.
9. P. Azaria, H. T. Diep and H. Giacomini, *Phys. Rev. Lett.* **59** (1987) 1629.
10. M. Debauche, H.T. Diep, P. Azaria, and H. Giacomini, *Phys. Rev. B* **44** (1991) 2369.
11. H.T. Diep, M. Debauche and H. Giacomini, *Phys. Rev. B* **43** (1991) 8759.
12. M. Debauche and H. T. Diep, *Phys. Rev. B* **46** (1992) 8214; H. T. Diep, M. Debauche and H. Giacomini, *J. of Mag. and Mag. Mater.* **104** (1992) 184.
13. H. Kitatani, S. Miyashita and M. Suzuki, *J. Phys. Soc. Jpn.* **55** (1986) 865; *Phys. Lett. A* **158** (1985) 45.
14. T. Horiguchi, *Physica A* **146** (1987) 613.
15. J. Villain, R. Bidaux, J.P. Carton, and R. Conte, *J. Physique* **41** (1980) 1263.
16. T. Oguchi, H. Nishimori, and Y. Taguchi, *J . Phys. Jpn.* **54** (1985) 4494.
17. C. Henley, *Phys. Rev. Lett.* **62** (1989) 2056.
18. A. Gaff and J. Hijmann, *Physica A* **80** (1975) 149.
19. J. E. Sacco and F. Y. Wu, *J. Phys. A* **8** (1975) 1780.
20. K. Binder and A. P. Young, *Rev. Mod. Phys.* **58** (1986) 801.

21. See the chapter by M. Gingras, this book.
22. T. Choy and R. Baxter, *Phys. Lett. A* **125** (1987) 365.
23. P. Azaria, H. T. Diep and H. Giacomini, *Phys. Rev. B* **39** (1989) 740.
24. D. Blankschtein, M. Ma and A. Berker, *Phys. Rev. B* **30** (1984) 1362.
25. H. T. Diep, P. Lallemand and O. Nagai, *J. Phys. C* **18** (1985) 1067.
26. J. Stephenson, *J. Math. Phys.* **11** (1970) 420; *Can. J. Phys.* **48** (1970) 2118; *Phys. Rev. B* **1** (1970) 4405.
27. J. Maillard , *Second conference on Statistical Mechanics*, California Davies (1986), unpublished.
28. F. Y. Wu and K.Y. Lin , *J. Phys. A* **20** (1987) 5737.
29. H. Giacomini , *J. Phys. A* **19** (1986) L335.
30. R. Baxter , *Proc. R. Soc. A* **404** (1986) 1.
31. P. Rujan , *J. Stat. Phys.* **49** (1987) 139.
32. M. Suzuki and M. Fisher, *J. Math. Phys.* **12** (1971) 235.
33. F. Y. Wu, *Solid Stat. Comm.* **10** (1972) 115.
34. L. Onsager, *Phys. Rev.* **65** (1944) 117.
35. G. H. Wannier, *Phys. Rev.* **79** (1950) 357; *Phys. Rev. B* **7** (1973) 5017 (E).
36. C. S. Hsue, K. Y. Lin, and F.Y. Wu, *Phys. Rev. B* **12** (1975) 429.
37. K. Y. Lin and I. P. Wang, *J. Phys. A* **10** (1977) 813.
38. J. Imbrie, *Phys. Rev. Lett.* **53** (1984) 1747.
39. D. Blankschtein, M. Ma , A. Nihat Berker, G. S. Grest, and C. M. Soukoulis, *Phys. Rev. B* **29** (1984) 5250.
40. O. Heinonen and R. B. Petschek, *Phys. Rev. B* **40** (1989) 9052.
41. M. L. Plumer, A. Mailhot, R. Ducharme, A. Caillé and H. T. Diep, *Phys. Rev. B* **47** (1993) 14312.
42. A. Bunker, B. D. Gaulin and C. Kallin, *Phys. Rev. B* **48** (1993) 15861.
43. See the chapter by O. Nagai, T. Horiguchi and S. Miyashita, this book.
44. P. Azaria, H. T. Diep and H. Giacomini, *Europhys. Lett.* **9** (1989) 755.

Magnetic System with Competing Interaction
Ed. H. T. Diep
©1994 World Scientific Publishing Co.

CHAPTER V

PROPERTIES AND PHASE TRANSITIONS

IN FRUSTRATED ISING SPIN SYSTEMS

Ojiro Nagai
Department of Physics, Faculty of Science, Kobe University
Kobe 657, Japan.

Tsuyoshi Horiguchi
Department of Computer and Mathematical Sciences, GSIS, Tohoku University
Sendai 980, Japan.

and

Seiji Miyashita
Graduate School of Human and Environmental Studies, Kyoto University
Kyoto 606, Japan.

1. Introduction

Frustration is one of the most interesting concepts in condensed matter physics. Historically, a striking affair concerning frustration was a discovery of helical spin structure in 1959.[1,2] Until that year, this structure has not been found in spite of the fact that there are many substances noticed later as helical magnets; the mathematical procedure to derive the helical spin structure is simple. The concept of frustration has a very important role on spin systems, including Ising spin systems, classical vector spin systems such as XY model, classical Heisenberg model and so on, and also quantum spin systems. The spin glasses are one of the most interesting problems in the spin systems, in which the frustration plays an important role. There exist experimentally various Ising magnets with hexagonal symmetry and also the ordered problem of adsorbates on a crystal plane as for the Ising spin systems. Note that the binary alloy problem is equivalent to Ising models. According to recent findings, the neural networks are deeply related with the Ising spin systems with frustration.[3]

The word "frustration" was first introduced by Toulouse[4,5] in 1977 for a plaquette shown in Fig.1 (see the next page). Here the plaquette is composed of three ferromagnetic bonds and one antiferromagnetic bond. An Ising spin, $\sigma_j = \pm 1$, is located at the jth vertex (or site). The Hamiltonian of this four-spin system is written as

$$H = J_{12}\sigma_1\sigma_2 + J_{23}\sigma_2\sigma_3 + J_{34}\sigma_3\sigma_4 + J_{41}\sigma_4\sigma_1. \tag{1}$$

The absolute value of J_{ij} is assumed to be equal to J. Then the degeneracy of the ground state for this system with Hamiltonian H is eight. A frustration parameter G defined by the sign of the product of interaction constants $\{J_{ij}\}$, namely

$G =$sign$(J_{12}J_{23}J_{34}J_{41})$, is negative for the plaquette shown in Fig.1.

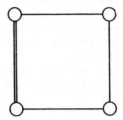

Fig. 1. A frustrated plaquette. A single bond and double bonds denote a ferromagnetic interaction and an antiferromagnetic interaction, respectively.

If G is negative (or positive), the plaquette is frustrated (or non-frustrated). When all the plaquettes of the square lattice are frustrated, the Ising model on the square lattice is called the odd model or Villain lattice (Fig.2).[6]

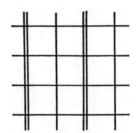

Fig. 2. The odd model or Villain lattice. The lattice is fully frustrated.

A spin model on a frustrated lattice is also called a frustrated spin model (or a spin model with frustration) on a lattice in this chapter. The partition function of the Ising model on the Villain lattice has been exactly calculated.[6,7] Zero-point entropy is of the order of N, where N denotes the total number of the lattice sites or spins. No long range order of spins exists in this system at any finite temperatures.

More generally, one may consider a polygon or plaquette with n vertices (sites); the jth vertex is labeled by j where $j \in \{1, 2, \cdots, n\}$. An Ising spin σ_j is located at the jth vertex. The neighboring spins, σ_j and $\sigma_{j'}$, interact with interaction constant $J_{jj'}$; $|J_{jj'}| = J$. The frustration parameter G is defined by

$$G = \text{sign}(\prod_{j=1}^{n} J_{jj+1}),$$ (2)

where $J_{n,n+1} = J_{n,1}$. If $G = -1$ (or $+1$), the polygon is frustrated (or non-frustrated).

The ground-state degeneracy of frustrated polygon is $2n$, although that of non-frustrated polygon is 2.

The simplest example of frustrated plaquettes is a triangle formed by three antiferromagnetic bonds. The Ising model on the antiferromagnetic triangular lattice was exactly solved by Wannier,[8] Houtappel,[9] Husimi and Syoji,[10,11] Stephenson[12,13] and so on.[14] Zero-point entropy of this system is again of the order of N and no long range order exists at any finite temperatures.

Consider the ferromagnetic Ising model on the linear chain, which is not frustrated. The partition function Z is a function of the reduced interaction constant K and the total number of lattice sites N:

$$Z(N, K) = \sum_{\sigma_1 = \pm 1} \sum_{\sigma_2 = \pm 1} \cdots \sum_{\sigma_N = \pm 1} \exp[K(\sigma_1\sigma_2 + \sigma_2\sigma_3 + \cdots + \sigma_{N-1}\sigma_N + \sigma_N\sigma_1)], \quad (3)$$

where K is βJ, J is the interaction constant, and β is $1/kT$; k is the Boltzmann constant and T the absolute temperature. If N is an even number and σ_{2j} is replaced by $-\sigma_{2j}$, we find

$$Z(N, K) = Z(N, -K). \quad (4)$$

Thus, the Ising model on the linear chain is local gauge invariant. In general, the Ising model on the hypercubic lattices, such as the square lattice and the simple cubic lattice and so on, is local gauge invariant. We write the Hamiltonian as follows:

$$H = -J \sum_{(ij)} \sigma_i \sigma_j, \quad (5)$$

where the summation with respect to (ij) means that the sum is taken over all nearest-neighbor pairs of lattice sites. The partition function is given by

$$Z(N, K) = \sum_{\{\sigma_i\}} \exp[K \sum_{(ij)} \sigma_i \sigma_j]. \quad (6)$$

Since the hypercubic lattice is a bipartite lattice and there are two equivalent sublattices, we replace all the spins, σ_i, on one of sublattices by $-\sigma_i$ and find Eq.(4). Hence, the critical temperature of the ferromagnetic Ising model on a hypercubic lattice is equal to that of antiferromagnetic Ising model on the hypercubic lattice. On the other hand, the antiferromagnetic Ising model on the triangular lattice is not gauge invariant. Therefore, the critical temperature of the ferromagnetic Ising model on the triangular lattice is different from that of the antiferromagnetic Ising model on the triangular lattice. The Ising model on the Villain lattice is not gauge invariant. In general, the Ising model with frustration on the hypercubic lattices is not gauge invariant, too. Since we can change the distribution of ferro- and antiferro-magnetic bonds but can not the distribution of the frustrated plaquettes by a local gauge transformation, the important point is the distribution of the frustrated plaquettes but not that of ferro- and antiferro-magnetic bonds.[15]

In recent years, the word "frustration" is used in a wide sense. For example, when there exist competing interactions among spins in a system, one may say that the system is frustrated. However, in this chapter, we concern with Ising models on the lattice with frustration in a narrow sense and mainly consider Ising models on the antiferromagnetic triangular (AFT) lattice and on the stacked AFT lattice. We explicitly demonstrate important formulae and develop full arguments leading to important results for these systems. A comprehensive review of frustrated Ising spin systems has been presented by Liebmann.[16] The articles and monographs until 1984 may be found in that textbook.

Frustrated Ising spin systems are characterized by a large degeneracy of the ground state. Since many·spin-states with various symmetries are included in the ground state, the realized spin-state is such that the number of states with some symmetry is largest. Since the number of degeneracy is strongly related to the spin ordering, the magnitude S of spin is an important physical parameter in the frustrated Ising spin systems. We show in this chapter that the symmetry of spin ordering strongly depends on the magnitude S of spin. There exist "free" spins or "free" linear-chains, on which internal fields are canceled out, in some frustrated Ising spin systems. These free spins and free linear-chains play an important role as for spin orderings. Existence of these free spins and free linear-chains is explicitly shown in this chapter and the role of them is discussed.

Another characteristic feature of frustrated Ising spin systems is that various metastable states exist in these systems. Existence of metastable states is closely related to the degeneracy of ground state and also to the excited states. Furthermore, the metastable states may give rise to a first order phase transition. Actually, some models introduced in this chapter show a first order phase transition. From this viewpoint, we try to understand the physics of frustrated Ising spin systems in this chapter. We notice terminology used in this chapter: We discuss Ising models with general spin S after section 5 in this chapter. But we use a spin variable $\sigma_i = S_i/S$ for spin S Ising spin and then the Ising model with spin $1/2$ is the same as the usual Ising model of ± 1. We discuss only the Ising model of ± 1 in sections 2, 3 and 4.

2. Ising model on two-dimensional frustrated lattice and on stacked frustrated lattice

The critical temperature of the Ising model on the Villain lattice is zero.[6] However, the critical temperature of the Ising model on the stacked Villain lattice is not zero.[17] We start with a brief explanation for this fact. The lattice structure of the stacked Villain lattice is shown in Fig.3 (see the next page). In this lattice, each xy-plane is frustrated but none of yz- nor zx-planes are frustrated. Each ferromagnetic yz-plane is connected by the ferro- and antiferro-magnetic bonds with its nearest yz-planes. Since the number of the ferromagnetic bonds and that of the antiferromagnetic bonds are same between two successive yz-planes, it is easily supposed that the critical temperature of this system is non-zero and close to that of the Ising model on the square lattice. Actually, according to the Monte Carlo (MC) simulations, the critical temperature T_c of the Ising model on the stacked Villain lattice is given by $kT_c/J \cong 2.9$.[17] Here we notice that the critical temperature of the Ising model on the square lattice is given by $kT_c/J \cong 2.27$.[18] These two values are in fact close.

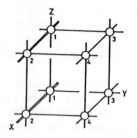

Fig. 3. The Stacked Villain lattice.

The lattice structure of the stacked AFT lattice is shown in Fig.4. We assume that the nearest-neighbor interaction constant between spins on each xy-plane is $-J(J > 0)$ and that between a spin on an xy-plane and one on its nearest xy-plane is $J_z > 0$.

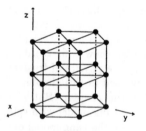

Fig. 4. The stacked AFT lattice.

In this system, none of zx- nor yz-planes are frustrated. The square lattices in the zx-plane or yz-plane are composed of the antiferromagnetic bond J and the ferromagnetic bond J_z. However, the effect of the J bonds cancels out each other on the ordering problem of those xy-planes. Therefore, the critical temperature of the Ising model on the stacked AFT lattice is supposed to be close to that of the Ising model on the square lattice composed of J and J_z. In order to see this, we calculate kT_c/J_z as a function of J/J_z. The critical temperature for the Ising model on the square lattice with J and J_z is calculated from the following equation:[18]

$$\sinh(2K_c)\sinh(2K_c^z) = 1, \qquad (7)$$

where $K_c = J/kT_c$ and $K_c^z = J_z/kT_c$. The critical temperature of the Ising model on the stacked AFT lattice is computed by the MC simulations.[17] These two critical temperatures are compared in Fig.5 (see the next page). We see that J/J_z dependencies of kT_c/J_z for these two systems show a similar behavior. This behavior supports our physical interpretation.

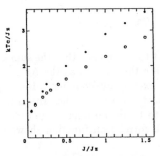

Fig. 5. J/J_z dependence of kT_c/J_z. The open circles denote the critical temperature for the Ising model on the square lattice and the solid circles for the Ising model on the stacked AFT lattice.

In the Ising model on two-dimensional fully-frustrated lattices such as the Villain lattice and the AFT lattice, there appear many free spins.[19] Corresponding to the free spins, there appear many free linear-chains in the Ising model on the stacked frustrated lattices.[20] The role of free linear-chains will be explicitly discussed in later sections.

3. Ising model on antiferromagnetic triangular lattice: Effect of ferromagnetic next-nearest-neighbor interaction

The Ising model is considered on the AFT lattice Λ shown in Fig.6; the set of lattice sites Λ is divided into three subsets of lattice sites denoted

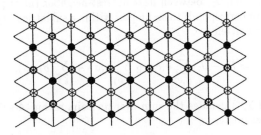

Fig. 6. The AFT lattice. The solid circles, the open circles and the double circles denote the lattice sites belonging to A-, B-, and C-sublattice, respectively.

by Λ_A, Λ_B and Λ_C, and we call these subsets as A-, B- and C-sublattice, respectively. We call spins on the A-sublattice A spins, those on the B-sublattice B spins and those on the C-sublattice C spins hereafter. If we assume the nearest-neighbor (nn) interaction, the Hamiltonian is written as

$$H_0 = J \sum_{(ij)} \sigma_i \sigma_j, \tag{8}$$

where the summation with respect to (ij) means that the sum is taken over all the nn pairs of lattice sites. We abbreviate this system as the AFT model in the following. The AFT model has been solved exactly by Wannier,[8] Houtappel,[9] and Husimi and Syoji[10,11] and also by Stephenson.[12,13] Recently, the critical indices are investigated by Horiguchi et al.[14] According to them, no long-range order exists in this system even at zero temperature but the correlation function shows not exponential decay but a power law decay:

$$< \sigma_i \sigma_j > \propto r_{ij}^{-\eta}, \tag{9}$$

where $\eta = 1/2$. r_{ij} denotes the distance between the sites i and j. Usually, we adopt a three-sublattice structure for the AFT model. The set of sublattice moments, M, is written as

$$\mathbf{M} = \frac{3}{N} (\sum_{i \in \Lambda_A} \sigma_i, \sum_{j \in \Lambda_B} \sigma_j, \sum_{k \in \Lambda_C} \sigma_k). \tag{10}$$

The ground-state degeneracy is $\exp(0.323N)$ and the ground state includes many spin-states with various symmetries. Then, if a very small perturbational energy with some symmetry is added to Eq.(8), the spin-state with the same symmetry may be chosen from the many spin-states in the ground state for the system of Eq.(8) as a new ground state. The new ground state may have a long-periodic symmetry which is not represented by the three-sublattice structure and a new state (or phase) may show a non-zero critical temperature.[21-26]

In this section, we show some interesting spin-states chosen by adding a small ferromagnetic interaction, J_2, between next-nearest-neighbor (nnn) pairs. We see in Fig.6 that an nnn pair corresponds to an nn pair within a sublattice. The Hamiltonian H_3 for the system is written as

$$H_3 = H_0 - J_2 (\sum_{(i,i')}^A \sigma_i \sigma_{i'} + \sum_{(j,j')}^B \sigma_j \sigma_{j'} + \sum_{(k,k')}^C \sigma_k \sigma_{k'}), \tag{11}$$

where $\sum^A$ ($\sum^B$ and $\sum^C$) denotes the sum over nn pairs within A sublattice (B and C sublattices): for example,

$$\sum_{(i,i')}^A \sigma_i \sigma_{i'} = \sum_{(i,i'), i \in \Lambda_A, i' \in \Lambda_A} \sigma_i \sigma_{i'} \tag{12}$$

and (i, i') denotes nn pairs of sites in the A-sublattice. We use this kind of notations throughout this chapter. We call this system given by H_3 S3. The effect of J_2 was previously investigated in the same or similar models by many people.[21-26] It has been known that there occurs a spin ordering within a sublattice at a temperature and a spin ordering between sublattices at a lower temperature than that temperature.[27]

The spin-state for this system at low temperatures is of the so-called ferrimag-netic (FR) structure. The FR structure, which is shown in Fig.7, is one of spin-states in the ground state occurring in the system with $J_2 = 0$. Here we show the MC data for $J_2 = 0.1J$ in Figs.8 and 9.

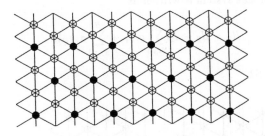

Fig. 7. The ferrimagnetic (FR) structure, $(1, -1, -1)$. The solid and open circles show up- and down-spins, respectively.

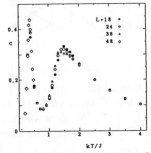

Fig. 8. Specific heat for S3 with $J_2 = 0.1J$.

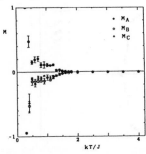

Fig. 9. Sublattice magnetization for S3 with $J_2 = 0.1J$.

The MC results of the specific heat, $c = C/Nk$, versus temperature, kT/J, are shown in Fig.8. Two peaks are observed in the specific heat. This figure is in accordance with the previous ones.[26] The temperature dependence of sublattice magnetizations is shown in Fig.9. The higher transition temperature is estimated to be $kT_1/J \cong 1.5$ and the lower transition temperature to be $kT_2/J \cong 0.4$.

Next, we consider the system in which the ferromagnetic interaction, J_2, exists between nn pairs within A-sublattice and also between those within B-sublattice. The Hamiltonian for this system is written as

$$H_2 = H_0 - J_2\left(\sum_{(i,i')}^{A} \sigma_i\sigma_{i'} + \sum_{(j,j')}^{B} \sigma_j\sigma_{j'}\right).\tag{13}$$

We call this system S2.

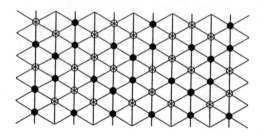

Fig. 10.   One of the PD structures, $(1, -1, 0)$.

At low temperatures, we assume that all the A spins take $+1$ and all the B spins take -1. Then the internal field to each of C spins vanishes. We call this spin-state the partial disorder (PD) structure $(1, -1, 0)$. We show an example of the PD structure in Fig.10. In other words, all the C spins are free. The number of spin-states for the case that the magnetic moment of C sublattice vanishes is $\binom{N/3}{N/6} \cong 2^{N/3}$.

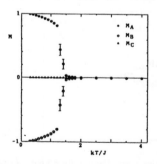

Fig. 11.   MC data of sublattice moment for S2.

Thus the number of spin-states for the PD structure, $(1, -1, 0)$, is extremely large compared to that of spin-states for the structure $(1, -1, d)$, where $d \neq 0$. The MC data on the sublattice moments for the case of $J_2 = 0.1J$ are shown in Fig.11. The temperature dependence of specific heat is shown in Fig.12. The critical temperature is estimated to be $kT_c/J \cong 1.4$ from these MC data. As noted in the above, the number of the spin-states in the ground state is $2^{N/3}$. These ground states belong to the Wannier state when $J_2 = 0$.

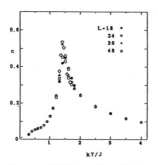

Fig. 12. MC data of specific heat for S2.

Finally, let us consider the system where the ferromagnetic interaction, J_2, exists only between nn pairs within the A-sublattice. The Hamiltonian for this system is written as

$$H = H_0 - J_2 \sum_{(i,i')}^{A} \sigma_i\sigma_{i'}. \tag{14}$$

We call this system S1. The spin-state in the ground state for this system is represented by $(1, -1/2, -1/2)$. This spin-state was proposed by Blankschtein et al[28] as a low temperature phase for the Ising model on the stacked AFT lattice (see Fig.4). We call this spin-state the BL structure. The reason why the system takes the BL structure at low temperatures will be explained as follows. Let us assume all the A spins take $+1$ and all the B and C spins take -1. The B spins or the C spins may be free. If one of B spins is reversed, then the directions of three C spins surrounding the B spin are fixed. Since the B- and C-sublattices are equivalent to each other, the ratio of the number of up-spins to that of down-spins is, in average, $1/3$ in the B- or C-sublattice. Hence, the spin state becomes the BL structure, $(1, -1/2, -1/2)$. Another explanation for the reason, why the BL structure is achieved, may be as follows. The number of the FR structure, W(FR), is equal to 1 and the number of PD structure, W(PD), is given by $2^{N/3}$. Therefore, the PD structure is achieved rather than the FR structure. However, the PD structure of $(1, 0, -1)$ and that of $(1, -1, 0)$ both appear with equal probability. Hence, the average of these two PD structure is given by the BL structure $(1, -1/2, -1/2)$. The MC results for the sublattice magnetizations for the case $J_2 = 0.1J$ are shown in Fig.13 (see the next page). The specific heat data are shown in Fig.14 (see the next page). The critical temperature is estimated to be $kT_c/J \cong 1.3$ from these data. One of the BL structures is shown in Fig.15 (see the next page). Since many free-spins exist in the BL structure, one may not observe the spin-state of the BL structure shown in Fig.15 on a snapshot in the MC simulations. The spin-state of the BL structure is included in the Wannier state for the system with $J_2 = 0$. This BL structure is realized as the ground state, when a uniform magnetic field is applied to the spins in one of sublattices, as can be easily supposed.

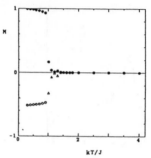

Fig. 13.   MC data of sublattice moment for S1.

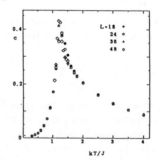

Fig. 14.   MC data of specific heat for S1.

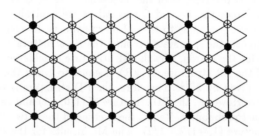

Fig. 15.   One of the BL structures.

We have seen that the FR, PD, and BL structures, respectively, appear for the S3, S2, and S1 at $T = 0$. If a perturbational interaction between far-neighbor spins is assumed, there may appear a multi-sublattice spin-structure as the ground state, which may reflect the symmetry of the far-neighbor interactions. Here, those multi-sublattice spin-structure should belong to the Wannier state for the unperturbed Hamiltonian of Eq.(8).

4. Ising model on stacked antiferromagnetic triangular lattice

The Hamiltonian of the Ising model on the stacked AFT lattice is written as

$$H = J \sum_{(i,j)}^{xy} \sigma_i \sigma_j - J_z \sum_{(i,j)}^{z} \sigma_i \sigma_j, \qquad (15)$$

where J denotes the nearest-neighbor (nn) interaction constant in the xy-plane and J_z the nn interaction constant along the z axis (see Fig.4). Here J and J_z are both positive. The sum runs over the nn pairs either in the xy-plane or along the z direction. It is noticed that a long-range order appears at a temperature $kT/J \cong 2.9$ when $J/J_z = 1$. Below the critical temperature, the spin-state is the PD structure, $(m, 0, -m)$.[29,30] At $T = 0$, since the spins along the z axis are parallel to each other, the spin-state in the xy-plane should be the same as that of the AFT lattice. That is, no long range order exists in the xy-plane at $T = 0$. Hence, there occurs some crossover from the PD structure to the Wannier state at a low temperature. In the MC simulations, two peaks are observed for the specific heat at $T \cong 2.9J/k(= T_1)$ and at $T \cong 0.9J/k(= T_2)$. The peak at T_1 is due to the phase transition. The reason that the second peak at T_2 appears is as follows. We know that there exist many free-spins in the AFT lattice at $T = 0$. Corresponding to these free-spins, there exist many free-linear-chains in the stacked AFT lattice. The free linear-chains in the system are parallel to the z axis. It is known that the specific heat of Ising model on the linear chain shows a Schottky-type peak at $kT/J \cong 0.9$. By using the MC simulations, we calculate the specific heat, c, for the Ising model on the stacked AFT lattice for various values of J/J_z; we use c given by

$$c = \frac{1}{N} \left(\frac{J_z}{kT}\right)^2 [\langle \left(\frac{H}{J_z}\right)^2 \rangle - \langle \frac{H}{J_z} \rangle^2]. \qquad (16)$$

In Fig.16 (see the next page), the MC data of the specific heat are shown as a function of kT/J_z. There is a small peak or bump in the specific heat. The position, the height, and the shape of the bump on the low temperature side are not affected by the value of the J/J_z, as can be seen in Fig.16. Hence, it is conjectured that this bump is due to the linear-chain-like excitations in the system, where the interaction constant of the "linear chain" is given by J_z.[20] Another interesting point is that the conventional single-spin-flip Monte Carlo simulation method may fail for this system at low temperatures. Let us consider the MC simulations for the Ising model on the linear chain. As is well-known, the internal energy, U, of the Ising model on the linear chain is given by

$$U = -LJ_z \tanh \frac{J_z}{kT} = -LJ_z[1 - 2\exp(-\frac{2J_z}{kT}) + \cdots], \qquad (17)$$

where L denotes the total number of spins in the linear chain. This expression indicates that the first excited energy is equal to $2J_z$ and also that the first excited state includes only one kink.

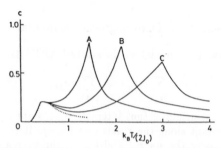

Fig. 16. MC data of the specific heat for various values of J/J_z: $J/J_z = 1$ (A), 2.0 (B), and 3.0 (C). The dotted line is explained in the text. Here J_0 is used in place of J_z.

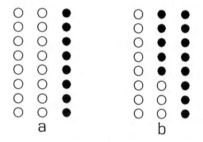

Fig. 17. (a) The ground state of a free linear-chain in the system. The solid and open circles denote the up- and down-spins, respectively. (b) The first exited state. The excitation energy is $2J_z$.

Thus, it is needed to adopt the cluster-flip Monte Carlo method, where the cluster in the system is a free linear-chain. We schematically show the ground state and the first excited state in the system in Fig.17(a) and (b). We take up the temperature dependence of long-range order-parameters in the whole temperature range. Let us define the order parameter X and the sublattice order-parameter Y by the following equations:

$$X = \frac{1}{N^2}\langle(\sum_i \sigma_i)^2\rangle, \tag{18}$$

$$Y = (\frac{3}{N})^2\langle(\sum_i^A \sigma_i)^2 + (\sum_j^B \sigma_j)^2 + (\sum_k^C \sigma_k)^2\rangle. \tag{19}$$

It is easily seen that $X = 0$ and $Y = 2$ for the PD structure, $(1, -1, 0)$, and $X = 1$ and $Y = 3$ for the FR structure, $(1, -1, -1)$. The temperature dependencies of X

and Y in the MC simulations are shown in Fig.18.

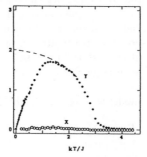

Fig. 18. MC data of X and Y for the Ising model on the stacked AFT lattice.

We see that $X = 0$ in the whole temperature range and the values of Y agree with those for the PD structure (dashed line) above $kT/J \cong 1.2$. The PD structure is obtained when we assume a ferromagnetic interaction, J_z, between nnn pairs in the xy-plane. Here these spin pairs are belonging to the A-sublattice or B-sublattice. Remember system S2 in the section 3. The dashed line in Fig.18 shows an extrapolation to the state with the PD structure. The dotted line shows an extrapolation to the Wannier state. We may assume that a temperature region between $kT/J = 0$ and 2.0 is the crossover region between a three-dimensional ordered state and the Wannier state.

The number of free linear-chains, which is denoted by $\gamma N^{2/3}$ at low temperatures, may be estimated from the MC data of the specific heat. The dotted line for the specific heat in Fig.16 is obtained by assuming $\gamma = 0.376$.[31] Then, in terms of the low-temperature-series-expansion (LTSE), the free energy F is calculated as

$$F = -N(J + J_z) - \gamma N \exp(-2J_z/kT) + \cdots. \qquad (20)$$

On the other hand, we have $\gamma = 1/3$ for the PD structure and $\gamma = 2/3$ for the FR structure. Thus, the free energy of a structure with the above $\gamma(= 0.376)$ is lower than that of the PD structure but higher than that of the FR structure. If we follow Eq.(20), the FR structure should be realized. However, the value of γ for the FR structure will be reduced by the flip of the free linear-chains. Then, the spin-state for the Ising model on the stacked AFT lattice will gradually fall into the Wannier state as the temperature decreases. This situation is shown by the extrapolation given by the dotted line in Fig.18.

5. Ising model with infinite-spin on antiferromagnetic triangular lattice

We know that there is no long-range order in the ground state of the Ising model with spin 1/2 on the AFT lattice. This property may change if the magnitude S of spin changes. For the Ising model with infinite-spin, there exists a long range order, as will be shown in this section, and hence there is an evidence of a phase transition at a finite temperature. The frustrated systems such as the Ising model with spin 1/2 on the AFT lattice are characterized by a large degeneracy of the ground state and also by other physical quantities, which may qualitatively change if the magnitude of spin changes. The effect of magnitude S of spin on physical properties of frustrated Ising spin systems has not been seriously investigated in previous researches except for a few examples.[32,33] It is interesting to show in this section that the nature of the spin ordering for the Ising model with large S is quite different from that with spin 1/2 on the AFT lattice.[34,37]

The Hamiltonian of the Ising model with spin S on the AFT lattice is given by

$$H = J \sum_{(i,j)} \sigma_i \sigma_j, \tag{21}$$

where $\sigma_i = S_i/S$. S_i denotes the Ising spin of magnitude S on the ith lattice site and takes one of the $2S + 1$ values, $S, S - 1. \cdots, -S$. We again assume the nn interaction in Eq.(21) and divide the triangular lattice into three sublattices.

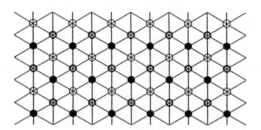

Fig. 19.   The partially disordered structure of the Ising model with spin S on the AFT lattice. The solid and open circles denote $+S$ spin and $-S$ spin, respectively. The free spin is represented by the double circles.

The spin with the limit of S infinity is also called the continuous Ising-spin since σ_i takes on $[-1, 1]$; we assume that the limit of S infinity is taken before the thermodynamic limit. The set of sublattice magnetizations at $T = 0$ is given by Eq.(10). We know that there are many free-spins, on which the internal fields are canceled out, in the ground state configurations for the AFT lattice. For the Ising model with spin S, each free-spin can take any value out of $2S + 1$ values $\{S, S - 1 \cdots, -S\}$. Hence, if n free-spins exist in the ground state, the degeneracy of the ground state is equal to $(2S + 1)^n$. The spin structure with maximum n is such that all the spins in one of sublattices, say the A-sublattice, take $+S$, and all the spins in the B-sublattice take $-S$, and all the spins in the C-sublattice are free. See Fig.19 for this situation. In this case, the set of the sublattice moments, M, is written as $(1, -1, 0)$ and the degeneracy of the spin-states is equal to $(2S + 1)^{N/3}$. We call this spin structure the partially disordered structure of $(1, -1, 0)$ type. Of course, there are six kinds of equivalent structures $(1, 0, -1)$, $(0, 1, -1)$ and so on.

We consider a spin on the A-sublattice (A spin) in the partially disordered structure of $(1, -1, 0)$ type. In order for the spin to become free, the neighboring three C spins must take $+S$. Thus, the number of free spins is reduced by 2. In Fig.20, these four spins, one A spin and three C spins, are enclosed by dashed lines.

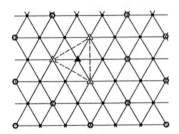

Fig. 20.   A misfit cluster in the partially disordered structure of $(1, -1, 0)$ type. The triangles represent the spins in the misfit cluster. The diamonds denote the free spins. The boundary of the misfit cluster is shown by the dashed lines.

The cluster composed of those four spins is called a misfit cluster of $(0, -1, 1)$ type. We define a boundary of the misfit cluster by the closed dashed lines in Fig.20. We may also consider a misfit cluster making a B spin free which is composed of one B spin and the neighboring three C spins. In order to demonstrate the stability of the partially disordered structure, we study the effect of misfit clusters. When n independent misfit clusters exist, the ground state degeneracy, W_n, is given by

$$W_n \cong (2S + 1)^{\frac{N}{3} - 2n} \left(\begin{array}{c} \frac{2N}{3} \\ n \end{array} \right). \tag{22}$$

The maximum of the distribution, or the maximum of the entropy, is given by solving the equation

$$\frac{\partial}{\partial n} \ln W_n = 0, \tag{23}$$

which yields

$$n = \frac{2N}{3} \frac{1}{(2S + 1)^2}. \tag{24}$$

Thus, we obtain $n = 0$ for $S = \infty$, since the limit of S infinity is taken before the thermodynamic limit is taken. If the density of misfits, n/N, is low enough, then a sublattice long-range order may exists in the system. We point out that this problem is similar to the percolation problem and that there might exist a critical value for the magnitude of spin, S_c, such that a sublattice long-range order exists for $S > S_c$.

Following an extension of Peierls' argument[38,39] for the spin ordering of the Ising model with spin 1/2 on the square lattice, we discuss a condition for the existence of long range order in the Ising model with spin S on the AFT lattice. Let us consider spin configurations in the system with a boundary condition which induces the order of $(1, -1, 0)$ type. We may consider various kinds of misfit clusters of larger size; an example is given in Fig.21.

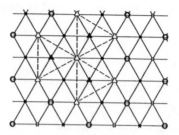

Fig. 21.   A misfit cluster of large size.

We define the boundary of misfit cluster, which is shown by closed dashed lines in Figs.20 and 21. If the length of boundary, f, is measured in unit of $\sqrt{3}\times$bond length, we have $f = 3$ for the misfit cluster in Fig.20 and $f = 9$ for the misfit cluster in Fig.21. If d denotes the reduction in the number of free spins due to the appearance of misfit cluster, we have $d = 2$ in the case of Fig.20 and $d = 4$ in the case of Fig.21. Generally, we have the following relation:

$$f \leq 3d \qquad (25)$$

between f and d. Consider a spin on the A-sublattice. Let p_+ be the probability that the spin is found in the cluster of $(1, -1, 0)$ type or $(1, 0, -1)$ type. Similarly, let p_- be the probability that the spin is found in the cluster of $(-1, 1, 0)$ type or $(-1, 0, 1)$ type and let p_0 be the probability that the spin is found in the cluster of $(0, 1, -1)$ type or $(0, -1, 1)$ type. Since $p_+ + p_- + p_0 = 1$, we have

$$\langle \sigma_A \rangle = \frac{3}{NS} \langle \sum_{i \in A} S_i \rangle = p_+ - p_- = 1 - p_0 - 2p_-, \qquad (26)$$

where the angular bracket denotes the average over all the spin-states in the ground state instead of the canonical weight in the case of Peierls' argument. We may classify the clusters by the boundary length. If we compare the number of degeneracies of the cases with and without the boundary of p_0 as in the usual Peierls' argument, then we have

$$p_0 < \sum_f \sum_k m_k (2S + 1)^{-d}, \qquad (27)$$

where $k = k(f)$ denotes the kth cluster and $d = d(k, f)$ the reduction in the number of free spins due to the appearance of the cluster. m_k denotes the number of sites of the A-sublattice in the kth cluster. If the number of clusters with the boundary length f is denoted by C_f, we have the following relation:

$$C_f \leq 2(z - 1)^f, \qquad (28)$$

where $z(= 6)$ is the number of nearest-neighbor sites. The factor 2 comes from the fact that there are two kinds of misfit clusters with free A spins, namely the clusters of $(0, 1, -1)$ type and those of $(0, -1, 1)$ type. Using Eq.(28) and the relation, $m_k < f^2$, we have

$$\sum_k m_k (2S + 1)^{-d} < 2f^2 \left(\frac{5^3}{2S + 1}\right)^{f/3}. \tag{29}$$

The series Eq.(29) converges if $2S + 1 > 5^3$. We can estimate p_- in the same way. Thus we have a finite value of $\langle \sigma_A \rangle$ for a large enough spin value.

It is known that the sublattice spin-pair correlation-function decays by a power law

$$\langle \sigma_i \sigma_{i+r} \rangle \sim r^\eta \tag{30}$$

where $\eta = 1/2$ for $S = 1/2$. If the sublattice long-range order exists, η is zero. Hence, it is interesting to see how η changes as S changes. For this purpose, we calculate the quantity A, which is defined by

$$A = \frac{1}{N} \langle (\sum_i^A \sigma_i)^2 + (\sum_i^B \sigma_i)^2 + (\sum_i^C \sigma_i)^2 \rangle \tag{31}$$

by the Monte Carlo simulations. The values of η are obtained from the size dependence of A:

$$A \sim L^{2-\eta} \tag{32}$$

The S dependence of η is shown in Fig.22.

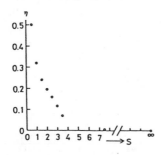

Fig. 22. S dependence of η; η is the exponent for the correlation function.

Though the value of η seems to be zero if $S = 15/2$, the precise value of S_c has not to yet been obtained.

It is interesting to study the magnetic properties of the Ising model with infinite-spin on the AFT lattice. We show the MC data of the specific heat for various values of S in Fig.23 (see the next page). The shape around the peak of specific heat, c, for the Ising model with infinite-spin is rather sharper than that for the Ising model with finite S, as can be seen in Fig.23. However, any size dependence was not observed in the peak of c for the Ising model with infinite-spin (see Fig.24).

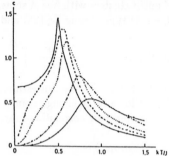

Fig. 23.   MC data of specific heat for the Ising model with spin S on the AFT lattice for various values of S. The solid line (A) is for $S = 1/2$, the dash-dot line for $S = 3/2$, the dotted line for $S = 7/2$, the dashed line for $S = 15/2$, and the solid line (B) for $S = \infty$.

The sublattice magnetizations of PD structure, $(M, 0, -M)$, were apparently observed up to the temperature $T = 0.1J/k$ for a fairly large lattice size ($L = 240$) and a long MC average time ($t_{MC} = 10^6$ MC steps per spin).[37] However, the temperature dependence of $2 - \eta$ decreases with respect to T and reaches at 1.75 at $T \sim 0.41J/k(= T_1)$ and suddenly decreases to zero as shown in Fig.25 (see the next page).[34] Here η denotes the exponent for correlation function and the value of η is 0.25 at the Kosterlitz-Thouless phase transition temperature. Since η should be zero in the spin-ordered region, no long range order exists at any finite temperatures except at very low temperatures for the Ising model of infinite-spin on the AFT lattice.

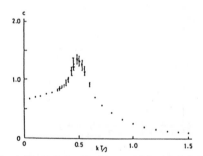

Fig. 24.   MC data of specific heat for the Ising model with infinite-spin for various lattice sizes. The data of $L = 30, 42$ and 72 are shown.

It is desired to be investigated in detail the behavior of η near $T = 0$ whether the long-range order exists or not at very low temperatures. On the other hand, η should be equal to 2 in the paramagnetic region. Thus, the MC data of η seem to indicate that the Kosterlitz-Thouless-type (KT-type) phase transition occurs at T_1. The microscopic understanding of magnetic properties of the system below T_1 is not obtained yet at present.

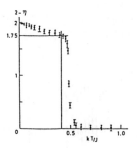

Fig. 25. MC data of $2 - \eta$ as a function of the temperature for the Ising model with infinite-spin. The KT value, 1.75, is found at $T = 0.41J/k$.

6. Ising model with infinite-spin on stacked antiferromagnetic triangular lattice

In the ground state, the Ising model with spin 1/2 on the AFT lattice does not show any long range order. While, the Ising model with infinite-spin on the AFT lattice shows a partial-disordered (PD) structure which is expressed as $(1, 0, -1)$, for example. Thus, in the Ising model on the AFT lattice, the magnitude S of spin, is an important physical parameter. The Ising model with spin 1/2 on the stacked AFT lattice shows a phase transition at a finite temperature, and the spin structure is expressed as a PD structure, $(m, 0, -m)$. However, the spin structure should fall into the Wannier state in the ground state, and hence it is expected to have a crossover from a three-dimensional (3d) ordered structure to a two-dimensional (2d) Wannier state. On the other hand, the spin structure for the Ising model with infinite-spin on the stacked AFT lattice is quite different from that for the Ising model with spin 1/2. The Ising model with infinite-spin shows a phase transition at a finite temperature. The spin structure in an intermediate temperature region is the PD structure but it is the ferrimagnetic (FR) structure, $(M, -m', -m')$, in a low temperature region. This FR structure is quite stable up to $T = 0$. This can be seen by the use of low-temperature-series-expansion (LTSE).[40-43]

The Hamiltonian of the present model is given by

$$H = J \sum_{(i,j)}^{xy} x_i x_j - J_z \sum_{(i,j)}^{z} x_i x_j, \tag{33}$$

where x_i takes on $[-1, 1]$. We assume $J = J_z$ in the present section. The MC data for the sublattice magnetizations are shown in Fig.26 (see the next page). As can be seen in Fig. 26, the system takes the PD structure in a temperature region, $T_1 < T < T_c$, where $T_1 \cong 0.8J/k$ and $T_c \cong 1.20J/k$. The PD structure is shown in Fig.10, where only a triangular plane is illustrated. The FR structure is observed in a temperature region, $0 < T < T_1$. The FR structure is illustrated in Fig.7. A sharp peak in the specific heat is observed at the temperature T_c. Although we think that the phase transition at $T = T_1$ is of the first order, a more precise study will be needed for this point.

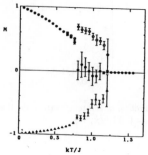

Fig. 26.   MC data of sublattice magnetizations. The spin structure is the FR structure for $T < T_1$ and it is the PD structure for $T_1 < T < T_c$.

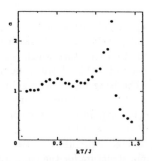

Fig. 27.   MC data of specific heat per spin.

Let us study theoretically the stability of the FR structure at low temperatures observed by the MC simulations. The partition function Z is written as

$$Z = \int_{-1}^{1} dx_1 \cdots \int_{-1}^{1} dx_N \exp(-\beta H). \tag{34}$$

We introduce new variables u and v for up- and down-spins, respectively: $x_i = 1 - u_i$ and $x_j = -1 + v_j$. Then Z is written as

$$Z = \int_{0}^{2} du_1 \cdots \int_{0}^{2} du_{N/3} \int_{0}^{2} dv_1 \cdots \int_{0}^{2} dv_{2N/3} \exp(-\beta H), \tag{35}$$

where

$$H = -N(J + J_z) + (6J + 2J_z) \sum_{i}^{A} u_i + 2J_z (\sum_{j}^{B} v_j + \sum_{k}^{C} v_k) + H'. \tag{36}$$

Here H' contains quadratic terms of u or v. A precise expression of H' will be omitted in this text. At low temperatures, we use the following replacement for the integrations

$$\int_0^2 du \to \int_0^\infty du \quad \text{and} \quad \int_0^2 dv \to \int_0^\infty dv. \tag{37}$$

Then, the free energy is obtained by the LTSE as

$$F = -N(J + J_z) - \frac{NkT}{3}\ln[\frac{kT}{6J + 2J_z}] - \frac{2NkT}{3}\ln\frac{kT}{2J} - \frac{N}{4}(\frac{kT}{J_z})^2 Q + \cdots, \tag{38}$$

where

$$Q = J(a^3 + 5a^2 + 6a - 9)/(a + 3)^2, \tag{39}$$

and $a = J_z/J$. The internal energy U is given by

$$U = -N(J + J_z) + NkT + \frac{N}{4}(\frac{kT}{J_z})^2 Q + \cdots. \tag{40}$$

The above expansions are not correct when $J_z < kT$, since free spins appear in the system and the above LTSE becomes invalid. When $J_z = J$, Eq.(40) agrees well with the MC data of U for $kT/J < 0.75$. Let us consider a FR structure with a single misfit which is defined in Fig.28.

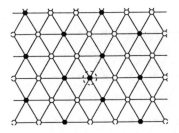

Fig. 28.   The FR structure with a single misfit. The misfit is marked by a dashed circle.

This structure is constructed from the FR structure in Fig.7 by replacing an open circle with a solid circle. Therefore there exist $L(2L^2/3 - 1)$ down-spins and $L(L^2/3 + 1)$ up-spins in the spin-state shown in Fig.28. Let us consider the difference between the free energy of FR structure and that of ferrimagnetic-with-misfit (FRm) structure. The difference is given as

$$\Delta F = F(\text{FR}) - F(\text{FRm})$$

$$= 3NkT\{\ln[\frac{2}{kT}(2J + J_z)] + \ln[\frac{2}{kT}(J + J_z)] - \ln[\frac{2}{kT}(3J + J_z)] - \ln\frac{2J_z}{kT} + \cdots\}. \tag{41}$$

The above equation is rewritten as

$$\Delta F = -3NkT \ln\left[\frac{(a+1)(a+2)}{a(a+3)}\right] + \cdots.$$ (42)

ΔF is negative for any positive value of a. Hence, the FR structure is more stable than the FRm structure. Furthermore, we notice that there are many metastable states with various symmetries. One of the possible metastable states will be the so-called (2×1) structure, which is shown in Fig.29. The free energy of this state is calculated by LTSE as

$$F(2 \times 1) = -N(J + J_z) + NkT \ln\left[\frac{2}{kT}(J + J_z)\right] + \cdots.$$ (43)

The difference between $F(\text{FR})$ and $F(2 \times 1)$ is as follow:

$$F(\text{FR}) - F(2 \times 1) = -\frac{NkT}{3} \ln\frac{(a+1)^3}{a^2(a+3)} + \cdots.$$ (44)

We see from Eq.(44) that $F(\text{FR}) - F(2 \times 1)$ is negative for any positive value of a. Thus, the FR structure is more stable than the 2×1 structure at low temperatures. In this way, it turns out that the FRm structure and the (2×1) structure are both metastable.[43]

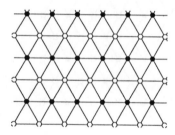

Fig. 29.   A 2×1 structure.

In a similar way, it turns out that the PD structure is also metastable. Furthermore, the spin-states in a "stacked" Wannier state are all metastable. Here a "stacked" Wannier state is defined by a set of spin-states each of which belongs to one of the Wannier state for the ground state of the Ising model on the AFT lattice in any xy-plane and is stacked in such way that all the spins on a column along the z axis have the same direction. Hence the number of metastable states in the system is of the order of $N^{2/3}$.

7.   Phase diagram in spin-magnitude versus temperature for Ising models with spin S on stacked antiferromagnetic triangular lattice

The magnetic property of the Ising model with infinite-spin on the stacked AFT lattice is well understood. That is, the system takes the PD structure in an intermediate temperature region and the FR structure in a low temperature region. Although the Wannier state is realized in the ground state, the FR structure is achieved in the limit of $T = 0$. On the other hand, the Ising model with spin 1/2 on the stacked AFT lattice takes the PD structure in an intermediate temperature region but the system falls into the Wannier state in the limit of $T = 0$. Because, the spins on a column along the z axis are each parallel to others at low temperatures and then the column may be regarded as a single Ising spin. Therefore, there is a crossover region in the temperature axis from the state of the three-dimensional PD structure to the two-dimensional Wannier state. Then, it is interesting to study the phase transition for the Ising model with general spin S.

We restrict our interest to the systems with $J = J_z$ in this section. The energy for the first excited state in the Ising model with spin S is of the order of J/S. Hence, if the temperature is lower than J/kS, the spins on a column along the z axis are each parallel to others and then the system will show the behavior of the Ising model on the AFT lattice. Hence the boundary of the temperature for the crossover region will depend on the value of J/S.[44]

In order to see the phase transitions in the systems, we calculate the sublattice magnetizations, X and Y defined by Eqs.(18) and (19), respectively, and also the specific heat.

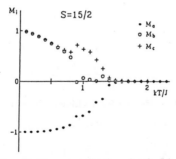

Fig. 30. MC data of the sublattice magnetizations for the Ising model with spin $15/2$. The magnetic structure is the PD structure in the intermediate temperature region and the FR structure in the low temperature region.

We show the MC data in Figs.30, 31 and 32 (for Figs 31 and 32, see the next page) for the Ising model with spin 15/2. We have $X = 0$ and $Y = 2$ for the PD structure, $(1, 0, -1)$, $X = 1$ and $Y = 3$ for the FR structure, $(1, -1, -1)$, and $Y = 0$ and $X = 0$ for the Wannier state. The sublattice magnetizations are shown as functions of the temperature in Fig.30. The temperature dependencies of Y and X are shown in Fig.31. T_1 denotes the critical temperature at which the sublattice magnetizations appear, Y becomes non-zero, and the peak of the specific heat is observed. T_2 denotes a transition temperature of the first order phase transition at which the transition from the state with the PD structure to that of the FR structure occurs, X becomes non-zero and the curvature of Y changes. Note that any appreciable anomaly is not observed in the specific heat at T_2. There is no crossover region from the three-dimensional ordered-state to the two-dimensional Wannier-state for $kT/J > 0.1$ in this case.

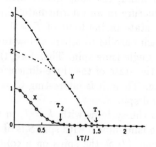

Fig. 31. MC data of Y and X. The dashed line denotes the extrapolation to the zero temperature. The arrows indicate the transition points.

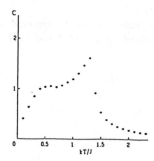

Fig. 32. MC data of specific heat for the Ising model with spin $15/2$.

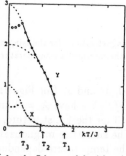

Fig. 33. MC data of Y and X for the Ising model with spin $5/2$. See the text for the explanation of T_1, T_2 and T_3.

Next we show the MC data for the Ising model with spin $5/2$. Since the statistical errors in the MC data of the sublattice magnetizations are very large, we show only the data of X and Y. The temperature dependencies of Y and X are shown in Fig.33 (see the previous page). In this figure, T_1 denotes the critical temperature of the phase transition from the paramagnetic state to the state with the PD structure. T_2 is a transition temperature of the first order from the state with the PD structure to that with the FR structure. T_3 denotes a turning point where the FR structure becomes unstable. The curve of Y between T_1 and T_2 tends towards the value 2 at $T = 0$ but the curve between T_2 and T_3 tends towards the value 3 at $T = 0$. However, the data of Y suddenly becomes small at T_3 and seem to tend towards zero at $T = 0$. Similarly the MC data of X between T_2 and T_3 tend towards the value 1 at $T = 0$ but suddenly becomes small at T_3 and tend towards zero at $T = 0$. We think that the temperature region between T_3 and zero is a crossover region for the system from the state with the FR structure to the Wannier state as the temperature decreases.

Thus, in this spin system there are two transition temperatures T_1 and T_2, and one turning point T_3 below which the system is in the crossover region. Similar analyses are made for the Ising models with various spin S on the stacked AFT lattice and an obtained $S - T$ phase diagram is shown in Fig.34.[44]

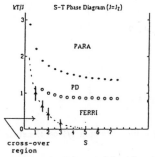

Fig. 34. $S - T$ phase diagram for the Ising model with spin S on the stacked AFT lattice. The dashed line shows an upper bound of the crossover region obtained by the MC simulations.

In Fig.34, the dashed line denotes the S dependence of T_3, which is an upper bound of the crossover region obtained by the MC simulations. As stated above, the dashed line behaves as a linear function of J/S.

8. Effect of antiferromagnetic interaction between next-nearest-neighbor spins in xy-plane

In this section, we discuss the effect of antiferromagnetic interaction J_2 in the Ising models with spin S and with infinite-spin on the AFT lattice and on the stacked AFT lattice. The Hamiltonian for both systems is given as follows:

$$H = J \sum_{(i,j)}^{xy} \sigma_i \sigma_j - J_z \sum_{(i,j)}^{z} \sigma_i \sigma_j + J_2 \sum_{(i,j)}^{xy} {}' \sigma_i \sigma_j. \tag{45}$$

Here $\sum'$ denotes the sum over the nnn pairs in the xy-plane. The effect of J_2 on the Ising model with spin 1/2 on the AFT lattice was previously studied by Saito.[45] Until recently, however, it has not been noticed that there exist long range orders in the Ising models with spin S, including the Ising model with infinite-spin, on the stacked AFT lattice.

The ground state shows the 2×1 structure given in Fig.29 if the antiferromagnetic interactions J_2 exist between the nnn pairs in the xy-plane. However, this 2×1 structure is metastable when $J_2 = 0$. In order to see this, we make the MC simulations for the case of $J_2 = 0$ by assuming the 2×1 structure as an initial spin configuration. We define the order parameter, Y_4, assuming a four-sublattice structure for a lattice in the xy-plane as follows:

$$Y_4 = (\frac{4}{N})^2 \left\langle (\sum_i^A \sigma_i)^2 + (\sum_j^B \sigma_j)^2 + (\sum_k^C \sigma_k)^2 + (\sum_l^D \sigma_l)^2 \right\rangle. \tag{46}$$

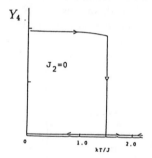

Fig. 35.   MC data of Y_4 for the Ising model with spin 1/2 on the stacked AFT lattice with $J_z = J$ and $J_2 = 0$.

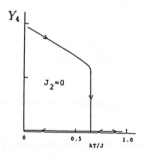

Fig. 36.   MC data of Y_4 for the Ising model with infinite-spin on the stacked AFT lattice with $J_z = J$ and $J_2 = 0$.

We show the temperature dependence of Y_4 for the Ising model with spin 1/2 on the stacked AFT lattice in Fig.35. The 2×1 structure is stable up to $T \cong 1.5J/k$, as can be seen in Fig.35. The temperature dependence of Y_4 for the Ising model with infinite-spin on the stacked AFT lattice is shown in Fig.36 (see the previous page) when $J_2 = 0$. We see the 2×1 structure is stable up to $T \cong 0.65 J/k$. The existence of "stable" metastable states suggests the appearance of a first order phase transition in the systems. Actually we find a first order phase transition in the systems.[46]

The energy for the 2×1 structure is given by

$$E_0 = -N(J + J_z + J_2). \qquad (47)$$

When the ground state is the 2×1 structure in the AFT lattice, the 2×1 structure also appears in each sublattice which is one of three sublattices in section 3. Hence, we may suppose that the phase diagram of T versus J is similar to that of T versus J_2. We found this similarity by the MC simulations as seen below. Hence we discuss only the case of $0 \leq J_2 \leq J$.

In Fig.37, we show the phase diagram obtained by the MC simulations for the Ising model with spin 1/2 on the AFT lattice.

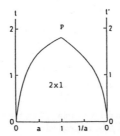

Fig. 37.   Phase diagram for the Ising model with spin 1/2 on the AFT lattice. a, t and t' denote $J_2/J, kT/J$ and kT/J_2, respectively.

In this system, there appear the paramagnetic state and the ordered state with the 2×1 structure. The paramagnetic state is denoted by P and the state with the 2×1 structure by 2×1 in Fig. 37 and also in the following figures.

In Fig.38 (see the next page), we show the phase diagram obtained by the MC simulations for the Ising model with spin 1/2 on the stacked AFT lattice. In the system, the spin structure is not clarified yet in an intermediate temperature region for small J_2 and even in the case of $J_2 = 0$ at low temperatures. In fact, the spin-state observed in the MC simulations shows some complex structure in this region and very sensitive to the values of parameters, J_z, J_2 and T. Hence we use the symbol C in Fig.38 to denote this complex spin-state. However, the phase boundaries are easily obtained by the MC simulations through Y_4 and other quantities. The spin structure at low temperatures is the 2×1 structure. We calculate the free energies in the LTSE for various spin-states. Since there is no free linear-chains in the system for the case of $J_2 \neq 0$, it is sufficient to take single spin flip-type excitations into account. The free energy for the 2×1 structure, $F(2 \times 1)$, is given as

$$F(2 \times 1) = -N(J + J_z + J_2) - NkT \exp[-\frac{4}{kT}(J + J_z + J_2)] + \cdots. \qquad (48)$$

The free energies for the FR structure and the PD structure are given as follows:

$$F(\text{FR}) = -N(J + J_z - 3J_2) - \frac{2NkT}{3} \exp[-\frac{4}{kT}(J_z - 3J_2)] + \cdots, \qquad (49)$$

$$F(\text{PD}) = -N(J + J_z - \frac{5}{3}J_2) - \frac{NkT}{3} \exp[-\frac{4}{kT}(J_z + J_2)] + \cdots. \qquad (50)$$

According to these equations, we have at $T = 0$

$$F(2 \times 1) < F(\text{PD}) < F(\text{FR}), \qquad (51)$$

while the PD or FR structure becomes stable at $T \neq 0$.

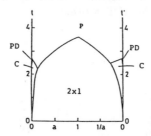

Fig. 38.   Phase diagram for the Ising model with spin 1/2 on the stacked AFT lattice with $J = J_z$. a, t and t' denote $J_2/J, kT/J$ and kT/J_2, respectively.

In Fig.39, we show the phase diagram obtained by the MC simulations for the Ising model with infinite-spin on the stacked AFT lattice.

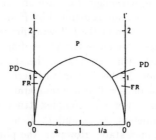

Fig. 39.   Phase diagram for the Ising model with infinite-spin on the stacked AFT lattice with $J = J_z$. a, t and t' denote $J_2/J, kT/J$ and kT/J_2, respectively.

The spin structures of the ordered states in this system are well-defined. For a very small value of J_2, we find that the spin structure is the 2×1 structure in a low temperature region, the FR structure in an intermediate low-temperature region and PD structure in an intermediate high-temperature region, as seen in Fig.39. In the

LTSE, the free energy for the 2×1 structure is given as

$$F(2 \times 1) = -N(J + J_z + J_2) + NkT\ln[\frac{2}{kT}(J + J_z + J_2)] + \cdots. \tag{52}$$

The free energy for the FR structure is given as

$$F(\text{FR}) = -N(J + J_z - 3J_2) + \frac{NkT}{3}\ln[(\frac{1}{kT})^3(6J + 2J_z - 6J_2)(2J_z - J_2)^2] + \cdots. \tag{53}$$

The free energies for the 2×1 and the FR structure for the system with $J_z = 0.3J$ and $J_2 = 0.05J$ are shown in Fig.40. In Fig.41, we show the results by the LTSE for the internal energy $u = U/NJ$ together with the results by the MC simulations for the system with $J_z = 0.3J$ and $J_2 = 0.05J$.

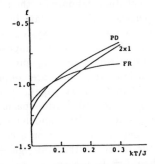

Fig. 40. The free energy by the low temperature expansion.

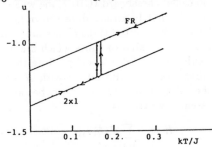

Fig. 41. The internal energy. The dots show the MC results and the solid lines the theoretical results. A first order phase transition is observed.

A first order phase transition occurs at $T = 0.165J/k$. A small hysteresis, $\Delta(kT/J) = 0.01$, is observed in this case, as shown in Fig.41.

9. Three-dimensional Ising paramagnet

Various frustrated Ising spin systems have been designed and studied to understand the effect of frustration in connection with the spin-glass problem.[47,48] It is well known that the frustration in general disturbs formation of long range order in spin systems. In spin systems on three dimensional lattices, however, some long range orders driven by thermal fluctuations can be found easily even in fully-frustrated Ising models. No long range order appears in the Ising model on the AFT lattice as a typical example, but the Ising model on the antiferromagnetic face-centered cubic lattice shows the phase transition from the paramagnetic state to a periodic long range order at a finite temperature.[16] The Ising model on the stacked Villain lattice, being different from the Ising model on the Villain lattice, also comes under this category. Then we have a following question. "Is there a 3d-Ising spin system which shows the paramagnetic behavior at all finite temperatures ?" We take up this problem in this section. It is possible in principle because we may construct such a 3d-Ising model from the Ising model on the AFT lattice or the Villain lattice by using a glue-Hamiltonian.[49] But we propose a realistic candidate for the 3d-Ising paramagnet and demonstrate some magnetic properties of the candidate by the MC simulations in order to understand the paramagnetic nature of the system.[50,51]

An Ising model proposed for the 3d-paramagnet is composed in the following way. Consider a cube with lattice-spacing a where an Ising spin of spin 1/2 is located at each corner of the cube and the neighboring spins are coupled by a ferromagnetic interaction J (see Figs. 42 and 43 in the next page). This cube is named a J-cube. Imagine a simple cubic (sc) lattice with lattice-spacing $2a$. Place the center of a J-cube on each lattice site of the sc lattice and let each of edges of J-cube parallel to that of the sc lattice. Let an Ising spin on a corner of J-cube interact with its nearest Ising spins on the other J-cubes through $+J_1$ or $-J_1$ interaction constant as shown in Fig.42. When $J = J_1$, the xy-, yz- or zx-plane in the resultant lattice corresponds to the so-called chessboard (CB) plane where the frustrated and non-frustrated plaquettes are alternately arranged, as shown in Fig.43. We call this resultant lattice the CB3 lattice since all the three planes are of the CB type.

The CB3 lattice contains another kind of cubes which are made up by $+J_1$ and $-J_1$ bonds. Each of these cubes is named a J_1-cube. Note that the J_1-cube is not frustrated. For the Ising model on the CB3 lattice with N sites, called the CB3 system hereafter, the number of both J-cubes and J_1-cubes amounts to $N/8$. The Hamiltonian of the CB3 system is written as

$$H = -\sum_{(i,j)} J_{ij}\sigma_i\sigma_j, \qquad (54)$$

where J_{ij} stands for the interaction constant between σ_i and σ_j and $\sum_{(ij)}$ indicates the sum over all the nn spin-pairs. It is easy to see that the ground state energy of CB3 system is $E_0 = -(1/2)NJ$ and the zero-point entropy is $S_0 = (N/8)\ln 2$ if $J = J_1$. Similarly, if $J < J_1$, we have $E_0 = -(1/2)NJ_1$ and $S_0 = (N/8)\ln 2$. For the case of $J = J_1$, the ferromagnetic state (F-state) is one of the spin states in the ground state, since $E_0 = -(1/2)NJ$. Because there are at least $N/4$ free-spins in this F-state, we have $S_0 \geq (N/4)\ln 2$ for the CB3 system when $J = J_1$. Let us make a thought experiment for the CB3 system according to the concept of the MC simulations. When $J_1 = 0$, the CB3 system is obviously paramagnetic at any finite temperatures.

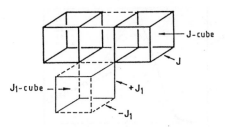

Fig. 42. A local structure of the CB3 system.

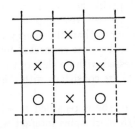

Fig. 43. A chessboard plane of alternating frustrated (marked by crosses) and non-frustrated (marked by circles) plaquettes.

However, if the conventional single-spin-flip (SSF) MC method is applied to the CB3 system, we are confronted with a crucial contradiction that a spin-glass order parameter tends to unity in the limit of temperature $T = 0$; the spin-glass order parameter, q, is defined by

$$q = \frac{1}{N} \sum_i |\langle \sigma_i \rangle|, \tag{55}$$

where the brackets denote the MC average. This discrepancy can be removed if we use the multi-spin-flip (MSF) MC method which updates all the spin configurations in a spin cluster at the same time.

We now take a J-cube as a spin cluster in the MC calculation for the CB3 system. 2^8 states of a spin-cluster are chosen as the achieved state of the cluster at each MC trial, according to the heat-bath importance sampling. We take advantage of the MSF as well as SSF methods for the present system. We calculate the specific heat, the susceptibility and the spin-glass order parameter. It is found that the spin-glass order parameter, q, vanishes at all temperatures for any J_1 value. In Fig.44(see the next page), the MC results of the specific heat, $c = C/Nk$, is shown for $J_1 = J$ (solid circles) and $J_1 = J/2$ (open circles). The exact result for $J_1 = 0$ is also plotted by the solid curve. No size dependence is observed in the specific heat data. Also the peak position at $T \cong 1.9 J/k$ is nearly fixed for different values of J_1.

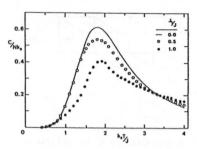

Fig. 44. MC data of the specific heat for the CB3 system.

In the case of $J_1 = 0$, the peak of specific heat is undoubtedly caused by the energy fluctuation of the paramagnetic J-cubes. We suppose that the peak, even for the case of $J = J_1$, is due to nothing more than the energy fluctuations of J-cubes or J_1-cubes. Since the J-cubes (or J_1-cubes) are connected by $\pm J_1$ bonds (or J bonds), the above fluctuation is now suppressed with no shift of the peak position in the case of $J_1 \neq J$. The MC results in Fig.44 are consistent with this tendency as the strength of J_1 increases.

In the limit of J infinity, the J-cube is regarded as a large "block spins" whose value is $+8$ or -8. Since these $N/8$ block-spins interact with each other via $\pm J_1$ in the CB3 system, the critical temperature should be zero in this limit.

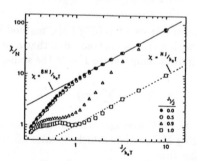

Fig. 45. MC data of the susceptibility for the CB3 system.

A similar conjecture holds for the case of $J > J_1$ at very low temperatures, because of the strong correlation between spins on the J-cube. Then, the susceptibility of the system must be written as

$$\chi = \frac{8NJ}{kT}. \tag{56}$$

Actually, the MC results of susceptibility show this paramagnetic behavior at low temperatures as seen in Fig.45. On the other hand, the low temperature behavior of

χ in the MC results behaves as the Curie law of $\chi = NJ/kT$ for the case of $J = J_1$. This does not indicate that there exist N free spins in the system, but should be read as a suggestion that the system may include a large number of not only free block spins but also free spins, induced by the frustration effect. The MC data of χ at low temperatures give another reason for us to support why we have to use the MSF MC method for this system: the excitations within the J-cube are more dominant in the system.

It is interesting that the correlation functions of the Ising model on the 2d chessboard lattice [52] decay exponentially at all temperatures, in contrast to those on the 2d purely frustrated lattice where a power-law decaying correlation appears in the ground state. In view of its paramagnetic behaviors, the CB3 system is expected to have the correlation functions of the exponential decay at all temperatures.

10. Concluding remarks

We discussed in this chapter several interesting features of the Ising spin systems with frustration. For the Ising model on the two-dimensional frustrated lattices such as the antiferromagnetic triangular (AFT) lattice and the Villain lattice, exact solutions were known. Based on those results, we focused our attention mainly to the following problems: spin orderings, effects of interactions between next-nearest-neighbor spin-pairs and effects of magnitude S of spin in Ising spin systems on the AFT lattice and on the stacked AFT lattice.

One of characteristic features of the Ising spin systems on frustrated lattices is the existence of "free" spins. As discussed in sections 2, 4, 7, and 9 in this chapter, however there may appear also "free" spin-clusters such as "free" linear-chains and "free" cubes. In those frustrated systems, cluster flip-type excitations exist in the low-lying energy states. Consequently, a cluster flip-type Monte Carlo method is needed in the MC simulations for the those frustrated systems.

The effects of the far-neighbor interactions in the Ising model on the AFT lattice and the stacked AFT lattice were clarified in this chapter. The ground state of the Ising model with only nearest-neighbor interaction on the AFT lattice is infinitely degenerate and is called the Wannier state. The Wannier state includes many states with various symmetries. Then if a perturbational interaction exists between spins on far-neighbor sites, the large degeneracy of the ground state is lifted and there appears a multi-sublattice spin structure corresponding to the symmetry of the far-neighbor interaction. It is intriguing to find the phase diagrams for those systems. This was discussed and clarified in section 3 and 8 to the Ising models with the perturbational interaction between spins on the next-nearest-neighbor sites.

Recently, it has been noticed that the magnitude S of spin can be an important physical parameter in both of classical spin systems and quantum spin systems. In the quantum spin systems, the Haldane problem is one of controversial problems. In frustrated Ising spin systems, the universality class may depend strongly on the magnitude S of spin, although no such S dependence is found in the Ising models on the non-frustrated lattices. We discussed S-dependencies of spin ordering for the

Ising spin systems on the frustrated lattices in sections 4, 5 and 6. We hope that more active investigations will be done in order to clarify S-dependent properties of the Ising spin systems on the frustrated lattices.

11. Acknowledgements

We thank M. Kang, Y. Yamada and M. Tanaka for their help in numerical computations and in preparing this chapter.

12. References

1. J. Villain, *Phys. Chem. Solids* 11 (1959) 303.
2. A. Yoshimori, *J. Phys. Soc. Jpn.* 14 (1959) 807.
3. D. J. Amit, *Modeling Brain Function: the world of attractor neural networks* (Cambridge University Press, Cambridge, 1989).
4. G. Toulouse, *Commun. Phys.* 2 (1977) 115.
5. J. Vannimenus and G. Toulouse, *J. Phys.* C10 (1977) L537.
6. J. Villain, *J. Phys.* C10 (1977) 1717.
7. G. Forgacs, *Phys. Rev.* B22 (1980) 4478.
8. G. H. Wannier, *Phys. Rev.* 79 (1950) 357.
9. M. Houtappel, *Physica* 16 (1950) 425.
10. K. Husimi and I. Syoji, *Prog. Theor. Phys.* 5 (1950) 177.
11. I. Syoji, *Prog. Theor. Phys.* 5 (1950) 341.
12. J. Stephenson, *Can. J. Phys.* 47 (1969) 2621.
13. J. Stephenson, *J. Math. Phys.* 11 (1970) 413.
14. T. Horiguchi, K. Tanaka and T. Morita, *J. Phys. Soc. Jpn.* 61 (1992) 64.
15. D. C. Mattis, *The Theory of Magnetism II* (Springer-Verlag, Berlin, 1985)
16. R. Liebmann, *Statistical Mechanics of Periodic Frustrated Ising Systems* (Springer-Verlag, Berlin, 1986).
17. O. Nagai, Y. Yamada and H. T. Diep, *Phys. Rev.* B32 (1985) 480.
18. L. Onsager, *Phys. Rev.* 65 (1944) 117.
19. O. Nagai, K. Nishino, J. J. Kim and Y. Yamada, *Phys. Rev.* B37 (1988) 5448.
20. J. J. Kim, Y. Yamada and O. Nagai, *Phys. Rev.* B41 (1990) 4760.
21. S. Miyashita, *J. Phys. Soc. Jpn.* 52 (1983) 780.
22. H. Takayama, K. Matsumoto, H. Kawahara and K. Wada, *J. Phys. Soc. Jpn.* 52 (1983) 2888.
23. M. Mekata, *J. Phys. Soc. Jpn.* 42(1977) 76.
24. S. Fujiki, K. Shutoh, Y. Abe and S. Katsura, *J. Phys. Soc. Jpn.* 52 (1983) 1531.
25. D. P. Landau, *Phys. Rev.*B27 (1983) 5604.
26. O. Nagai, M. Kang and S. Miyashita, to be published.
27. S. Miyashita, H. Kitatani and Y. Kanada, *J. Phys. Soc. Jpn.* 60 (1991) 1523.
28. D. Blankschtein, M. Ma, A. N. Berker, G. S. Grest and C. M. Soukoulis, *Phys.*

Rev. **B29** (1984) 5250.

29. A. N. Berker, G. S. Grest, C. M. Soukoulis, D. Blankschtein and M. Ma, *J. Appl. Phys.* **55** (1984) 2416.
30. F. Matsubara and S. Inawashiro, *J. Phys. Soc. Jpn.* **56** (1987) 2666.
31. S. N. Coppersmith, *Phys. Rev.* **B32** (1985) 1584.
32. S. Miyashita, *Prog. Theor. Phys. Suppl.* **87** (1976) 112.
33. D. C. Mattis, *Phys. Rev. Lett.* **42** (1979) 1503.
34. O. Nagai, T. Horiguchi and S. Miyashita, to be published.
35. O. Nagai, S. Miyashita and T. Horiguchi, *Phys. Rev.* **B47** (1993) 202.
36. T. Horiguchi, O. Nagai and S. Miyashita, *J. Phys. Soc. Jpn.* **60** (1991) 1513.
37. T. Horiguchi, O. Nagai, S. Miyashita, Y. Miyatake and Y. Seo, *J. Phys. Soc. Jpn.* **61** (1992) 3114.
38. R. Peierls, *Proc. Cambridge Phil. Soc.* **32** (1936) 477.
39. R. B. Griffiths, in *Phase Transitions and Critical Phenomena*, ed. C. Domb and M. S. Green (Academic Press, London, 1972), Vol.1, p.1.
40. C. Domb, in *Phase Transitions and Critical Phenomena*, ed. C. Domb and M. S. Green (Academic Press, London, 1974), Vol.3, p.1.
41. T. Horiguchi, *J. Phys. Soc. Jpn.* **59** (1990) 3142.
42. T. Horiguchi, O. Nagai and S. Miyashita, *J. Phys. Soc. Jpn.* **61** (1992) 308.
43. O. Nagai, M. Kang, T. Horiguchi and H. T. Diep, to be published.
44. O. Nagai, Y. Yamada, M. Tanaka, M. Kang, T. Horiguchi and S. Miyashita, to be published.
45. Y. Saito, *Phys. Rev.* **B24** (1981) 6652.
46. O. Nagai, M. Kang, T. Horiguchi and H. T. Diep, to be published.
47. K. Binder and A. P. Young, *Rev. Mod. Phys.* **58** (1986) 801.
48. S. Kirkpatrick, in *Disordered Systems and Localization*, ed. C. Castellani, C. Di Castro and L. Pelti (Springer-Verlag, Berlin, 1981), p.280.
49. T. Horiguchi, *Physica* **146A** (1987) 613.
50. H. T. Diep, P. Lallemand and O. Nagai, *J. Phys. C.* **18** (1985) 1067.
51. J. J. Kim, K. Nishino, Y. Yamada and O. Nagai, in *Proceeding of Fourth Asia Pacific Physics Conference* , ed. S. H. Ahn, S. H. Choh, H. T. Cheon and C. Lee (World Scientific, Singapore, 1990).
52. G. André, R. Bidaux, J. -P. Carton, R. Conte and L. de Seze, *J. Physique* **40** (1979) 479.

Magnetic System with Competing Interaction
Ed. H. T. Diep
©1994 World Scientific Publishing Co.

CHAPTER VI

COMPETITION BETWEEN FERROMAGNETIC AND

SPIN GLASS ORDER IN RANDOM MAGNETS:

THE PROBLEM OF REENTRANT SPIN GLASSES

Michel J. P. Gingras

TRIUMF, Theory Group, 4004 Wesbrook Mall
Vancouver, B.C., V6T-2A3, Canada.

1. Introduction

Simple magnetic systems have been used in the past to benchmark concepts developed to investigate collective phenomena. The experimental availability of a large class of magnets, the development of the renormalization group methods, and the notion of universality class have led to a generally consistent understanding of most classical collinear three-dimensional ($3D$) magnets. [1]

Motivated by striking successes in understanding the collective behaviour of conventional magnets, researchers then turned their attention to materials that exhibit novel types of ordering as a consequence of "frustrated" interactions. [2] These interactions can arise due to the combination of simple antiferromagnetism and certain lattice symmetries, such as triangular and tetrahedral, and consequently are common in nature. Bond frustration arises when a geometrical unit (plaquette) made of spins at its corner cannot minimize its total energy by minimizing each energy bond individually. Bond frustration has been shown to lead to novel and interesting effects that can manifest themselves either at finite and/or zero temperature. The previous chapters of this book have largely concentrated on how frustration can affect, if not dictate, the nature of the ordering in geometrically frustrated magnetic systems. Bond frustration also frequently arises in random disordered magnets, but with its presence having even more dramatic effects than in nonrandom magnets.

Until the mid seventies, most of the condensed matter research was focused towards perfect disorder-free systems. That was true for both experimental and theoretical studies. However, the observation that all real systems in nature contain a certain amount of randomly frozen (quenched) disorder eventually led researchers to investigate the effects of disorder in their prefered systems. Again, it became clear that magnets with almost perfectly homogeneous microscopic disorder would be the ideal systems to investigate the influence of disorder on collective phenomena. It was hoped that, here too, a handful of relevant parameters would dictate the universality classes of disordered systems, with the universal features being most easily unravelled by studies of "simple" random magnets.

The introduction of random disorder in a magnetic system can generate frustration via a competition between ferromagnetic and antiferromagnetic exchange couplings. The simplest example is magnetic impurities in a non-magnetic metallic host (e.g AuFe). Here, the magnetic Fe impurities are coupled to each other via the RKKY exchange interaction, J_{RKKY}, mediated by the Au conduction electrons. Due to the oscillatory dependence of J_{RKKY} on the distance r_{ij} separating the Fe atoms ($J_{RKKY} \propto \cos\{2k_F r_{ij}\}$, k_F is the wavevector at the Fermi surface), some Fe moments are coupled ferromagnetically while others are coupled antiferromagnetically. Randomly located ferromagnetic and antiferromagnetic bonds results in random bond frustration.

In ferromagnets with weak random frustration, the Curie temperature, T_c, decreases with increasing disorder up to a certain level where the random bond frustration is so strong that the system cannot accomodate a percolating ferromagnetic ground state. However, such strongly disordered tuned systems often exhibit a *spin-glass* transition: the spins freeze cooperatively in a nontrivial pattern that is random in space with no delta function magnetic Bragg peaks. The reader interested in the spin glass problem should refer to the reviews of Binder and Young, [3] Mézard et al., [4] and Fischer and Hertz. [5] In this chapter we focus specifically on the behavior of random magnets close to the point in the disorder-temperature phase diagram where the paramagnetic (PM), ferromagnetic (FM) and spin glass (SG) phase meet and where the so-called *reentrant spin glass* behavior is observed.

There have been some recent experimental reviews of the problem of reentrant spin glasses. [6,7,8,9,10] In this chapter we present a review of both the current theoretical and experimental status on this problem. Our purpose here is to review the mainstream research directions in the problem of reentrant spin glasses, and the generic features that have emerged over the years. Since the previous reviews have largely focused on the experimental investigations, we will put here a slight emphasis on the theoretical and numerical work on reentrant spin glasses.

The rest of this chapter is divided as follows: In the next section we introduce the problem of reentrant spin glasses. In Section 3 we discuss the mean mean-field theory of spin glasses. This theory shows ferromagnetic to spin glass reentrance within an inaccurate replica-symmetric treatment, but *not* in the presumably exact non-replica-symmetric theory of Parisi. In Section 4 we review some of the numerical studies searching for reentrance in models of two-dimensional ($2D$) and $3D$ random ferromagnets. Sections 5 discusses the most recent experimental investigations of reentrant spin glass materials. Recent progress on the rigorous theoretical constraints on the phase diagram of spin glass models are presented in Section 6. Sections 4, 5 and 6 discuss how reentrance does not occur in most cases of randomly frustrated ferromagnets. However, some hints suggesting the possibility of true reentrance in the most frustrated ferromagnets are mentioned in Section 5. We close this chapter by a short conclusion in Section 7.

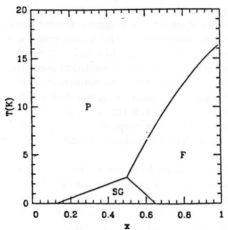

Fig. 1. Concentration, x, temperature, T, phase diagram of the insulating Heisenberg ferromagnet $Eu_xSr_{1-x}S$ which shows reentrant behavior in the range $0.5 < x < 0.65$. [11] $Eu_xSr_{1-x}S$ can be described by a Heisenberg spin Hamiltonian, H, with $H = \sum_{(i,j)} -J_{ij}(\vec{S}_i \cdot \vec{S}_j)$, where the sum runs over the nearest and second-nearest neighbors of a regular FCC lattice whose sites are randomly occupied by either the Eu or Sr atoms. The exchange couplings J_{ij} are ferromagnetic and antiferromagnetic for the nearest (J_1) and second-nearest (J_2) neighbors respectively, with $J_2/J_1 \approx -1/2$.

2. Reentrant Spin Glasses

2.1. Magnetization Measurements

A large number of disordered ferromagnets tuned close to their paramagnetic-ferromagnetic-spin-glass (PM-FM-SG) multicritical point display history dependence below a temperature $T_g < T_c$. In particular, the magnetization, M_z, measured during warming starting from a state reached at $T < T_g$ in a zero-field cooling procedure (ZFC) differs from the magnetization measured upon cooling in a field (FC). Such behavior has been observed in diluted insulating magnets (e.g. $Eu_xSr_{1-x}S$), metallic alloys (e.g. AuFe), amorphous $Fe-TM$ metallic glasses (TM is a transition metal; e.g. a-$Fe_{1-x}Mn_x \equiv (Fe_{1-x}Mn_x)_{75}P_{16}B_6Al_3$), and iron-rich $Fe-ETM$ (ETM is an early transition metal; e.g. Fe_xZr_{1-x}). It has become general practice to refer to random magnets displaying such behavior as *reentrant spin glasses*.

The insulating Heisenberg ferromagnetic system $Eu_xSr_{1-x}S$ is probably the best known example that displays reentrant spin glass behavior for a range of concentration $0.5 < x < 0.65$. Eu is magnetic with $|\vec{S}| = 7/2$ and Sr and S are nonmagnetic. Figure 1 shows the phase diagram for this material. [11]

It is very important to realize that in this phase diagram the spin glass phase (SG) labels the *whole* $x-T$ region where the system shows history dependence with a value of the FC magnetization larger than the ZFC one. The negative slope of the ferromagnetic to spin glass boundary in Figure 1 implies that upon lowering the

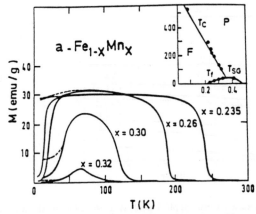

Fig. 2. Low field magnetization versus temperature in the a-$Fe_{1-x}Mn_x$ system. The applied field is 20 Oe. The magnetization is measured in ZFC state (solid line) and in the FC state (dashed line). The inset shows the phase diagram deduced from these measurements. We notice that for x close to $x_c = 0.35$ ($x = 0.30$ and $x = 0.32$), the FC measurements reach a maximum upon cooling, and then a noticeable decrease down to a temperature where the FC and ZFC values depart from each other. (From Ref. 12).

temperature, and for a range of concentration of the magnetic species, x, the following sequence of two transitions is observed:

$$paramagnetic \xrightarrow{T_{PF}} ferromagnetic \xrightarrow{T_{FG}} spin\,glass \quad . \tag{1}$$

The original name "reentrant spin glasses" is directly motivated by the above generic phase diagram topology of all reentrant spin glass materials where the system *would appear to reenter* into a less magnetically ordered state, the spin glass phase, as the temperature is decreased. The terminology "reentrant spin glasses" is somewhat misleading, since the system does not reenter a spin glass phase at low temperatures, as it was a paramagnet, not a spin glass, at high temperatures. However, the system does reenter into a less magnetically ordered state at low temperatures, the spin glass phase. The above sequence of phase transitions implies that the spin glass phase has the lowest energy while the ferromagnetic phase has more entropy, which is rather counterintuitive. This is probably the main motivation for the research efforts devoted to understanding the problem of reentrant spin glasses in the past twenty years.

From a rigorous point of view, one would like the definition of a reentrant spin glass to be that of a random magnet which shows a loss of collinear long-range ferromagnetic (or antiferromagnetic) order *upon cooling* and appearance of a spin glass phase. However, as emphasized above, the ZFC magnetization branch is obtained during warming, starting from a state reached from cooling in zero field. There is no decrease in the FC magnetization in systems with low level of frustration (see Figure 2).

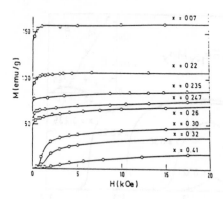

Fig. 3. High field magnetization versus applied field at $T = 11K$ in the a-Fe$_{1-x}$Mn$_x$ system. (From Ref. 12).

It is interesting to look at the magnetization measurements performed in high magnetic field and low temperature ($H \leq 15kOe, T \sim 11K$). Figure 3 for weakly frustrated samples of a-Fe$_{1-x}$Mn$_x$ shows a sharp increase of the $M(H)$ curve followed by a saturation plateau. This is similar to what is seen in conventional ferromagnets in which the magnetic field aligns large domains, suppressing the Bloch walls. However, a slope persists at high fields in $M(H)$ for the most frustrated (large x) alloys. When increasing the frustration, the knee in the $M(H)$ curve smears out, and finally disappears in the true spin glass sample ($x = 0.41$).

It appears from the simplest magnetization measurements that the behavior of reentrant spin glasses is not the same for all systems. In particular, in some systems (e.g. a-Fe$_{1-x}$Mn$_x$), the most frustrated samples, with their disorder tuned close to the PM-FM-SG multicritical point, seem to display a decrease in the FC magnetization unlike what is seen in the less frustrated samples (see Figure 2). One well defined question that has been the main focus of the numerical work and the most recent experiments presented in Sections 4 and 5 is: "Are there indications of reentrant-like behavior in weakly frustrated systems?" Before discussing these theoretical and experimental studies, it is useful to consider the behavior of frustrated magnets within mean-field theory.

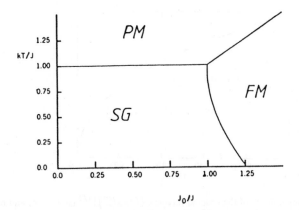

Fig. 4. Replica-symmetric mean-field phase diagram of the Gaussian Ising spin glass model. J_0 and J are the mean and width of the Gaussian probability distribution $\mathcal{P}(J_{ij})$ distribution. Notice the reentrant ferromagnetic (FM) to spin glass (SG) transition for $1 < J_0/J < 5/4$. (From Ref. 13).

3. Mean-Field Theory of Spin Glasses

The experimental discovery of reentrant spin glasses and the orginal assumptions assuming a collapse of the ferromagnetic order did not really come as a surprise. Indeed, the replica-symmetric mean-field theory of the Ising spin glass did predict such behavior for a ratio of the mean and width of the Gaussian exchange coupling distribution within a certain range. [13] Figure 4 shows the phase diagram of the Ising spin glass model with a Gaussian distribution of exchange couplings J_{ij} as obtained from the replica-symmetric mean-field theory. [13] The Hamiltonian for this model is

$$H = - \sum_{(i,j)} J_{ij}\, S_i S_j \tag{2}$$

where the sum runs over all pairs (i,j) of Ising spins S_i and S_j. J_0 and J are the mean and width of the Gaussian probability distribution $\mathcal{P}(J_{ij})$.

$$\mathcal{P}(J_{ij}) \;=\; \frac{1}{\sqrt{2\pi J^2}} \exp\{(J_{ij} - J_0)^2/2J^2\} \tag{3}$$

The system is in the paramagnetic phase at high temperatures (PM), and $[\langle S_i \rangle]_d = 0$ for all spins S_i. Here, $\langle ... \rangle$ refers to a Boltzmann (thermal) average and $[...]_d$ is an average over different realizations of the disorder (disorder average). The system is ferromagnetic (FM) at low temperatures for sufficiently weak disorder $(J_0/J > 5/4)$. In the ferromagnetic phase, the primary order parameter is the average magnetization per spin, M_z, with $M_z \equiv 1/N \sum_i [\langle S_i \rangle]_d > 0$. For large disorder $(J_0/J < 1)$ and low temperatures, the system is in the spin glass phase (SG) where

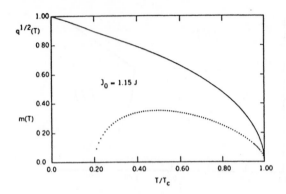

Fig. 5. Temperature dependence of the ordered moment $\{Q_{EA}(T)\}^{1/2}$ (solid line) and magnetization $M_z(T)$ (dotted line) for $J_0/J = 1.15$. (From Ref. 13).

the spins are frozen in a random spatial pattern with zero magnetization ($M_z = 0$). In the replica-symmetric mean-field theory, the Edwards-Anderson order parameter $Q_{EA}(T) = 1/N \sum_i [\langle S_i \rangle^2]_d > 0$ is nonzero and characterizes the extent of the static frozen random spin order. The most interesting feature of this phase diagram is the reentrant ferromagnetic (FM) to spin glass (SG) transition for $1 < J_0/J < 5/4$. This can be made more explicit by plotting the static frozen ordered moment $Q_{EA}^{1/2}$ and the magnetization $M_z(T)$ in Figure 5. This clearly shows "true" reentrant behavior in the rigorous sense, since the equilibrium value of $M_z(T)$, as obtained from the replica-symmetric mean field theory, does diminish for $T < 0.5$ and eventually vanishes at $T \approx 0.20$ for $J_0/J = 1.15$.

However, the replica-symmetric solution is incorrect (unstable) below the so-called Almeida-Thouless (AT) line. [3,4,5] Indeed, for all values $J_0/J > 1$, the AT line resides *above* the FM-SG boundary predicted by the replica-symmetric theory. It is now known that the correct non-replica-symmetric mean-field solution of Parisi does not show reentrance. [3,4,5] Indeed, the presumably exact Parisi solution give a strictly vertical SG-FM phase boundary. [14]

Most experimental systems have vector spins, for which Ising models are likely to be a poor approximation since they do not allow for the possibility of ordering transitions in different spin directions. The replica-symmetric solution to the mean-field theory of the Heisenberg model was published shortly after the one for the Ising model (see Ref. [15] for studies of spin glass order in antiferromagnets). This replica-symmetic solution shows three magnetic transitions in the weakly frustrated regime ($J_0/J > 1$) as shown in Figure 7. [16]

The first transition, PM → FM is an ordinary paramagnetic to ferromagnetic transition. The second transition, FM → M1, is from a ferromagnetic phase to a

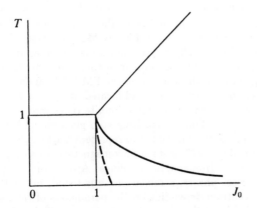

Fig. 6. Mean-field phase diagram of the Gaussian Ising spin glass model as predicted by the correct non-replica-symmetric Parisi solution. The AT line (boldface line) lies everywhere above the reentrant FM-SG boundary (dashed line) predicted by the incorrect replica-symmetric solution. The true FM-SG boundary is vertical and located at $J_0/J = 1$ (solid line).

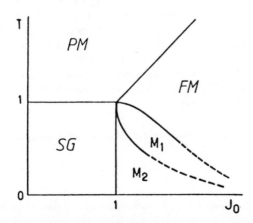

Fig. 7. Mean-field phase diagram of the Gaussian Heisenberg spin glass model. J_0 and J are the mean and width of the Gaussian probability distribution $\mathcal{P}(J_{ij})$ distribution. There is a spin glass freezing of the XY components at the M1 phase boundary. The transition from ferromagnetic to spin glass order occurs at $J_0/J = 1$. As in the Ising model, there is no FM—SG reentrant transition and the boundary between these two phases is vertical. (From Ref. 16).

mixed state characterized by long-range ferromagnetic order in M_z, but spin-glass order in the transverse components of S_{ix} and S_{iy}. The third transition, M1 → M2 marks the onset of strong irreversibility in the z component of $\vec{S}_i$. However, the replica-symmetric solution fails *on* the FM → M1 transition. This is easy to understand, as the FM → M1 is really a spin-glass transition in an $m-1$ component spin glass model (where m is the total number of spin components; $m=3$ in the Heisenberg model). Again, as in the Ising case, the FM-SG boundary beyond which no average magnetization occurs is vertical and located at $J_0/J = 1$.

Hence, there is no reentrance in non-diluted infinite range models of spin glasses. However, a generalized *dilute* infinite range Ising spin glass model has been studied by Viana and Bray. [17] The model is intended to represent $Eu_xSr_{1-x}S$ by taking ferromagnetic bonds with probability 2/3, and antiferromagnetic ones with probability 1/3 and half the strength of the ferromagnetic couplings. Apart from the paramagnetic, ferromagnetic and spin glass phases, a mixed phase is found, but with the FM-SG phase boundary being reentrant depending on the parameters of the model.

Reentrant behavior has been observed in binary mixtures, [18] liquid crystals, [19] and superconducting systems. [20] More recently, a reentrant melting transition has been seen in a polymer glass [21] while the wrinkling of partially polymerized membranes upon a decrease of temperature could be interpreted as the reentrance of a crumpled phase. [22] The presence of some "hidden" interactions which are not necessary to explain the thermodynamic stability (existence) of the intermediate ordered phase are usually the origin of a reentrant phenomenon. [23] The existence of such interactions and of a related disordering mechanism has not yet been identified in reentrant spin glasses. Hence, the physical origin of reentrance in realistic $2D$ and $3D$ spin glass systems, if it does indeed occur, is unclear, since short range models of spin glasses are still poorly understood. For this reason, several numerical studies investigating the possible occurence of reentrance in randomly frustrated $2D$ and $3D$ ferromagnetic models have been performed in the past few years. The following section reviews the results obtained in some of these studies.

4. Search for Reentrance in Numerical Studies of Random Ferromagnets

4.1. Methods

The exact temperature-disorder phase diagram of infinite-range Ising and vector (XY and Heisenberg) spin glass models is now fairly well understood within the Parisi mean-field solution. However, very few rigorous results exist for $2D$ and $3D$ spin glass models. In fact, there is an ongoing debate as to whether or not the predictions of mean-field theory apply at all to $2D$ and $3D$ spin glasses. [24] In this context, one of the key issues is whether or not a true thermodynamic spin-glass transition occurs in the presence of an applied field.[24-30] Most of what we know about $2D$ and $3D$ spin glasses is due to various types of numerical calculations. We can identify four types of methods: (1) Monte Carlo simulations, (2) high temperature series expansions, (3) defect-wall energy calculations, and (4) approximate real-space renormalization calculations based on the Migdal-Kadanoff scheme. These four methods have all been used to investigate reentrant spin glasses. Before presenting results that were obtained using these methods, we briefly describe each of them.

- *Monte Carlo Method*

  Computer simulation based on the Monte Carlo method is now a standard technique used in the study of the statistical mechanics of strongly interacting systems. [31] The main features of Monte Carlo simulations have already been reviewed in the Chapter *"Critical Properties of Frustrated Vector Spin Systems"* by Plumer et al. in this book. Basically, the method generates a set of phase-space configurations via a Markov chain which uses transition probabilities within the phase-space states that guarantees ergodicity, detailed balance, and that the states are distributed according to a Boltzman distribution. Once it has been verified that the system has reached equilibrium, [3,32] one can start to calculate thermodynamic quantities. [31] In the context of spin glasses, Monte Carlo simulations became a most useful tool right from the birth of the field. [13] We refer the reader to references[29,30,32−60] for applications of Monte Carlo simulations to spin glass models.

  It has been suggested that it is possible to speed up the relaxation of classical vector (non-Ising) spin systems by using a "hybrid" Monte Carlo technique. This method combines true deterministic dynamics with stochastics Monte Carlo moves. [43] For classical vector (non-Ising) spin systems, one can explicitly write down Newton's equations of motion for the spins. From the torque (local field) acting on a spin at a time t_1, one integrates Newton's equation to find the orientation of the spin at a time $t_2 > t_1$. This is the spin analogue version of the well known molecular dynamics technique used in numerical study of systems with particles having a center of mass motion. [31] The key feature of this method is that it conserves the total average energy (kinetic plus potential). Hence, the kinetic energy taken alone fluctuates, and so does the temperature of the system. By combining sequences of deterministic spin dynamics and stochastic Monte Carlo moves (hence the name hybrid Monte Carlo) one can effectively stabilize the temperature of the system at a desired nominal value. [43] This method is analogous to the constant temperature molecular dynamics method in which the velocities are rescaled during the course of the simulation such that the average kinetic energy does correspond to the desired average temperature. [31]

  It is possible to implement ideas of real-space renormalization within the Monte Carlo method to obtain accurate values of critical coupling constants and exponents. [61] This method has also been applied to the problem of the Ising spin glass. [62] New ideas on how to reduce the number of steps needed to reach equilibrium in numerical simulations of spin glasses have been put forward. [57,58,59]

A temperature scan in a Monte Carlo simulation of spin glass models (or any other random systems) is done for a fixed realization of the disorder (e.g. values of the exchange couplings). Due to the limited sample size available in computer simulations ($10^2 - 10^3$ spins), once must repeat this temperature scan for a very large number of realizations of disorder in order to perform the disorder average (quench average) of the various thermodynamic quantities. It is most important in simulations of random systems to make sure that a sufficiently large number of disorder realizations are considered in the disorder averaging, as some quantities are not self-averaging, [3,4,5] and this can drastically affect the final results. [3,56]

- *High temperature series expansion*

In the context of critical phenomena, the high temperature series expansion technique was developed several years before the renormalization-group methods were invented. As it was made available even before the advent of computers, it provided the very first way to extract non mean-field critical exponents for $3D$ spin systems.

As the name indicates, the high-temperature expansion method is based on a (Taylor) expansion of thermodynamics quantities (e.g. the uniform susceptibility in a ferromagnet) in powers of the (small) inverse temperature parameter $\beta = 1/T$. Once a series with a "large" number of powers of β has been derived, one then uses the Padé approximant method to extract from this Taylor expansion the "best" form of the singularity describing the thermodynamic quantity close to the critical point. This procedure then allows an approximate determination of the underlying critical exponent characterizing the singularity. This method has been successfully used in the context of spin glasses where, for example, one can calculate a high-temperature series expansion of the spin-glass susceptibility χ_{SG}. [63,64]

Low-temperature expansions can also be done in systems where both the ground state and the elementary excitations are known. One example is the spin-wave expansion in Heisenberg ferromagnets. However, this method cannot be applied to spin glasses since one does not know, a priori, the nontrivial ground state(s) of the system. In fact, it still unclear whether or not $2D$ and $3D$ spin glasses have a unique ground state (apart from trivial global symmetry operation on $\vec{S}_i$) or, as predicted by mean-field theory, a macroscopic number of ground states. [24-30,65]

- *Defect-wall energy method*

The defect-wall energy method is a numerical technique which allows to determine how the energy (at zero temperature) or the free-energy (at nonzero temperature) of a spin system changes upon introduction of a small perturbation (defect) and, in particular, how this change scales with the system size.

The defect-wall energy, E_{def}, is obtained by calculating the difference in the ground state energy for periodic and antiperiodic boundary conditions in one direction, while keeping periodic boundary conditions in the remaining $d-1$

directions (d is the spatial dimension).[66-79] This quantity fluctuates between samples, and has a distribution with zero mean, but with a standard deviation which is proportional to the disorder-average of the absolute value of E_{def}, $[|E_{def}(L)|]_d$, and which scales with the linear size L of the system as

$$[|E_{def}(L)|]_d \propto L^\theta .$$ (4)

If the exponent θ is negative, the width of the distribution vanishes on large length scales and the system does not have a spin glass phase at positive temperature. Conversely, if θ is positive, the system is expected to have a spin glass transition at nonzero temperature.[76] The marginal case, $\theta = 0$, occurs when the system is at its lower critical dimension (LCD).

It is easy to show that below the LCD the spin-glass correlation length ξ diverges at $T = 0$ with the critical exponent $\nu = -1/\theta$.[67] Also, for systems with a nondegenerate ground state (apart from the trivial two-fold degeneracy arising from the global spin inversion symmetry) the spin-glass correlation function does not decay at $T = 0$. Hence, the correlation function exponent η is equal to $2 - d$, and ν is the only independent non-trivial static critical exponent for a system below the LCD. It is therefore possible to use the defect-wall method to determine the universality class of spin glass models with nondegenerate ground states below their LCD. The advantage of this method is that one can use results obtained at $T = 0$ to extract the singular dependence on T as $T \to 0$, avoiding the extrapolation on temperature needed to get ν from Monte Carlo simulations.[67,71,78,79]

- *Migdal-Kadanoff renormalization-group method*

The Migdal-Kadanoff bond-moving scheme is a simple real-space renormalization group method which, despite its simplicity, has proved to be a useful tool in investigating spin glasses.[80] For example, it appears to be able to predict the lower critical dimension and the spin-stiffness exponent of Ising[81-85] and XY[73,86-89] spin glasses correctly. It also predicts that XY gauge glasses are most likely to be in a different universality class than other isotropic m-component spin glasses.[78,90,91] The method has severe limitations, as it is not able to give good estimates of the critical temperature in spin glasses and gives inaccurate critical exponents. It is often argued, however, that it can give the correct topology of the phase diagram. The Migdal-Kadanoff scheme follows the distribution of couplings under the renormalization-group iterations and is able to locate the paramagnetic, ferromagnetic and the spin glass phases. It is therefore a well suited tool to examine reentrant behavior. We refer the reader to Refs.[88,89,92] on how to carry the Migdal-Kadanoff procedure for generic Q-state clock spin glasses ($Q = 2$ for Ising and $Q = \infty$ for XY spin glasses).

4.2. Spin Hamiltonians and Effects of Randomness

The simplest Hamiltonian of a disordered spin system is:

$$H = \sum_{(i,j)} -J_{ij}\, \vec{S}_i \cdot \vec{S}_j \; . \tag{5}$$

Here the sum is taken over pairs of spins (i,j) interacting via a distance dependent exchange coupling J_{ij} (as in the RKKY interaction for example). Random frustration arises in this model when there are random competition between positive (ferromagnetic) and negative (antiferromagnet) J_{ij} couplings in the above Hamiltonian. [3,5]

As described in the previous chapters of this book, nonrandom exchange frustration can possibly lead to "new" universality classes, but not to spin glass order. Interestingly, some uniformly frustrated systems such as Josephson junction arrays with irrational magnetic flux per plaquette, [93] and antiferromagnetically coupled Heisenberg spins on stacked kagomé planes with single-ion anisotropy, [94] have been shown to display disorder-free spin-glass-like behavior. A "small" amount of *positive* random J_{ij}, which does not introduce exchange frustration, can lead to a change of the universality class. [95,96] In some cases, the randomness can also manifest itself in the form of random fields via an added H_{RF} term to Eq. 5 with

$$H_{RF} = \sum_i \vec{h}_i \cdot \vec{S}_i \; . \tag{6}$$

The presence of weak random fields are detrimental to the magnetic ordering since they increase the lower critical dimension for the occurence of long-range ferromagnetic order at nonzero temperature. [97] However, both in non-frustrated random bonds and in random field systems, the primary order parameter is a Fourier component of the magnetization $\vec{M}(\vec{R})$, rather than an Edwards-Anderson spin glass order parameter, and these systems are not spin glasses. Still, some features of non-frustrated random bonds and random field systems are very similar to those found in spin glasses. One example is the extremely slow dynamics of all random spin systems at low temperatures. [98,99,100] Such slow dynamics have been explicitly demonstrated in charge [101,102] and spin [102,103] density wave systems, which are analogous to random field XY systems, and where ageing effects similar to those found in spin glasses have been observed.

In non-Ising systems, where $\vec{S}_i$ is either a two (XY) or three (Heisenberg) component vector, random anisotropic exchange couplings can also occur. These interactions, D_{ij}^{ab}, couple the cartesian a component of spin $\vec{S}_i$ with the b component of spin $\vec{S}_j$ (as well as $S_{i,b}$ with $S_{j,a}$). The random matrix D_{ij}^{ab} can be either symmetric $(D_{ij}^{ab} = D_{ij}^{ba})$ or antisymmetric $(D_{ij}^{ab} = -D_{ij}^{ba})$. Physical realization of these two cases are dipole-dipole and Dzyaloshinkii-Moriya interactions respectively. [3,5] Another physical realization is in amorphous materials where there can be a random easy-axis anisotropy which enters in Eq. 5 as

$$H_{RA} = -\sum_i D_i(\hat{n}_i \cdot \vec{S}_i)^2 \tag{7}$$

where D_i is a random positive coupling, and $\hat{n}_i$ is a m-component random vector. Random anisotropy destroys long-range ferromagnetic order in spatial dimensions less than or equal to four [104] as for random fields in systems with continous symmetry. [97] However, unlike in the case of random fields, it has been argued that random anisotropy not only destroys ferromagnetic order at large length scale in less then four dimensions, but may also lead to a spin glass transition in the same universality class as the Ising spin glass. [104,105] Some numerical studies in $2D$ are compatible with an Ising spin glass transition at zero temperature in vector spin glasses with random axis [71] or random bond [79] anisotropy. However, the situation appears more complicated in $3D$, with the possibility of quasi long-range order in the XY random axis model. [106] Much work remains to be done to understand the effects of random anisotropy and random fields in XY ferromagnets.

4.3. Edwards-Anderson Model

In their seminal paper Edwards and Anderson argued that quenched randomness *and* frustration are the only relevant features leading to spin glass behavior in systems with J_{ij} decaying sufficiently fast with distance (i.e. short range systems). [107] In practical terms, this means that the main ingredient to model spin glass order is the presence of a large concentration of both positive (ferromagnetic) and negative (antiferromagnet) J_{ij}s leading to considerable random frustration. The Edwards-Anderson's postulate means that the universality class (lower critical dimension and critical exponents) can be investigated on simple d-dimensional cubic where the details leading to frozen random frustration is swamped into a quenched probability distribution of J_{ij}s, $\mathcal{P}(J_{ij})$, given apriori. The two distributions $\mathcal{P}(J_{ij})$ that have been mostly used in numerical studies are the bimodal and Gaussian probability distributions. The ferromagnetically biased bimodal distribution is given by

$$\mathcal{P}_B(J_{ij}) = x\delta(J_{ij} - J_0) + (1 - x)\delta(J_{ij} + \lambda J_0) \ . \tag{8}$$

A bond between sites i and j has a probability x to be ferromagnetic and of strength J_0, and a probability $1 - x$ of being antiferromagnetic and of strength $-\lambda J_0$. The Gaussian distribution is given by

$$\mathcal{P}_G(J_{ij}) = \frac{1}{\sqrt{2\pi J^2}} \exp\{-(J_{ij} - J_0)^2/(2J^2)\} \ , \tag{9}$$

where J_0 and J are the mean and width of the distribution, respectively. Other distributions have also been considered in an infinite range model [17] and for a $3D$ cubic lattice. [108]

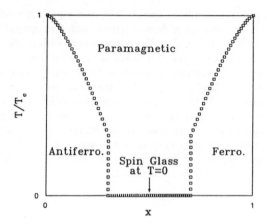

Fig. 8. Schematic temperature-disorder phase diagram of the $2D$ bimodal Ising spin glass with $\lambda = 1$ on a square lattice.

4.4. Ising Models

The Ising model is obtained by restricting the vector $\vec{S}_i$ to be in either the $+z$ or $-z$ direction in Eq. 5.

After several years of numerical studies, it is now generally accepted that the temperature-disorder phase diagram of the bimodal Ising spin glass has the topology shown in Figures 8 and 9 in $2D$ and $3D$ respectively for $\lambda = 1$. The topology of the phase diagram for the Gaussian Ising spin glass is essentially identical to the bimodal version, with J_0/J replacing x.

As found by direct transfer matrix calculations, [3,109] Monte Carlo simulations, [36,60] series expansion, [63,64] defect-wall, [66-69,79,83] and MKRG calculations, [81-85] the $2D$ Ising spin glass model exhibits a zero temperature spin glass transition only for $x < x_c(\lambda)$. However, the nature of this $T = 0$ transition is interesting in the case of the bimodal model as the ground state of this system is believed to be extensively degenerate. [110] This is presumably due to the combination of the discrete nature of J_{ij} and of S_i leading to decoupled spin-bloks with vanishing local field acting at their boundaries at zero temperature. This means that the critical exponent η characterizing the decay of the spin-glass correlation function in the ground state is $\eta > 2 - d$. We say *presumably* since the origin of degenerate ground states in spin glass models is not well understood. [51,83,87] For example, a recent Monte Carlo simulations of the $2D$ bimodal XY spin-glass model (XY spin glass) suggests a $T = 0$ transition to a degenerate ground state *even* in this system with continuous symmetry. [51,87]

In both $2D$ and $3D$, the system supports long range ferromagnetic order for small disorder. Uncorrelated random bond disorder is relevant at the paramagnetic to ferromagnetic transition since the specific heat exponent, $\alpha = 2 - d\nu$, is positive in $3D$ (ν is the ferromagnetic correlation length critical exponent). [95,96] Hence, the critical

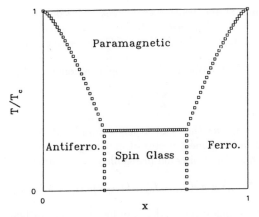

Fig. 9. Schematic temperature-disorder phase diagram of the $3D$ bimodal Ising spin glass with $\lambda = 1$ on a cubic lattice.

exponents for the PM-FM transition are different than those of the pure $3D$ Ising ferromagnet and controlled by a "new" disordered fixed point. In two dimensions, the Onsager solution of the Ising ferromagnet gives a logarithmic divergence of the specific heat ($\nu = 1$ in $2D$). In this interesting case, Harris' criterion cannot predict whether or not weak disorder is relevant or not at the pure PM-FM transition fixed point. Recent theoretical work suggests that weak random bond disorder is marginally relevant in the $2D$ Ising ferromagnet with a $\log(\log|T - T_c|)$ singularity rather than the $\log|T - T_c|$ singularity of the pure model. [111]

The main feature of the $3D$ Ising spin glass is that a spin-glass transition at $T > 0$ is generally believed to occur,[3,36,53,63,64,66-69,79,81-85] while there is only spin-glass order at $T = 0$ in $2D$. [110] However, it is worth mentioning some recent extensive Monte Carlo work by Marinari et al. [60] where the question of spin-glass order at $T > 0$ has been questioned.

As in the mean-field theory, there is no reentrant behavior in either $2D$ (with a FM→PM transition) or in $3D$ (with a FM→SG transition).

4.5. N-Component Vector Models

As described above, there appears to be no reentrant spin-glass transition from either a ferromagnet to a spin glass (in $3D$) or from a ferromagnet to a paramagnet (in $2D$) upon a decrease of the temperature in the Ising models. These results agree with the experimental findings that Ising systems do not show a reentrant transition. [112] This failure to find reentrance in Ising models might suggest that some relevant ingredients leading to reentrance are missing in the Ising models (see, however Ref. [17]) One possibility is that reentrance occurs in isotropic vector spin systems because of the not-yet-understood behavior of transverse fluctuations. The second is that couplings between the longitudinal and transverse degrees of freedom, such as those present in dipole-dipole interactions, are needed to induce reentrance. In this section we review some of the numerical work on $2D$ and $3D$ vector spin glasses where these two hypothesis were tested.

The first model we describe is the isotropic m-component bimodal spin glass model with Hamiltonian

$$H = \sum_{<i,j>} -J_{ij}\, \vec{S}_i \cdot \vec{S}_j \ , \tag{10}$$

where the sum is taken over the nearest-neighbors of a d-dimensional cubic lattice and $\vec{S}_i$ is a classical m-component vector of unit length. The exchange interactions, J_{ij}, are random with a biased bimodal probability distribution, $\mathcal{P}(J_{ij})$, given by

$$\mathcal{P}(J_{ij}) = x\delta(J_{ij} - J_0) + (1 - x)\delta(J_{ij} + \lambda J_0) \ . \tag{11}$$

This is one of the simplest model that can be used to investigate the competition between paramagnetic, ferromagnetic, and spin glass order in a vector system.

The case of the three-component ($m = 3$, Heisenberg) model has attracted much attention recently, in particular in the context of the slightly doped antiferromagnetic cuprate high-T_c superconductors. [113,114,115] In these systems, the localization of holes in the slightly doped insulating materials generates effective ferromagnetic bonds giving rise to random frustration and to the spin glass properties observed in the intermediate region between the antiferromagnetic and the superconducting portion of the phase diagram. [116,117,118,119]

The XY model ($m = 2$) has also been much studied in recent years.[33,50,51,120−125] Results on a $2D$ XY model similar to the one above, and suggestive of a reentrant behavior from an algebraically ordered phase to a paramagnemetic phase, have been observed by using an iteration of local mean-field equations. [33] Also, the possible existence of a chiral glass [50,75,77,125] showing a true spin glass transition in $3D$ at finite temperature has been investigated, but no concensus has yet emerged. Still, the critical exponents for the thermodynamic quantities probing the chiral order are different then those associated to the spin variables, even in $2D$, where both the spins and chiral variable freeze at $T = 0$. [51] Nevertheless, most numerical studies (Monte Carlo,[34,35,46−48,53] defect-wall energy,[70,73,79] and Migdal-Kadanoff calculations[73,86−89]) agree that there is no freezing transition in the spin degrees of freedom at $T > 0$ in $2D$ and $3D$ XY and Heisenberg spin glasses.

The temperature disorder phase diagram of the $3D$ bimodal XY ($m = 2$) model has been studied using the Migdal-Kadanoff method, with the phase diagram presented in Figure 10. [89] No spin glass transition occurs at $T > 0$ and there is no ferromagnetic to paramagnetic reentrance.

The temperature disorder phase diagram of $3D$ Heisenberg ($m = 3$) model has been studied using Monte-Carlo simulations. [49,48,52] The phase diagram is shown in Figure 11. Unlike in the Migdal-Kadanoff method, the Monte Carlo method allows investigation of the behavior of the spin components perpendicular to the direction of the magnetization. It is observed that for any amount of disorder, there is a

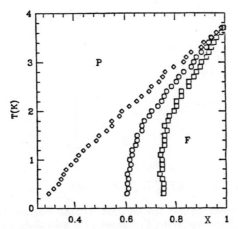

Fig. 10.   Temperature (T) versus impurity concentration (x) phase diagram of the $3D$ bimodal XY model as obtained by the Migdal-Kadanoff method (Eqs. (10) and (11)). The temperature, T, is measured in units of J_0/k_B. The boundaries separate the paramagnetic phase (P) at high temperatures from the low temperature ferromagnetic phase (F). The diamonds, circles and squares are for $\lambda = 0.0$, 0.3 and 0.7 respectively. No spin glass phase at $T > 0$ occurs in this $3D$ model, and no ferromagnetic to paramagnetic reentrance occurs either. (From Ref. 89).

dynamic freezing transition observed in the transverse spin components and not a true thermodynamic transition. This is qualitatively similar to what happens in mean-field theory at the true thermodynamic FM-M1 transition discussed in Section 2.

As we will discuss in Section 5, a large number of amorphous ferromagnets do not show a transverse spin freezing for an amount of disorder less than some threshold value. In a recent paper, Nielsen et al. [126] argue that this could be due to site frustration rather than bond frustration as present in a model described by Eqs. (10) and (11). In the site frustrated model described in Ref. 126, some spins are coupled to their neighbors via an even number of antiferromagnetic couplings only, with no ferromagnetic ones. In this case, the frustrated sites "flip down" in the opposite directions of the percolating ferromagnetic cluster, hence interacting minimally with the ferromagnetic network. Transverse spin freezing appears only for a critical concentration of frustrated sites when these latter start to interact among themselves. [126]

Another model we consider applies to a metallic (nonmagnetic) system doped with magnetic impurities with classical magnetic moment $\vec{S}_i$ whose orientation is constrained in the xy plane. The Hamiltonian is

$$H_{DM} = - \sum_{\langle i,j \rangle} J_{ij} \vec{S}_i \cdot \vec{S}_j - \sum_{\langle i,j \rangle} D_{ij} \; \hat{z} \cdot \vec{S}_i \times \vec{S}_j \; , \tag{12}$$

where $\hat{z}$ is a unit vector perpendicular to the XY plane and $\vec{S}_i$ is of unit length. The sum is taken over nearest-neighbor sites of a d-dimensional cubic lattice. The presence

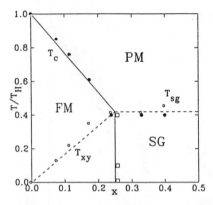

Fig. 11. Temperature (T) versus impurity concentration (x) phase diagram of the $3D$ bimodal Heisenberg spin glass model. The temperature, T, is measured in units of J_0/k_B. The boundaries separate the paramagnetic phase (PM) at high temperatures from the low temperature ferromagnetic phase (FM). The PM to spin glass (SG) transition here is a dynamical transition rather than a true thermodynamical transition. The spin glass phase in this model is believed to occur at $T = 0$ only. [35,73] Also, there is a dynamical freezing transition in the transverse components of the spins within the ferromagnetic phase. (From Ref. 48).

of additional randomly positioned spin-orbit impurity scatterers are responsible for the random Dzyaloshinskii-Moriya interactions, D_{ij}. [3,5] We want to emphasize that H_{DM} is rotationally invariant. This is not the case for the equivalent Heisenberg Hamiltonian where $D_{ij}\hat{z}$ becomes a three-component vector, $\vec{D}_{ij}$, with random orientation. We take D_{ij} to be given by a Gaussian probability distribution, $\mathcal{P}(D_{ij})$, with zero mean and width D

$$\mathcal{P}(D_{ij}) = \frac{1}{\sqrt{2\pi D^2}} \exp\{-(D_{ij})^2/(2D^2)\} \quad . \tag{13}$$

If the magnetic impurities are randomly distributed, one would also expect the exchange interaction, J_{ij}, to vary in sign due to the usual oscillatory RKKY interaction. Here we take the average distance between the magnetic atoms to be given by the lattice spacing and adjusted such that J_{ij} are ferromagnetic ($J_{ij} = J_0 > 0$). This choice is not essential as random J_{ij} are generated under renormalization. [89,90] It is useful to introduce the angle ϕ_i that $\vec{S}_i$ makes with respect to a fixed axis and rewrite Eq. (12) in the following form

$$H_{DM} = \sum_{<i,j>} -K_{ij} \cos(\phi_i - \phi_j - A_{ij}) \tag{14}$$

where $K_{ij} = \{J_0^2 + D_{ij}^2\}^{1/2}$ and $\tan(A_{ij}) = D_{ij}/J_0$. Written in this form, H_{DM} is now a lattice model whose Hamiltonian describes a random array of superconducting grains in a magnetic field and coupled via the Josephson coupling K_{ij}. Within this interpretation of the model, A_{ij} is the line integral of the vector potential, $\vec{A}$, from one grain to the other $(A_{ij} = 2\pi\Phi_0^{-1} \int_i^j \vec{A}(\vec{r}) \cdot dl)$, and Φ_0 is the elementary flux quantum, $\Phi_0 = hc/2e$. This model for the random Josephson junctions has recently been the subject of an intense study in the limit of strong disorder. [40,78,90,91,127,128,129] In the limit of extreme disorder, where A_{ij} undergoes large variations from one pair of grains to the other, Eq. (14) is believed to be an adequate model to describe the vortex glass phase in type-II superconductors with point disorder. [127,128] The paramagnetic, ferromagnetic and spin glass phases in the magnetic problem translate into the normal, superconducting and vortex-glass phases respectively for the superconducting problem. Here we concentrate on the regime of small disorder where $D < J_0$ in order to investigate the competition between ferromagnetic (superconducting) and spin glass (vortex glass) order, and the eventual existence of a ferromagnetic to spin glass reentrance in this regime.

A $2D$ version of Eq. (12) has been studied by Rubinstein, Shraiman and Nelson [130] using Kosterlitz-Thouless-like renormalization-group equations. [92,131] More recently, Paczuski and Kardar [132] have extended the work of Rubinstein et al. by studying the effect of nonrandom symmetry-breaking anisotropies. The random $2D$ Josephson-junction formulation of the model, Eq. (14), has been investigated by Granato and Kosterlitz. [133] The main interesting result which comes out of the above theoretical studies based on Kosterlitz-Thouless renormalization-group equations is the prediction of a reentrant transition from an ordered phase to a paramagnetic phase for any, but nonzero, value of the disorder D. By ordered ferromagnetic phase in the $2D$ XY model we mean a phase with quasi-long-range order with a spin-spin correlation function decaying as a power law with distance. For D larger than some critical value, D_c, the ferromagnetic phase is absent and the system is paramagnetic at all temperatures. See Figure 12 for a schematic representation of the predicted $D - T$ phase diagram for the $2D$ Dzyaloshinskii-Moriya XY model.

The Dzyaloshinskii-Moriya XY model is an interesting model to study for the following three reasons. Firstly, this model contains random couplings between the longitudinal and transverse spin components due to the random D_{ij} or A_{ij} quantities. Hence, this model allows us to investigate the proposal that such random longitudinal-transverse couplings may be responsible for reentrant behavior in vector spin systems. Secondly, we have at our disposal a theoretical prediction of reentrance from the rather rigourous Kosterlitz-Thouless renormalization-group framework. [130,132,133] Furthermore, a $2 + \epsilon$ expansion of the $2D$ renormalization-group equations of Ref. [130] predicts reentrance for $\epsilon > 0$. This suggests that reentrance may indeed occur for model H_{DM} in $3D$. Finally, and perhaps most importantly, the Josephson-junction formulation of the problem is attractive since experiments on such systems with a controllable amount of disorder can be performed with the theory being checked accurately. [134]

The phase diagram for the $3D$ Dzyaloshinskii-Moriyia XY model, as obtained by the Migdal-Kadanoff method, is shown in Figure 13, [89] where now, unlike in the bimodal $3D$ XY model, a spin glass phase at nonzero temperature is observed. [40,78,129]

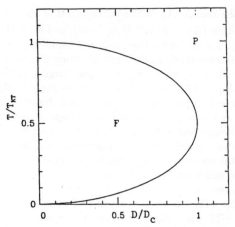

Fig. 12. Schematic representation of the disorder, D, temperature, T, phase diagram for the two-dimensional XY ferromagnet with random Dzyaloshinskii-Moriya interactions. T_{KT} and D_c are the Kosterlitz-Thouless transition temperature in the absence of disorder and the critical amount of disorder to observe an ordered phase. (Phase diagram adapted from Ref. 130).

The slope of the phase boundary between the ferromagnetic and spin glass phase is, within its width, slightly negative, suggesting a possible transition from a spin glass to a ferromagnetic phase upon decreasing the temperature. However, as we will discuss in Section 6, this is probably an artifact of the method, as such a spin glass to ferromagnetic transition is formally forbidden in most gauge glasses like the bimodal Ising spin glass with $\lambda = 1$ and a slight generalization of H_{DM}. The reentrant transition predicted for this model in $2D$ is expected to occur for any nonzero D. [130,132,133] Our results are certainly not compatible with such behavior in $3D$.

We are not aware of Monte Carlo results on this $3D$ model in the weakly frustrated regime $D/J_0 \ll 1$. However, in the limit of strong disorder ($\mathcal{P}(A_{ij}) \in [0, 2\pi]$), Monte Carlo, [40,129] and defect-wall calculations [78] suggest the possibility of a thermodynamics transition at $T > 0$ in $3D$. However, experimental results [134] and Monte Carlo simulations [41,42] on random two-dimensional Josephson junction systems in a magnetic field have failed to observe the reentrant behavior predicted in the theoretical studies. [130,132,133]

Thus, the $2D$ Monte Carlo phase diagram has the same topology as in Figure 13, with the spin glass phase absent. Although explanations for the absence of reentrance based on too small system sizes in the simulations and on disorder-induced vortex pinning in both experiments and simulations have been proposed, [42] a convincing argument for the discrepancy between the simulations and the theory is lacking. It has recently been proposed that the $2D$ version of the model described by Eqs. (12) and (13) may be paramagnetic at all temperatures, except for D strictly equal to zero. [135]

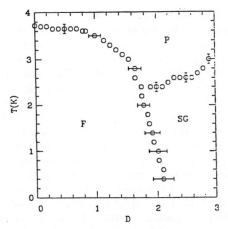

Fig. 13. Temperature (T) versus disorder D) phase diagram of the $3D$ XY ferromagnet with random Gaussian Dzyaloshinkii-Moriya interactions (H_{DM}) as obtained by the Migdal-Kadanoff method (Eqs. (12) and (13)). The temperature, T, is measured in units of J_0/k_B. The uncertainty on the location of the phase boundaries is shown by the error bars. Paramagnetic (P), ferromagnetic (F) and spin glass (SG) phases are found. (Taken from Ref. 89).

We conclude that there does not presently appear to be any experimental or numerical evidence supporting the existence of reentrant behavior in the $3D$ version of an XY ferromagnet with random Dzyaloshinskii-Moriya interactions. In two dimensions the situation is less clear. Indeed, a recent work by Korshunov claims that the $2D$ XY magnet is unstable to the presence of any random Dzyaloshinskii-Moriya interactions, with short-range magnetic order for all $T \geq 0$. [135] The failure of the Kosterlitz-Thouless-like theory, if Korshunov's claim is indeed correct, is that the Kosterlitz-Thouless equations used to predict reentrance are derived within an expansion to the lowest order in a *nonrandom* vortex fugacity, y. However, the microscopic disorder does lead to a microscopic random $y(\vec{r})$ that should be considered within the theory. Such random contribution to the bare $y(\vec{r})$ may generate an infinite number of dangerously relevant operators destroying the algebraically ordered phase for any $D > 0$. Another effect which has not been taken into account in any theoretical work is the effect of the spatial correlations between the randomness in $y(\vec{r})$ and the random D_{ij}. It might be that the strongest D_{ij} bonds are "screened" by localized vortices due to the local increase of $y(\vec{r})$ correlated to the strong D_{ij}'s, and that the destruction of the algebraically ordered phase is prevented via these nontrivial correlations. A defect-wall energy study of H_{DM} and calculations of the spin stiffness, spin-glass stiffness, ground state chaoticity level [74] and vortex density suggest an ordered ground state at small disorder D, with a $T = 0$ transition to a $2D$ gauge glass beyond a critical value D_c. [136] This result would argue against the Kosterlitz-Thouless theories finding reentrance, [130,132,133] and also against the claim for the absence of an ordered state at any nonzero $D > 0$. [135] Clearly, more analytical and numerical work is needed to further our understanding of this interesting model.

We now comment on the relevance of the type of disorder considered for the paramagnetic to ferromagnetic transition in the bimodal vector spin glass and the Dzyaloshinskii-Moriya XY spin glass. The continuum description of the bimodal XY model is known as the random-T_c XY model. The Harris criterion [95,96] states that a small amount of disorder of this type should be irrelevant at the PM−FM transition, since the specific heat exponent, α, is slightly negative for the pure XY model ($\alpha \approx -0.01$). [137] Thompson et al. have, using a Monte Carlo simulation, confirmed the previous statement for a bimodal Heisenberg model for which they find the same critical exponents as for the pure (nonrandom) Heisenberg ferromagnet which has $\alpha \approx -0.115 < 0$. [48] In the case of the $3D$ Dzyaloshinkii-Moriya XY model, Hertz has shown that the kind of disorder considered in Eq. (14) is irrelevant near the usual Gaussian fixed point in four dimension. [138] From these results, we expect the $3D$ bimodal XY model and $3D$ Dzyaloshinkii-Moriya XY model to have a stable ferromagnetic phase at finite temperature in three dimensions, and the paramagnetic to ferromagnetic phase transition to lie in the universality class of the pure ferromagnetic XY models for small disorder ($x \approx 1$ and $D \ll 1$). However, recent large scale simulations of a classical $3D$ Heisenberg model describing the dilute insulating frustrated ferromagnet $Eu_xSr_{1-x}S$ suggest that for strong disorder the critical behavior crosses over to a new type of critical behavior. This result is different from that of the pure model, in contrast to the prediction of the Harris criterion. [139] Similar results have been found in Migdal-Kadanoff calculations of random Potts models. [140] It is interesting to note that such behavior has been observed in the heat capacity measurements of $Eu_xSr_{1-x}S$. [141] One should note that T_c for pure EuS is $\approx 16K$, and since the spin value of Eu^{2+} is $S = 7/2$, dipole-dipole interactions in that system are nonnegligible compared to T_c. The system is therefore not truly Heisenberg like. It would be interesting to investigate the effects of the important dipolar interactions on the critical behavior of $Eu_xSr_{1-x}S$.

5. "Recent" Experimental Investigations of Reentrant Spin Glasses

In this section we review results from various recent experimental studies of reentrant spin glasses. We refer the reader to previous experimental reviews. [6–10] As discussed briefly in Section 2, a large number of disordered ferromagnets exhibit the reentrant spin-glass syndrome below some transition temperature T_g. At $T < T_g$ the system develops history dependence in magnetization experiments, and, which is accompanied by anomalies in neutron diffraction. The first detailed experiments on these systems were done by Coles and the Imperial College group at the end of the 1970s. [142–146]

The results on reentrant spin glasses were first interpreted in terms of a model in which the system splits into a percolating ferromagnetic cluster of spins below the Curie temperature T_c, plus small "decoupled" paramagnetic spin clusters. At a lower temperature, $T_g < T_c$, the small spin clusters freeze and affect the ferromagnetic order of the nearby ferromagnetic network. This description has been widely used, even in recent neutron diffraction experiments of amorphous FeMn [147] and muon spin resonance spectroscopy of $Fe_{80-x}Ni_x(Cr_x)_{20}$. [148]

A very different picture comes out of the mean-field theory for Heisenberg spin-glasses as presented in Section 3. In that description, the system develops collinear ferromagnetic order characterized by a magnetization below T_c, but at T_{xy}, the transverse (XY) components of *each* individual spin freeze in a random transverse direction. In other words, the moments become locally canted with respect to the original magnetization direction. The system can be viewed as a longitudinal ferromagnet in the $\hat{z}$ direction with spin-glass-like order in the XY directions. Monte Carlo simulations of the $3D$ bimodal Heisenberg model (Section 4.5) suggest that such behavior, at least in the sense of a dynamical freezing of the transverse components, also occurs. [48,126]

In the following subsections we review recent experimental results which suggest that in the majority of cases, the behavior of the system is well accounted for by the transverse spin freezing picture when slightly modified to account for short-range effects and random anisotropy. In other words, the glass transition temperature, T_g, occuring at $T_g < T_c$, corresponds to the freezing temperature, T_{xy}, of the transverse spin components:

$$T_g = T_{xy} \ .$$

This is essentially the mean-field picture described in Section 3. However, as discussed in the following subsections, there are some interesting cases where the system does not appear to behave as in the transverse spin freezing mean-field picture. Indeed, there are possibly some conceptual difficulties with the transverse spin freezing picture, even at the qualitative level, when the system is tuned close to its multicritical paramagnetic-ferromagnetic-spin-glass (PM-FM-SG) multicritical point.

5.1. Neutron Depolarization Experiments

In an unsaturated ferromagnet, in zero or small magnetic field, one observes the formation of micron size domains due to dipolar interactions: the system breaks into domains in order to decrease the demagnetization energy. The spatial arrangement of the domains is a complicated problem and results from the competition of all the large *and* small energies in the system such as the exchange, magnetostatic, magnetocrystalline, anisotropy and magnetostriction.

The neutron depolarization technique allows a study of domains in well ordered ferromagnet. In such experiments, a polarized neutron beam passes through the sample and becomes partially depolarized due to the Larmor precession experienced by the neutron as it follows the nonadiabatic change of the internal magnetic field when passing from one domain to the other. [8,12]

In a paramagnet or spin glass, the neutrons suffer no depolarization as the temporal (in the paramagnet) or spatial (in the spin glass) variations of the internal magnetic field are too fast to depolarize the neutrons. In a multi-domain ferromagnet

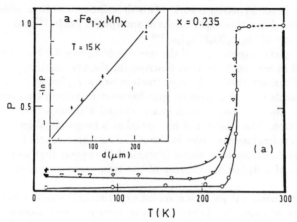

Fig. 14. Polarization versus temperature in a-Fe$_{1-x}$Mn$_x$ ($x = 0.235$) measured at several thicknesses, d. The crosses, inverted triangles, and circles are for $d = 25$, 100 and 200 μm respectively. (Taken from Ref. 12).

or an inhomogeneous ferromagnet (e.g. a ferromagnetic-like clusters embeded in spin glass regions), the neutron beam becomes depolarized and emerges from the sample with a reduced polarization, P. This exit polarization depends on (1) the value of the components of the magnetic field within a doman parallel and perpendicular to the incident neutron polarization, (2) the doman size, (3) the neutron wavelength and (4) the thickness of the sample. By studying the dependence of the exit polarization on the above parameters, and comparing with simple theoretical predictions, one can develop a physical picture of the nature of the internal magnetic domain structure in the material under study. For more details, we refer the reader to the recent review by Mirebeau et al. [8]

In weakly frustrated systems (far from the PM-FM-SG point), the polarization versus temperature of reentrant systems typically resembles Figure 14. The polarization decreases monotonically with decreasing temperature, indicating an increase of the intradomain magnetization. Furthermore, at fixed temperature, P decreases exponentially with the thickness of the sample (see inset of Figure 14). As there is no sudden increase of P upon a decrease of T, we conclude that no "true" reentrance occurs in the sense of either a decrease of internal domain magnetization or an increase of the number of domains. Interestingly, the temperature dependence of the polarization in low applied fields also shows irreversabilities and different results are obtained depending whether the experiment is a FC or ZFC one. [8,12]

One finds that the neutron depolarization does not behave the same for all strongly frustrated reentrant systems when the disorder x is tuned close to the PM-FM-SG point at $x = x_c$. For example, the material Ni$_{1-x}$Mn$_x$ shows a monotonously decreasing polarization with decreasing temperature for all concentration above the true spin glass regime as shown in Figure 15. [149,150] However, some systems such as

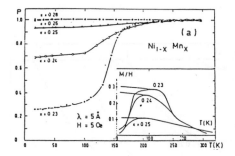

Fig. 15. Temperature dependence of the polarization in a-Ni$_{1-x}$Mn$_x$ ($x_c = 0.25$). (Taken from Ref. 8).

Au$_{1-x}$Fe$_x$ ($x_c = 0.15$) and a-Fe$_{1-x}$Mn$_x$ ($x_c = 0.35$) show a minimum in $P(T)$ as shown for in Figure 16. In these systems it would appear that the canted state is associated with a decrease of the intradomain magnetization.

Also, recent neutron depolarization measurements on the alloy Fe$_{82-x}$Ni$_x$Cr$_{18}$ (see phase diagram in Figure 17) coupled to muon spin resonance measurements of the distribution of internal field (see section 5.4) have allowed determination of the average size R_d of the magnetic clusters. [148] These results are shown in Figure 18 for $x = 26$. The domain structure ($R_d \approx 1000\text{Å} - 2000\text{Å}$) formed in the ferromagnetic region below T_c are argued to disintegrate into smaller clusters with a radius $R_d \approx 100 - 200\text{Å}$. [148]

An interesting system is Fe$_{70}$Al$_{30}$ where a strong increase of P is observed. From the neutron depolarization measurements it has been suggested that this material undergoes a transition onto a superparamagnetic (SP) phase before entering the spin glass phase as shown in Figure 19. [154]

5.2. Neutron Scattering Experiments

Several neutron studies have been performed on reentrant spin glasses. Previous small angle neutron scattering (SANS) experiments on strongly frustrated systems (close to x_c), and performed as a function of decreasing temperature, showed an increase in the scattered intensity at temperatures below the high temperature anomaly associated with the critical scattering peak at the PM-FM transition. The occurence of this intensity rise was first attributed to the broadening of the Bragg peak at the zone center (longitudinal correlations). Consequently, the inverse of this width, the correlation length, was found first to increase for $T < T_c$, up to some maximum value, and then to dimisnish upon further decreasing the temperature below T_g. This behavior was interpreted in terms of a rapid collapse of the average domain size. [147]

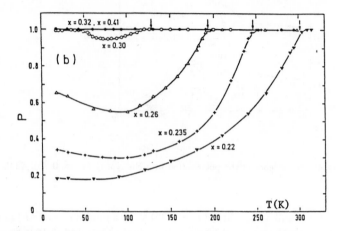

Fig. 16. Polarization versus temperature in a-$Fe_{1-x}Mn_x$ measured at several values of the disorder x ($x_c = 0.32$). (Taken from Ref. 12).

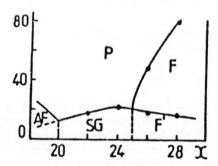

Fig. 17. Phase diagram of the alloy $Fe_{82-x}Ni_xCr_{18}$. (Taken from Ref. 148).

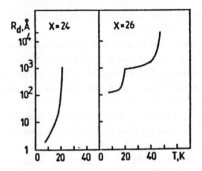

Fig. 18. Temperature dependence of the domain radius R_d in $Fe_{82-x}Ni_xCr_{18}$ in the spin glass phase ($x = 0.24$) and in the reentrant regime ($x = 0.26$). (Taken from Ref. 148).

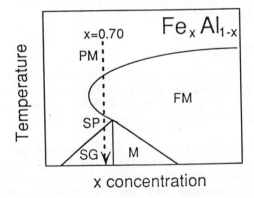

Fig. 19. Schematic phase diagram of Fe_xAl_{1-x} where the location of the sample studied in Ref. 154 is indicated by an arrow. The symbols PM, FM, SG and SP refer to paramagnetic, ferromagnetic, spin glass and superparamagnetic, respectively. (Taken from Ref. 154).

This behavior is similar to the one seen in neutron depolarization experiments mentioned above for some materials close to the multicritical point. The problem with this interpretation is that the two length scales extracted from the SANS and neutron depolarization experiments differ by two order of magnitudes, $10\mathring{A}$ against $1000\mathring{A}$ in SANS and neutron depolarization respectively.

The first neutron experiments on reentrant spin glasses were done in zero applied magnetic field. In such a case, it is impossible to investigate the possibility that the temperature dependence of the scattered intensity is due to a freezing of the transverse spin components rather than a modification in the correlation of the longitudinal components as has been assumed in other studies. [147]

A study of the anisotropy of the scattering under an applied magnetic fields demonstrates that the intensity is largely due to the freezing of the transverse spin components. [8,151] Furthermore, unlike what is found in systems in the true spin glass regime, spin-waves have now been observed in several reentrant spin glass systems. [8,151,152,153] Well-defined magnon-like excitations remain present *below* T_g. However, the temperature dependence of the magnon stiffness is not well understood and appears to vary from system to system. This implies that above T_g the magnetization is uniform within domains with no transverse order. There is only dynamical canting on very short time scales of the spin-wave type. Below T_g, there is a freezing of the transverse components with ferromagnetic short-range ordering on the scale of a few or tens of angstöms. This short-range scale is believed to be related to the typical distance between vortices in the spin configuration, and has been shown to vary with the strength of the applied field. [8,151]

5.3. Mössbauer Spectroscopy

The Mössbauer spectroscopy method is based on γ-ray absorption of the nuclear moments in the samples. By varying the velocity (in the lab-frame) of the γ source $(O\{10^1 mm/s\})$, one can Doppler shift the emitted photon to achieve a resonant absorption with the various nuclear transitions of the nuclear moments of the material under study.

For all the systems studied there are no signs of an absorption line around zero velocity below T_c and which would be there if there were decoupled paramagnetic clusters. Below T_g, new hyperfine transitions with $\Delta m_I = 0$ appear, signalling the freezing of the transverse spin components, as suggested by the neutron depolarization and neutron scattering experiments. Figure 20 shows that the average hyperfine field in $(Fe_{65}Ni_{55})_{1-x}Mn_x$ $(x = 0.113)$ measured by Mössbauer spectroscopy increases abruptly below $T_g \approx 60K$. This abrupt increase of the average hyperfine field is due to the freezing of the transverse spin components at T_g. [7]

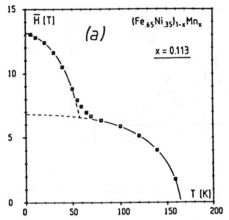

Fig. 20. Temperature dependence of the average hyperfine field in $(Fe_{65}Ni_{35})_{0.887}Mn_{0.113}$ showing the abrupt increase at $T_g \sim 60K$. (Taken from Ref. 7).

5.4. Muon Spin Resonance Experiments

In recent years the muon spin resonance (μSR) technique has become a very powerful tool to investigate the dynamics of the spin freezing in various types of magnetic systems. [155] This method consist of injecting a muon beam into a material. The (positive) muon has a half-life time of $\approx 2\mu s$ and disintegrates into a positron, an electron neutrino and a muon antineutrino. In the presence of a local magnetic field, the muon precesses at the muon Larmor frequency with respect to a direction determined by the total local magnetic field. Since the positron is ejected preferentially in the instantaneous direction of the muon spin at the moment of disintegration, a detection of the direction of the outgoing positron with scintillation counters allows characterization of the probability distribution of the local magnetic field at the site(s) occupied by the muon. By modelling the time dependence of the asymmetry in the number of positrons emitted in the direction of the incoming beam and opposite to it, one can extract the temperature dependence of the muon depolarization relaxation rate, Λ, as well as the mean value B_0 and the dispersion Δ of the local magnetic fields. The μSR technique has been used in the study of spin glasses [156,157] and reentrant spin glasses. [148,158,159]

Figures 21 and 22 show the temperature dependence of Λ, B_0 and Δ for the alloy $Fe_{82-x}Ni_xCr_{18}$ with $x = 26$ (phase diagram in Figure 17). The dynamical relaxation rate, Λ shows two distinct peaks, which are associated with two well separated spin freezing transitions. There is a rise of B_0 at T_c, followed by a saturation once deep inside the ferromagnetic regime. At temperatures below T_g, B_0 increases rapidly again as for the hyperfine field measured in Mössbauer. The temperature trend of the dispersion of the local field, $\Delta(T)$, is the same as for B_0. Similar behavior has very recently been found in the strongly frustrated system a-$Fe_{1-x}Mn_x$ at $x = 0.30$

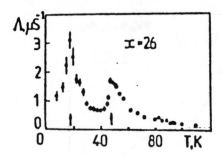

Fig. 21. Temperature dependence of the dynamics relaxation rate $\Lambda(T)$ in $Fe_{82-x}Ni_xCr_{18}$ with $x = 26$. (Taken from Ref. 148).

$(x_c = 0.35)$. [159]

It is not immediately obvious from these μSR results that the quality of the data truly allows a quantitatively accurate determination of both the mean local field B_0 and the dispersion Δ. Indeed, we see that at $T = T_c$ for $Fe_{82-x}Ni_xCr_{18}$, there is a rise of Δ which is not accompanied by a rise in B_0. This should not be the case if the system does develop partial collinear order at T_c. This possibly reflects on the difficulty associated with the data analysis in that temperature regime. Once in the ferromagnetic regime $T/T_c < 0.8$, we see that $\Delta \sim B_0$. This shows that this system has a fair amount of dispersion in the local magnetization, which also takes into account the various locations at which the muon can reside in this alloy. Below T_g, the rise of B_0 suggests that even though the transition under investigation is a transverse spin freezing, there is not complete cancellation of the superposition of the (dipolar) field generated by frozen transverse spin components of the neaby spins at the muon sites. This possibly reflects the short-range correlations existing between the transverse spin components below the freezing temperature T_g as shown in SANS in an applied magnetic field.

5.5. Lorentz Electron Microscopy

Most of the above experiments use a microscopic probe of some sort. These give experimental data which provide the desired information about the type of magnetic order only after a sophisticated data analysis (e.g. μSR). The Lorentz electron microscopy (LEM), however, provides direct visual observation of the domain structure. [160] Senoussi and coworkers used LEM to show that for various weakly frustrated reentrant alloys (i.e. far from critical disorder x_c), the domain structure in small applied field does not change when the temperature is lowered below T_g. [161] The average

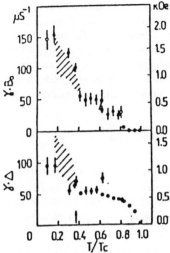

Fig. 22. Temperature dependence of the mean value of the internal field, B_0, and the dispersion, Δ in in $Fe_{82-x}Ni_xCr_{18}$ with $x = 26$. (Taken from Ref. 148).

domain magnetization does not go down for $T < T_g$. The domains are large, typically 10μm, with no signs of spin-glass cluster formation below T_g. These experiments do show directly that weakly frustrated reentrant spin glasses are ferromagnets at low-temperatures ($T < T_g$). Unfortunately, the LEM method cannot be used for systems close to the PM-FM-SG multicritical point since the domains become invisible when the domain size, R_d, becomes of the order of the domain wall thickness ($\sim 10^3 \mathring{A}$). This prevents a direct visualization of the domain structure in the interesting cases such as a-$Fe_{1-x}Mn_x$ ($x = 0.30$), $Fe_{70}Al_{30}$ or $Fe_{82-x}Ni_xCr_{18}$ ($x = 0.26$, Figure 18) where there are indications from other methods of a possible decrease of the intradomain magnetization and of the domain size.

5.6. Hysteresis and Influence of Anisotropy

As mentioned in Section (2), a key signature of reentrant behavior is the appearance of history dependence with different low-field responses in field-cooled (FC) and zero-field-cooled (ZFC) susceptibility measurements at low temperatures. In ferromagnets, the low-field response is due to domain wall movements, and the magnetic permeability is reduced if this motion is inhibited. The transverse spin freezing discussed above is thought to lead to this increased coercivity.

In amorphous metallic alloys, the Dzyaloshinsky-Moriya (DM) interaction is an important energy correction to the exchange Hamiltonian. The DM energy reads:

$$H_{DM} = C_{so} \sum_{(i,j)} \vec{D}_{ij} \cdot \vec{S}_i \times \vec{S}_j \; . \tag{15}$$

Here $\vec{D}_{ij}$ is a random vector in real space and C_{so} is a constant which depends on the spin-orbit coupling of the alloy. This interaction for the case of a $2D$ XY model (two components spins) was discussed in Section 4.5. To understand the role of the DM anisotropy and its effect below the transverse spin freezing, we decompose the vector spins into longitudinal components S_i^l parallel to the intradomain magnetization, and transverse components $\vec{S}_i^t$, and isolate the term in H_{DM} which couples directly within a mean-field approximation to the intradomain magnetization M_d: [160]

$$H_{DM} = -C_{so} \sum_i \vec{S}_i^l \cdot \left[\sum_j \vec{D}_{ij} \times \vec{S}_j^t \right] + H_{as}^t \ . \tag{16}$$

H_{as}^t is the antisymmetric contribution (in the spin components) to H_{DM}, arising from the coupling between the transverse components of $\vec{S}_i$ and $\vec{S}_j$. We absorb this term into the random exchange part of the Hamiltonian for the transverse components.

Within a mean-field approximation, S_i^l is virtually constant inside a domain, and is proportional to $\vec{M}_d$. We have

$$E_{DM} \propto C_{so} \sum_{domains} \sum_j \vec{h}_{dj} \cdot \vec{M}_d \tag{17}$$

where $\vec{h}_{dj} = \left[\vec{D}_{ij} \times \vec{S}_j^t \right]$, is a local anisotropy field.

In the ferromagnetic phase, the transverse components, $\vec{S}_i^t$, precess very rapidly so that on average $\vec{h}_d \approx 0$. However, as the temperature is lowered below $T_g = T_{xy}$, the transverse components freeze out, locked to the longitudinal components via the DM interactions. Because of the DM anisotropy, the frozen $\vec{S}_i^t$ are not randomly distributed, the DM anisotropy is not zero, and $\vec{h}_{dj}$ can be decomposed as an average unidirectional anisotropy, $\vec{H}_a$, parallel to the intradomain magnetization, and a spatially random term $\Delta \vec{h}_{dj}$:

$$\vec{h}_{dj} = \vec{H}_a + \Delta \vec{h}_{dj} \ . \tag{18}$$

Both $\vec{H}_a$ and $\Delta \vec{h}_{dj}$ are responsible for the increased coercivity phenomena below T_{xy}. The microscopic transverse spin freezing transition thus leads to the hysteresis observed at low temperatures in weakly frustrated reentrant spin glasses.

5.7. Discussion

In this section, we have presented results from several experiments which strongly suggest that a large proportion of ferromagnets exhibiting reentrant spin glass behavior display a spin freezing of the transverse spin components perpendicular to the direction of the intradomain magnetization established at $T = T_c$. Such behavior is compatible with the predictions of mean-field theory for the Heisenberg spin glass (Section 3). When the system under consideration is weakly frustrated (deep in the ferromagnetic regime and far from the PM-FM-SG multicritical point), the transverse spin freezing leads to zero or extremely small change of either the intradomain magnetization or of the domain size. The small random anisotropy, as introduced by the Dzyaloshinsky-Moriya (DM) interactions, leads to an intradomain uniaxial anisotropy below the transverse freezing which results in a strong increase in the domain wall mobility below T_{xy}. This explains the hysteresis that develops at low temperatures in these systems. It is also possible that dipolar effects alone, without invoking DM interactions, could explain the increased coercitivity below the transverse spin freezing.

The above scenario breaks down even qualitatively for a system sufficiently close to the PM-FM-SG multicritical point. In this case, one cannot disregard the strong coupling between the longitudinal and transverse spin components, so the decomposition done in Eqs. (15), (16) and (17) is invalid. In amorphous systems with Dzyaloshinskii-Moriya interactions, there cannot be true long-range order in less than four dimensions. This is because the Dzyaloshinskii-Moriya interactions are not rotationally invariant and lead to tranverse fluctuations that are infrared divergent in less than four dimensions. In other words, the Dzyaloshinskii-Moriya interactions introduce an Imry-Ma length scale beyond which ferromagnetic order is destroyed. [97] If the system is weakly frustrated, and the mean ferromagnetic exchange is large, the Imry-Ma scale is larger than the mean demagnetization domain size. However, it may be possible that for strongly frustrated systems with a large spin orbit constant, the Imry-Ma scale is larger than the demagnetization domain size for $T_g < T < T_c$, but becomes smaller for $T < T_g = T_{xy}$. This could explain the decrease of the intradomain magnetization observed in neutron depolarization experiments on systems with a large spin orbit constant.

Although our discussion has focused on random bond systems, there are other systems, such as amorphous random axis magnets, and which also exhibit reentrance. [105,106,162] Another interesting series of materials that has recently attracted much attention are the oxyde based manganese series $Y_2Mn_2O_7$ and $Lu_2Mn_2O_7$. These systems exhibit a behavior very similar to the random axis magnets, even though they are crystalline and the nominal off-stoichiometric disorder is extremely low. [163]

6. Theoretical Constraints on the Phase Diagram of Gauge Glass Models

In Sections 3 and 4 we showed that the ferromagnetic to reentrant transition has not been found in mean-field theory or in numerical simulations of $2D$ or $3D$ spin glass models. In Section 5 we presented results from various experiments suggesting that a large proportion of the systems investigated do not show true reentrance, that is a spontaneous loss of collinear feromagnetic order upon a decrease of temperature.

These results raise the question whether or not reentrance is indeed allowed in realistic short range $2D$ and $3D$ spin glass models. Recently, there has been some interesting progress made in that direction. In this section we briefly review the rigorous theoretical constraints that exist on the allowed topology for the phase diagram of spin glass models with frozen gauge invariance. We shall refer to this generic class of spin glasses as "gauge glasses". This section is somewhat more technical than the previous ones.

6.1. Nishimori Line and Phase Diagram of Gauge Glasses

Few exact results are available for spin glasses in finite dimensions. However, Nishimori found that for the bimodal Ising model with a probability distribution of exchange J_{ij}

$$\mathcal{P}(J_{ij}) = x\delta(J_{ij} - J) + (1 - x)\delta(J_{ij} + J) , \qquad (19)$$

there is a line in the $x - T$ phase diagram where the energy of the system can be calculated *exactly*, and is regular despite the fact that it crosses the ferromagnetic boundary. [164] The equation of the "Nishimori line" is

$$2x - 1 = \tanh(J/kT) \qquad (20)$$

where x is the concentration of ferromagnetic bonds.

It is the special choice of $\mathcal{P}(J_{ij})$ in Eq. 19 which allows the existence of a gauge invariance in the bimodal Ising spin glass system and consequently, permits the calculation of the the energy exactly on the Nishimori line in *any dimension*. This is a remarquable result given that one cannot even calculate the energy of simpler non-random ferromagnetic models except at zero temperature! By a suitable choice of a random bond probability distribution, one can construct other spin glass models with a different spin symmetry that also have a Nishimori line. [165,166] For example, discrete clock spin glass models [165] and XY gauge glass models (as in Eq. 12, but with a particular choice of random exchange J_{ij} and Dzyaloshinskii-Moriya couplings D_{ij}) can accomodate a Nishimori line, while the XY and Heisenberg spin glasses do not. [165,166]

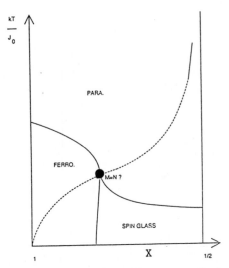

Fig. 23. Allowed phase diagram in a gauge glass with paramagnetic, ferromagnetic and spin glass phases. The system displays ferromagnetic to spin glass reentrance. The Nishimori line (dashed line) cannot enter the spin glass phase.

The existence of the underlying gauge invariance in gauge glasses allows for the construction of rigorous inequality. The most important ones state:

- The Nishimori line *cannot* lie within the spin glass phase of a gauge glass.

- The magnetization has to vanish for all temperatures (at constant disorder) whenever that disorder-temperature coordinate on the Nishimori line lies in the paramagnetic phase.

- The magnetization is nonzero on the Nishimori line whenever the line is not in the paramagnetic phase.

These constraints, imposed by the existence of the Nishimori line, imply that *the ferromagnetic boundary has to be either reentrant or strictly vertical* in all gauge glasses. Notice, however, that these constraints do not allow one to predict whether or not a spin glass phase occurs at $T > 0$ in a specific model, or what the critical exponents that characterize the various transitions are. The topology of some phase diagrams that are allowed in gauge glasses as well as some of those that are not are illustrated in Figures 23, 24, 25, and 26.

Obviously, the Nishimori line must intercept the paramagnetic-ferromagnetic boundary at some point, $\mathcal{N}$, corresponding to some temperature-disorder coordinate. It was suggested by Georges et al. that the location of the point $\mathcal{N}$ corresponds

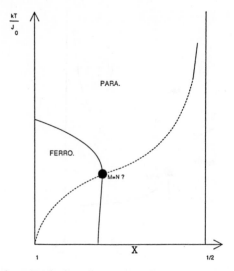

Fig. 24. Allowed phase diagram in a gauge glass with paramagnetic and ferromagnetic phases, but no spin glass phase. The system displays ferromagnetic to paramagnetic reentrance.

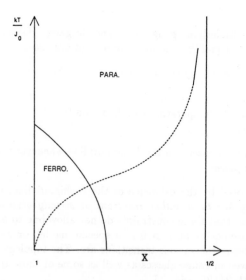

Fig. 25. Phase diagram that is not allowed for a gauge glass. There is a range of disorder x for which the system is ferromagnetic at low temperature while it is paramagnetic at higher temperature on the Nishimori line.

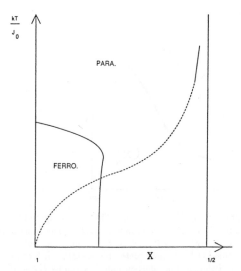

Fig. 26. Phase diagram that is not allowed for a gauge glass. There is a range of disorder px for which the system is ferromagnetic at high temperatures while it is paramagnetic at lower temperatures on the Nishimori line.

exactly to the location of the multicritical point, $\mathcal{M}$, where the paramagnetic, ferromagnetic and spin glass phases meet. [167] Later, Le Doussal and Georges, [165] and Le Doussal and Harris, [168] presented compelling renormalization-group and symmetry arguments for the multicritical PM-FM-SG point $(\mathcal{M})$ to lie on the Nishimori line. This was confirmed by Singh within a high-temperature series expansion of the $3D$ bimodal Ising spin glass. [64] One of the important results that came out of Le Doussal and Georges, [165] Le Doussal and Harris, [168] and Singh [64] studies is that the Nishimori line is one of the scaling axis of the renormalization-group flow, while the second scaling axis is parallel to the temperature axis. This result made plausible the suggestion that the whole ferromagnetic−spin-glass boundary in gauge glasses would be strictly vertical, emerging from $\mathcal{N} = \mathcal{M}$.

In two very recent papers, Kitani [169] and Ozeki and Nishimori [166] further pursued the use of gauge invariance ideas in gauge glasses. The most outstanding result coming from these two pieces of work is that *reentrance from a ferromagnetic to a paramagnetic, or from a ferromagnetic to a spin glass phase is excluded in all gauge glass models that admit a Nishimori line.*

We shall not present here a full discussion of the work of Kitani and Ozeki and Nishimori which involves somewhat technical gauge transformations. However, we briefly discuss the essence of their results from a physical point of view.

The proof obtained by Kitani [169] and Ozeki and Nishimori [166] relies solely on gauge invariance in gauge glasses admitting a Nishimori line, and without appealing

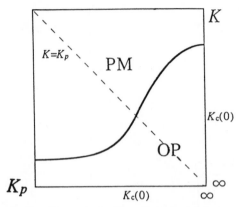

Fig. 27. Schematic phase digram in the $K_p - K$ plane. The bottom right of the figure $(K_p, K) = (\infty, \infty)$ is the zero temperature limit of the nonrandom system. The value of K_p increases from left to right while the value of K increases from top to bottom. The bold line is the boundary separating the paramagnetic phase (PM) to the ordered phases OP, which can be the ferromagnetic and/or spin glass phases. The Nishimori line (dashed line) is given by $K_p = K$.

to *any* explicit results from numerical studies.

Consider a gauge glass model where K_p is a single-valued function of the variable p that controls the amount of randomness. For example, take the bimodal Ising spin glass for which $p \equiv x$. In this case,

$$\exp\{-2K_p(x)\} = (1-x)/x .\tag{21}$$

A function K_p can be defined in all gauge glasses. Define also a variable $K = J_0/T$ where J_0 is some coupling constant setting the temperature scale T. Instead of drawing a $p - T$ phase diagram, one can draw a $K_p - K$ phase diagram as shown in Figure 27. This is convenient, because in this "representation" the condition for the existence of a Nishimori line in a gauge glass is

$$K_p = K .\tag{22}$$

The disordered phase is the paramagnetic phase (PM), and both the ferromagnetic and spin glass phases (if a spin glass phase does indeed occur at $T > 0$ in the model considered) are the ordered phases (OP) located below the bold curve. The Nishimori line (dashed line) is given by $K_p = K$. The value of K_p at which the Nishimori line intercepts the PM-OP boundary (the point $\mathcal{N}$) is labelled $K_c(0)$. However, recall that one does not know (except in mean-field theory) where the PM-OP phase boundary is without performing some kind of numerical investigation.

One introduces a parameter a which parametrizes the line

$$K = K_p + a\tag{23}$$

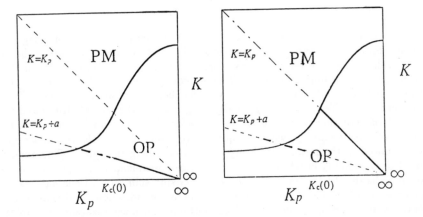

Fig. 28. Schematic phase diagram in the $K_p - K$ plane for the original model in (a) and the modified model in (b) with the parameter $a > 0$. The system cannot be ferromagnetic on the left of $K_p = K_c(0)$ since it would contradict the second constraint on the topology of gauge glasses described above. The system is ferromagnetic on the line $K = K_p + a$ with the magnetization identical to the one on $K = K_p$ in the modified model.

and which lies below the Nishimori line $K = K_p$. See Figure 28.

The important part of the argument is that all the points on the line $K = K_p + a$ from $K_c(0) < K_p \leq \infty$ in the original model lie (map) on the line $K = K_p$ of the "modified model with a", which has nonzero magnetization by gauge invariance *for all the points* $K_p > K_c(0)$. Since this argument applies for any value of the parameter a, and making the intuitively plausible assumption that no spin glass phase occurs *above* the line $K_p = K$ in the modified model, if follows by induction that a gauge glass cannot display reentrant behavior.

Ozeki and Nishimori also considered a slightly modified version of the $2D$ XY Dzyaloshinskii-Moriya problem discussed in Section 4.5, but which admits a Nishimori line. [166] Using the same type of gauge invariance arguments discussed above, they concluded that the occurence of reentrance was excluded in their version of the $2D$ XY Dzyaloshinskii-Moriya model. However, even if correct, this result still does not explain why the theoretical renormalization-group calculations fail. [130,132,133] Also, it does not enable one to discriminate between a situation in which the algebraically ordered phase extends for finite disorder, or whether the system is paramagnetic at all nonzero temperature for $D > 0$ as suggested by Korshunov. [135]

Finally, we mention an interesting paper of Nishimori [170] in which he conjectures that all bimodal spin glass models (e.g. Eqs. (10) and (11) with $\lambda = 1$) have a zero temperature ferromagnetic.to spin glass transition at a *universal* value of the disorder x in Eq. (11) that is independent of the number of spin components m. Numerical calculations based on the Migdal-Kadanoff scheme on hierarchical lattices appear to support this claim at the $1-2\%$ confidence level. [171]

7. Conclusion

In this chapter we have reviewed the problem of reentrance in random magnetic systems. From a theoretical and numerical point of view, there appears to be no strong evidence for true reentrant behavior, that is a disappearance of the equilibrium magnetization during cooling. The only counter example that we know of is the work of Viana and Bray. [17] Further work on their model would be useful. From the experimental side, it appears generally accepted now that weakly frustrated ferromagnets do not show a decrease of the intradomain magnetization, and that the behavior of the Heisenberg systems is qualitatively very close to the transverse spin freezing found in the mean-field theory of the Heisenberg spin glass. The appearance of irreversability effects below $T_g = T_{xy}$ can be explained by a pinning of the domain walls due to the anisotropic off-diagonal spin-spin interactions and via the transverse spin freezing transition. However, there are indications that some experimental systems display a decrease of the intradomain magnetization close to the paramagnetic-ferromagnetic-spin glass multicritical point.

8. Acknowledgments

Much of my own work reported in Section (4.5) was done with E. S. Sørensen; I thank him for a stimulating collaboration. I also thank A.J. Berlinsky, C. Kallin, K. Kojima, G. Morris, A. C. Shi, T. Uemura and W.D. Wu for other interesting collaborations on reentrant spin glasses. I am indebted to M. Hennion, I. Mirebeau, D.H. Ryan, D.J. Sellmyer, M.J. O'Shea, and A.P. Young for instructive and useful discussions. I wish to thank G. Aeppli, A. Aharony, D. Bensimon, R.J. Birgeneau, M. Cieplak, M. Grant, M. Kardar, J.M. Kosterlitz, D.R. Nelson, F. Nori, H. Orland, A.-M. Tremblay, and J. Vannimenus for useful discussions or correspendence. This work was supported by the Natural Sciences and Engineering Research Council of Canada and by the French Direction des Etudes et Recherches Techniques.

9. References

1. However, some collinear $3D$ antiferromagnets behave somewhat mysteriously. One famous current example is the weak moment heavy fermion magnet URu_2Si_2. See M.B. Walker et al., Phys. Rev. Lett., **71** (1993) 2630 and references therein.

2. G. Toulouse, Commun. Phys. **2** (1977) 115.

3. K. Binder and A.P. Young, Rev. Mod. Phys., **58** (1986) 801.

4. M. Mézard, G. Parisi and M.A. Virasoro, *Spin Glass Theory and Beyond*, (World Scientific, Singapore, 1987).

5. K H. Fischer and J.A. Hertz, *Spin Glasses*, (Cambridge University Press, Cambridge, 1991).

6. B.R. Coles, Phil. Mag. B, **49** (1984) L21.

7. D.H. Ryan, *Recent Progess in Random Magnets*, ed. D.H. Ryan (World Scientific, Singapore, 1992), p. 1.

8. I. Mirebeau, M. Hennion, S. Mitsuda, and Y. Endoh, *Recent Progess in Random Magnets*, ed. D.H. Ryan (World Scientific, Singapore, 1992) p. 41.

9. I.A. Campbell and S. Senoussi, Phil. Mag. B, **65** (1992) 1267.

10. B.R. Coles and S.B. Roy, *Selected Topics in Magnetism*, eds. L.C. Gupta and M.S. Multani (World Scientific, Singapore, 1993), p. 363.

11. H. Maletta and P. Convert, Phys. Rev. Lett., **42** (1979) 108; H. Maletta, G. Aeppli and S.M. Shapiro, Phys. Rev. Lett., **48** (1982) 1490.

12. I. Mirebeau, S. Itoh, S. Mitsuda, T.Watanabe, Y. Endoh, M. Hennion, and R. Papoular, Phys. Rev. B, **41** (1990) 11405.

13. D. Sherrington and S. Kirkpatrick, Phys. Rev. Lett., **35** (1975) 1792; S. Kirkpatrick and D. Sherrington, Phys. Rev. B, **17** (1978) 4384.

14. G. Toulouse, J. Phys. (Paris) Lett., **41** (1980) L447.

15. Yu A. Fedorov, I. Ya Korenblit and E.F. Shender, J. Phys. Cond. Matter, **2** (1990 1669 and references therein.

16. M. Gabay and G. Toulouse, Phys. Rev. B, **B5** (1981) 8454.

17. L. Viana and A.J. Bray, J. Phys. C, **18** (1985) 3037.

18. J.S. Walker and C.A. Vause; Phys. Lett., **79A** (1980) 421; ibid. Scientific American, May (1987).

19. J.O. Indekeu and A.N. Berker, Phys. Rev. A, **33** (1986) 1158.

20. T.H. Lin, X.Y. Shao, M.K. Wu, P.H. Hor, X.C. Jin, C.W. Chu, N. Evans and R. Bayuzick, Phys. Rev. B, **29** (1984) 1493.

21. S. Rastogi, M. Newman and A. Keller, Nature, **353** (1991) 55.

22. M. Mutz, D. Bensimon and M.J. Brienne, Phys. Rev. Lett., **67** (1991) 923.

23. For example, the presence of dipolar forces are not necessary to explain the existence of the layered smectic-A phase in liquid crystals (see H.N.W. Lekkerkerker, D. Frenkel and A. Stroobants, Nature, **332** (1988) 822). Such interactions are believed, however, to play an essential role in the reentrant transition from a smectic-A phase to an (apparently) less ordered nematic phase in reentrant liquid crystal systems. [19]

24. D.S. Fisher and D.A. Huse, Phys. Rev. B **38** (1988) 373 and 386.

25. A. Georges, M. Mézard, J.S. Yedidia, Phys. Rev. Lett. **64** (1990) 2937.

26. S. Caracciolo, G. Parisi, S. Patarnello and N. Sourlas, Europhys. Lett., **11** (1990) 1859; ibid, J. Phys. France, **49** (1988) 429.

27. D.A. Huse and D.S. Fisher, J. Phys. I France, **1** (1991) 621.

28. R.R.P. Singh and D.A. Huse, J. Appl. Phys. **69** (1991) 5225.

29. E.R. Grannan, R.E. Hetzel, and R.R.P. Singh, Phys. Rev. Lett. **67** (1991) 907.

30. R.E. Hetzel and R.N. Bhatt, Europhys. Lett. **22** (1993) 383.

31. K. Binder, *Monte Carlo Methods in Statistical Mechanics*, (2nd ed. Springer-Verlag, 1986).

32. A.P. Young, *Disorder in Condensed Matter Physics*; eds. J.A. Blackman and J. Tagüeña (Oxford University Press, Oxford, 1991).

33. W.M. Saslow and G. Parker, Phys. Rev. Lett., **56** (1986) 1074.

34. S. Jain and A.P. Young, J. Phys. C, **19** (1986) 3913.

35. J.A. Olive, A.P. Young and D. Sherrington, Phys. Rev. B, **34** (1986) 6341.

36. R.N. Bhatt and A.P. Young, Phys. Rev. B, **37** (1988) 5606.

37. R.N. Bhatt and A.P. Young, J. Phys. Condens. Matter, **1** (1989) 2997.

38. J.D. Reger, R.N. Bhatt, and A.P. Young, Phys. Rev. Lett., **64** (1990) 1859.

39. M.Scheucher, J.D. Reger, K. Binder, and A.P. Young, Phys. Rev. B, **42** (1990) 6881.

40. J.D. Reger, T.A. Tokuyasu, A.P. Young and M.P.A. Fisher, Phys. Rev. B, **44** (1991) 7147.

41. A. Chakrabarti and C. Dasgupta, Phys. Rev. B, **37** (1988) 7557.

42. M.G. Forrester, S.P. Benz, and C.J. Lobb, Phys. Rev. B, **41** (1990) 8749.

43. F. Matsubara, T. Iyota and S. Inawashiro, J. Phys. Soc. Jpn., **60** (1991) 41.

44. F. Matsubara, T. Iyota and S. Inawashiro, Phys. Rev. Lett., **67** (1991) 1458.

45. F. Matsubara, T. Iyota and S. Inawashiro, Phys. Rev. B, **46** (1992) 8282.

46. A. Chakrabarti and C. Dasgupta, Phys. Rev. Lett., **56** (1986) 1404.

47. J.D. Reger and A.P. Young, Phys. Rev. B **37** (1988) 5493.

48. J.R. Thompson, H. Guo, D.H. Ryan, M.J. Zuckermann, and M. Grant, Phys. Rev. B, **45** (1992) 3129.

49. A. Ghazali, P. Lallemand, H.T. Diep, Physica, **134A** (1986) 628.

50. H. Kawamura and M. Tanemura, Phys. Rev. B, **36** (1987) 7177.

51. P. Ray and M. Moore, Phys. Rev. B, **45** (1992) 5361.

52. F. Matsubara, T. Iyota and S. Inawashiro, J. Phys. Soc. Jpn., **60** (1991) 4022.

53. F. Matsubara and M. Iguchi, Phys. Rev. Lett., **68** (1992) 3781.

54. The data analysis reported in Ref. [53] has recently been criticized; see A. Chakrabarti and C. Dasgupta, Phys. Rev. Lett., **70** (1993) 3179 and F. Matsubara and M. Iguchi, ibid, **70** (1993) 3180.

55. I.A. Campbell, Phys. Rev. Lett., **68** (1992) 3351.

56. It has recently been pointed out that it is very important to consider a large number of samples to perform the disorder averaging in Monte Carlo simulations of spin glasses. The above work of Campbell [55] has recently been the subject of some criticism; see R.N. Bhatt and A.P. Young, Phys. Rev. Lett., **69** (1992) 3130, and I.A. Campbell, ibid, **69** (1992) 3131.

57. S. Liang, Phys. Rev. Lett., **69** (1992) 2145.

58. E. Marinari and G. Parisi, Europhys. Lett. **19** (1992) 451.

59. B.A. Berg and T. Celik, Phys. Rev. Lett., **69** (1992) 2292.

60. E. Marinari, G. Parisi and F. Ritori (submitted to J. Phys. A, 1993).

61. R.H. Swendsen, J. Stat. Phys., **34** (1984) 963; G.S. Pawley, R.H. Swendsen, D.J. Wallace and K.G. Wilson, Phys. Rev. B, **29** (1984) 4030.

62. J.S. Wang and R.H. Swendsen, Phys. Rev. B, **37** (1988) 7745.

63. R.R.P. Singh, Comment Condens. Matt., **13** (1988) 275.

64. R.R.P Singh, Phys. Rev. Lett., **67** (1991) 899.

65. C.M. Newman and D.L. Stein, Phys. Rev. Lett., **72** (1994) 2286; M. Cieplak, A. Maritan, and J.R. Banavar, Phys. Rev. Lett., **72** (1994) 2320.

66. J.R. Banavar and M. Cieplak, Phys. Rev. Lett., **48** (1982) 432.

67. A.J. Bray and M.A. Moore, J. Phys. C, **17** (1984) L463.

68. W.L. McMillan, Phys. Rev. B, **29** (1984) 4026.

69. W.L. McMillan, Phys. Rev. B, **30** (1984) 476.

70. W.L. McMillan, Phys. Rev. B, **31** (1985) 342.

71. A.J. Bray and M.A. Moore, J. Phys. C, **18** (1985) L139.

72. A.J. Bray and M.A. Moore, Phys. Rev. B, **31** (1985) 631.

73. B.W. Morris, S.G. Colborne, M.A. Moore, A.J. Bray, and J. Canisius, J. Phys. C, **19** (1986) 1157.

74. A.J. Bray and M.A. Moore, Phys. Rev. Lett., **58** (1987) 57.

75. H. Kawamura and M. Tanemura, J. Phys. Soc. Jpn., **60** (1991) 608.

76. A.J. Bray, Comments Cond. Mat. Phys., **14** (1988) 21.

77. H. Kawamura, Phys. Rev. Lett., **68** (1992) 3785.

78. M.J.P. Gingras, Phys. Rev. B, **45** (1992) 7547.

79. M.J.P. Gingras, Phys. Rev. Lett., **71** (1993) 1637.

80. P.W. Anderson ad C.M. Pond, Phys. Rev. Lett., **40** (1978) 903.

81. B.W. Southern and A.P. Young, J. Phys. C, **10** (1977) 2179.

82. M.A. Zaluska-Kotur, M. Cieplak and P. Cieplak, J. Phys. C, **20** (1987) 3741.

83. M. Cieplak and J.R. Banavar, J. Phys. A, **23** (1990) 4385 (1990) and references therein.

84. J.R. Banavar and A.J. Bray, Phys. Rev. B **35** (1987) 8888.

85. J.D. Reger and A.P. Young, J. Phys. Condens. Matter, **1** (1989) 915.

86. J.R. Banavar and A.J. Bray, Phys. Rev. B, **38** (1988) 2564.

87. M.J.P. Gingras, Phys. Rev. B, **46**, 14900 (1992) and references therein.

88. M. Cieplak, J.R. Banavar, M.S. Li, and A. Khurana, Phys. Rev. B, **45** (1992) 786.
89. M.J.P. Gingras and E.S. Sørensen, Phys. Rev. B, **46** (1992) 3441.
90. M.J.P. Gingras, Phys. Rev. B, **43** (1991) 13747.
91. M.J.P. Gingras, Phys. Rev. B, **44** (1991) 7139.
92. J.V. José, L.P. Kadanoff, S. Kirkpatrick and D.R. Nelson, Phys. Rev. B, **16** (1977) 1217.
93. T.C. Halsey, Phys. Rev. Lett., **55** (1985) 1018.
94. M.J.P. Gingras and S.T. Bramwell (TRIUMF report, 1993).
95. A.B. Harris, J. Phys. C, **7** (1974) 1671.
96. J.T. Chayes, L. Chayes, D.S. Fisher and T. Spencer, Phys. Rev. Lett., **57** (1986) 2999.
97. Y. Imry and S.-K. Ma, Phys. Rev. Lett., **35** (1975) 1399.
98. T. Nattermann and J. Villain, Phase Transitions, **11** (1988) 5; T. Nattermann and P. Rujan, Int'l. Jour. Mod. Phys. B, **3** (1989) 1597.
99. D.A. Huse and C.L. Henley, Phys. Rev. Lett., **54** (1985) 2708.
100. Y. Shapir, *Recent Progess in Random Magnets*, ed. D.H. Ryan (World Scientific, Singapore, 1992), p. 309.
101. K. Biljakovíc, J.C. Lasjaunias, P. Monceau, and F. Lévy, Phys. Rev. Lett., **67** (1991) 1902.
102. K. Biljakovíc, in *Phase Transitions and Relaxation in Systems with Competing Energy Scales*, NATO Advanced Study Institute, Geilo, Norway (1993), eds. T. Riste and D. Sherrington, Kluwer Academic Publishers, Dordrecht (1993).
103. J.C. Lasjaunias, K. Biljakovíc, F. Nad, P. Monceau, and K. Bechgaard, Rev. Lett., **72** (1994) 1283.
104. A. Aharony, Sol. St. Comm., **28** (1978) 667.
105. Y.Y. Goldschmidt, *Recent Progess in Random Magnets*, ed. D.H. Ryan (World Scientific, Singapore, 1992), p. 151.
106. R. Fisch and A.B. Harris, Phys. Rev. B, **41** (1990) 11305; R. Fisch, Phys. Rev. B, **39** (1989) 873; Phys. Rev. B, **41** (1990) 11705; Phys. Rev. B, **42** (1990) 540; Phys. Rev. Lett., **66** (1991) 2041.
107. S.F. Edwards and P.W. Anderson, J. Phys. F, **7** (1975) 965.
108. H.T. Diep and O. Nagai, J. Phys. C, **18** (1985) 369.
109. Y. Ozeki and H. Nishimori, J. Phys. Soc. Jpn., **56** (1987) 3265.
110. L. Saul and M. Kardar, Phys. Rev. B, **48** (1993) 3221 and references therein.
111. R. Shankar, Phys. Rev. Lett., **58** (1987) 2466; J.-K. Kim and A. Patrascioiu, Phys. Rev. Lett., **72** (1994) 2785.
112. P.Z. Wong, S. von Molnar, T.T.M. Palstra, J.A. Mydosh, H. Yoshizawa, S.M. Shapiro, and A. Ito, Phys. Rev. Lett., **55** (1985) 2043.
113. A. Aharony, R.J. Birgeneau, A. Coniglio, M.A. Kastner and H.E. Stanley; Phys. Rev. Lett., **60** (1988) 1330; R.J. Birgeneau, M.A. Kastner and A. Aharony, Z. Phys. B, **71** (1988) 57.
114. I. Morgenstern, Z. Phys. B, **80** (1990) 7.

115. A.W. Sandvik and D.J. Scalapino, Phys. Rev. Lett., **72** (1994) 2777 and references therein.
116. D.R.Harshman, G. Aeppli, G.P. Espinosa, A.S. Cooper, J.P. Remeika, E.J. Ansaldo, T.M. Riseman, D.Ll Williams, D.R. Noakes, B. Ellman and T.F. Rosenbaum, Phys. Rev. B, **38** (1988) 852.
117. B.J. Sternlieb, G.M. Luke, Y.J. Uemera, T.M. Riseman, J.H. Brewer, P.M. Gehring, K. Yamada, Y. Hidaka, T. Murakami, T.R. Thurston and R.J. Birgeneau, Phys. Rev. B, **41** (1990) 8866.
118. I. Watanabe, K. Kumagai, Y. Nakamura, T. Kimura, Y. Nakamichi and H. Nakajima, J. Phys. Soc. Jpn., **56** (1987) 3028.
119. Y.Endoh, K. Yamada, R.J. Birgeneau, D.R. Gabbe, H.P. Jenssen, M.A. Kastner, C.J. Peters, P.J. Picone, T.R. Thurston, J.M. Tranquada, G. Shirane, Y. Hidaka, M. Oda, Y. Enomato, M. Suzuki and T. Murakami, Phys. Rev. B, **37** (1988) 7443; J. Rossat-Mignot, L.P. Regnault, J.M. Jurgens, P. Burlet, J.Y. Henry, G. Lapertot and C. Vettier, in *Dynamics of Magnetic Fluctuations in High-Temperature Superconductors*, eds. G. Reiter, P. Horsch and G.C. Psaltakis, Plenum Press, New-York (1991). These last authors suggest that the reentrant behavior seen in $YBa_2CU_3O_{6+\delta}$ is not of spin glass origin.
120. J.V. José, Phys. Rev B, **20** (1979) 2167.
121. H. Kawamura and M. Tanemura, J. Phys. Soc. Jpn. **54**, 4479 (1985); ibid **55** (1986) 1802.
122. G.N. Parker and W.M. Saslow, Phys. Rev. B, **38** (1988) 11718; ibid. Phys. Rev. B, **38** (1988) 11733.
123. J. Vannimenus, S. Kirkpatrick, F.D.M. Haldane and C. Jayaprakash, Phys. Rev. B, **39** (1989) 4634.
124. P. Gawiec and D.R. Grempel, Phys. Rev. B, **44** (1991) 2613.
125. J. Villain, J. Phys. C, **10** (1977) 1717; ibid., **10** (1977) 4793.
126. M. Nielsen, D.H. Ryan, H. Guo, M. Zuckermann, to appear in J. Appl. Phys., (1995).
127. M.P.A. Fisher, D.S. Fisher and D. A. Huse, Phys. Rev. B, **43** (1991) 130.
128. G. Blatter, M.V. Feigel'man, V.B. Geshkenbein, A.I. Larkin, and V.M. Vinokur, (preprint 1993).
129. D.A. Huse and H.S. Seung, Phys. Rev. B, **42** (1990) 1059.
130. M. Rubinstein, B. Shraiman and D.R. Nelson, Phys. Rev. B, **27** (1983) 1800.
131. J.M. Kosterlitz and D.J. Thouless, J. Phys. C, **6** (1973) 1181; J.M. Kosterlitz, J. Phys. C, **7** (1974) 1046.
132. M. Paczuski and M. Kardar, Phys. Rev. B, **43** (1991) 8331.
133. E. Granato and J.M. Kosterlitz, Phys. Rev. B, **33** (1986) 6533; ibid., Phys. Rev. Lett., **62** (1989) 823.
134. M.G. Forrester, H. Jong-Lee, M. Tinkham and C.J. Lobb, Phys. Rev. B, **37** (1988) 5966; S.P. Benz, M.G. Forrester, M. Tinkham and C.J. Lobb, Phys. Rev. B, **38** (1988) 2869.
135. S.E. Korshunov, Phys. Rev. B, **48** (1993) 1124.
136. A.J. Berlinsky, M.J.P. Gingras, C. Kallin, and A.C. Shi (unpublished).

137. J.C. Le Gouillou and J. Zinn-Justin, Phys. Rev. B, **21** (1980) 3976.
138. J.A. Hertz, Phys. Rev. B, **18** (1978) 4875.
139. M. D'Onorio De Meo, J.D. Reger and K. Binder in *Computer Simulation Studies in Condensed-Matter Physics VI*, eds. D.P. Landau, K.K. Mon and H.-B. Schüttler, Springer Proceedings in Physics ¡bf 76 (1993) 193.
140. J. Machta and M.S. Cao, J. Phys. A, **25** (1992) 529.
141. J. Wosnitza and H.V. Löhneysen, J. Phys. (Paris), **49** (1988) 1203.
142. A.P. Murani, J. Phys. F, **4** (1974) 757; Phys. Rev. B, **28** (1983) 432.
143. A.P. Murani, S. Roth, P. Radhakrishnan, B.D. Rainford, B.R. Coles, K. Ibel, G. Goetz, and F. Mezei, J. Phys. F, **6** (1976) 425.
144. B.R. Coles, R.H. Taylor, B.V.B. Sarkissian, J.A. Kahn, and M.H. Bennett, Physica B, **98-88** (1977) 275.
145. B.R. Coles, B.V.B. Sarkissian, and R.H. Taylor, Phil. Mag. B, **37** (1978) 489.
146. B.V.B. Sarkissian, Phil. Mag. B, **39** (1979) 413; J. Phys. F, **11** (1981) 2191.
147. G. Aeppli, S.M. Shapiro, R.J. Birgeneau and H.S. Chen, Phys. Rev. B, **28** (1983) 5160; ibid., **29** (1984) 2589.
148. S.G. Barsov et al., Hyp. Inter., **64** (1990) 415.
149. I. Mirebeau, S. Itoh, S. Mitsuda, T.Watanabe, Y. Endoh, M. Hennion, and P. Calmettes, Phys. Rev. B, **44** (1991) 5120.
150. T. Sato, T. Ando, T. Watanabe, S. Itoh, Y. Endoh, and M. Furusaka, Phys. Rev. B, **48** (1993) 6074.
151. M. Hennion, B. Hennion, I. Mirebeau, S. Lequien, F. Hippert, J. Appl. Phys., **63** (1988) 4071.
152. B. Hennion, M. Hennion, I. Mirebeau and F. Hippert, Physica, **136B** (1986) 49.
153. S. Lequien, B. Hennion, and S.M. Shapiro, Phys. Rev. B, **38** 2669 (1988).
154. S. Mitsuda, H. Yoshizawa, and Y. Endoh, Phys. Rev. B, **45** (1992) 9788.
155. A. Schenck, *Muon Spin Rotation Spectroscopy*, (Adam Hilger, Bristol, 1985).
156. Y.J. Uemura, T. Yamazaki, O.R. Harshman, M. Senba, and E.J. Ansaldo, Phys. Rev. B, **31** (1985) 546.
157. A. Keren et al., (in press, Hyp. Int. (1994)).
158. C. Boekema et al., Hyp. Inter., **31** (1986) 369.
159. M.J.P. Gingras, W.D. Wu, K. Kojima, G.D. Morris, S.R. Dunsiger, Y.J. Uemura (unpublished).
160. S. Hadjouji, S. Senoussi, and I. Mirebeau, J. Mag. Mag. Mat., **93** (1991) 136.
161. S. Senoussi, S. Hadjouji, R. Fourmeaux, Phys. Rev. Lett., **61** (1988) 1013.
162. D.J. Sellmyer and M.J. O'Shea, *Recent Progess in Random Magnets*, ed. D.H. Ryan (World Scientific, Singapore, 1992), p. 71.
163. J.N. Reimers, J.E. Greedan, R.K. Kremer, E. Gmelin, and M.A. Subramanian, Phys. Rev. B, **43** (1991) 3387; A. Maignan, Ch. Simon, J.E. Greedan, J. S. Pedersen, and M.A. Subramanian, (unpublished, submitted to Phys. Rev. B, (1993)).
164. H. Nishimori, Prog. Theor. Phys., **66** (1981) 1169; ibid, **76** (1986) 305; ibid J. Phys. Soc. Jpn., **55** (1986) 3305.

165. P. Le Doussal and A. Georges, (unpublished).
166. Y. Ozeki and H. Nishimori, J. Phys. A, **26** (1993) 339
167. A. Georges, D. Hansel, P. Le Doussal, and J.P. Bouchaud, J. Phys. (Paris), **46** (1985) 1827.
168. P. Le Doussal and A.B. Harris, Phys. Rev. Lett., **61** (1988) 625; ibid., Phys. Rev. B, **40** (1989) 9249.
169. H. Kitani, J. Phys. Soc. Jpn., **61** (1992) 4049
170. H. Nishimori, J. Phys. Soc. Jpn., **61** (1992) 1011.
171. M.J.P. Gingras and E.S. Sørensen, (unpublished).

Magnetic System with Competing Interaction
Ed. H. T. Diep
©1994 World Scientific Publishing Co.

CHAPTER VII

EXPERIMENTAL STUDIES OF

GEOMETRICALLY-FRUSTRATED MAGNETIC SYSTEMS

Bruce D. Gaulin

Department of Physics and Astronomy, McMaster University
Hamilton, ON, Canada, L8S 4M1

1. Introduction

Materials containing antiferromagnetically-coupled magnetic moments which reside on geometrical units that can inhibit the formation of a collinear, magnetically ordered state often exhibit unusual behaviour at low temperatures. This phenomena is known broadly as frustration[1]. The materials of interest may undergo phase transitions with novel properties to an unusual ordered state, or they may not undergo a conventional phase transition at all, entering a glass-like state as the temperature is lowered. This chapter reviews recent experimental progress in understanding the phases and phase transitions displayed by several such magnetic materials. Much of it will focus on neutron-scattering studies, as such studies have played a central role in characterizing the properties of these antiferromagnets and their phase transitions. The review presented here is not intended to be exhaustive in nature, but rather to give an overview of several materials with which the author is familiar, and to discuss some of the experimental challenges which must be dealt with in such studies.

The experimental realizations of such frustrated systems are typically comprised of two types of geometric building blocks: triangles and tetrahedra. The frustration inherent in the local ordering of these structures is illustrated in Fig. 1: (a) shows a trio of Ising spins with nearest-neighbour-only antiferromagnetic interactions on a triangle, and (b), a quartet of similar spins on an isolated tetrahedron. As can be seen, arranging any two of the neighbouring spins into an up-down pair does not allow the remaining spin(s) to align either up or down without frustrating some of the interactions. The ground state for such a lattice therefore possesses a greater spin degeneracy than would exist on a comparable unfrustrated lattice. The consequences of the frustration are most severe for Ising spins (the case shown in Fig. 1), as such spins have the fewest degrees of freedom with which to resolve their dilemma. However, frustration has interesting manifestations for vector spins as well.

Two-dimensional (2D) and three-dimensional (3D) crystal structures can be constructed by assembling the triangles or tetrahedra such that neighbouring geometrical units share either a common edge or a common corner. In two dimensions, a network of edge-sharing triangles forms the familiar triangular lattice[2], which is very common in nature as it forms the basal plane of all crystal structures with hexagonal symmetry. A network of corner-sharing triangles forms the kagomé lattice[3]. In three dimensions, edge-sharing tetrahedra give rise to the face-centred cubic lattice, while a network of corner-sharing tetrahedra form structures found in spinel, Laves phase, and pyrochlore crystals[4].

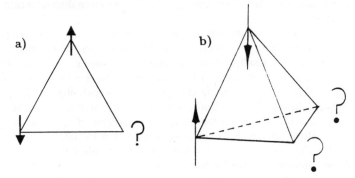

Figure 1. The frustration inherent in the combination of antiferromagnetically-coupled nearest-neighbour spins and a) triangular or b) tetrahedral coordination.

2. Neutron Scattering and Magnetic Phase Transitions

Magnetic materials have been the physical systems of choice for the study of cooperative phenomena and phase transitions. This is due to the fact that the microscopic interactions in magnetic insulators, tend to be predominantly nearest-neighbour-only, and are well described by a variety of simple Hamiltonians. Experimentally, the most powerful technique for elucidating the behaviour of magnetic systems has been neutron scattering[5-8], although other techniques do play important roles. As the neutron is electrically neutral, its only interaction with the electronic degrees of freedom in matter is via the magnetic dipole moment it carries, which allows it to couple to magnetic moments in solids. In addition, the energy of neutrons, produced in copious quantities in nuclear reactors and now at spallation sources, are very well matched to the energies of typical elementary excitations in magnetic materials. The consequence of these attributes is that scattering experiments can probe the behaviour of the magnetism on different and controllable length scales (through the wavevector transfer, Q, of the scattering event) as well as on different and controllable time scales (through the energy transfer, $E=\hbar\omega = h\nu$, of the scattering event). This is particularly important for antiferromagnets in which there is no net magnetic moment present in the absence of a magnetic field, and the ordering wavevector is non-zero. Most magnetic materials found in nature, and all of the magnets to be discussed in this chapter, are indeed antiferromagnets.

2.1. Critical Phenomena

A phase transition from a disordered to an ordered magnetic state is characterized by the appearance of an order parameter, $M(Q_{ord})$, which signifies that a symmetry has been broken. In a so-called continuous transition the order parameter rises continuously from zero at some critical temperature T_N, while in a discontinuous transition there is a discontinuity (a "jump") in $M(Q_{ord})$ at T_N. In an antiferromagnet, the order parameter is the sublattice magnetization appropriate to its ordered magnetic structure, and Q_{ord}, is the corresponding ordering wavevector for this structure.

Continuous phase transitions are also characterized by the buildup of fluctuations near the ordering wavevector, and the subsequent divergence of these fluctuations at T_N. Physically, these fluctuations correspond to short-range ordered droplets of the ordered structure which are out of phase with respect to each other. They are described by a diverging Q-dependent susceptibility, $\chi(Q_{ord})$, and correlation length, ξ. In addition the heat capacity of the material, C_P, either diverges, or exhibits a cusp at T_N. Taken together, the behaviour of these physical quantities define the critical phenomena exhibited by a given material.

Near T_N these quantities are usually assumed to obey power laws governed by the critical exponents defined below:

$$M(Q_{ord}) \sim \frac{(T_N - T)^\beta}{T_N} \tag{1}$$

$$\chi(Q_{ord}) = \chi_0^\pm \frac{|T_N - T|^{-\gamma}}{T_N} \tag{2}$$

$$\xi = \xi_0^\pm \frac{|T_N - T|^{-\nu}}{T_N} \tag{3}$$

$$C_P = C_0^\pm \frac{|T_N - T|^{-\alpha}}{T_N} \tag{4}$$

Equations (2-4) are relevant both above and below T_N, and the "+" or "-" superscript on the amplitude denotes on which side of the transition the expression applies. The exponents are universal in the sense that a wide range of seemingly disparate materials exhibit power-law divergences with the *same* exponents in the vicinity of their phase transitions. As such, the exponents in large part define a universality class. A material belongs to a universality class on the basis of a few quite general properties, such as the symmetry of its order parameter and its spatial dimensionality. In contrast, the amplitudes, $\chi_0^\pm$ etc., are material-dependent, non-universal quantities. However the ratio of the amplitudes above and below T_N, e.g. χ_0^+/χ_0^-, are again universal quantities, and as such, are of considerable interest.

These equations describe the asymptotic critical behaviour for a material, i.e. the behaviour very close to T_N for which the correlation length dominates all other length scales of relevance to the material. This will only occur when ξ is sufficiently large, and therefore when the reduced temperature, $t = |T_N - T|/T_N$, is sufficiently small. If t is not in this asymptotic regime, the temperature dependence will exhibit corrections to the simple power-law behaviour of Eqs. (1-4). These correction-to-scaling terms give rise to relations of the form:

$$A = A_0^{\pm} t^a [1 + Bt^x + ..]$$ (5)

where x has the value $\frac{1}{2}$ in mean-field theory.

Theory predicts that the critical exponents within a given universality class should satisfy certain scaling relations. Two of these are:

$$\gamma + \alpha + 2\beta = 2$$ (6)

$$D\nu = 2 - \alpha$$ (7)

where D is the spatial dimensionality of the system. Scaling relations involving the spatial dimensionality, D, are referred to as hyperscaling relations. It is important to point out that most data can be reasonably fit to power laws even when the system is not at sufficiently small t that the asymptotic behaviour is present. An analysis of this data will yield "effective" exponents. Theoretical work[9] has shown that the "effective" exponents will obey scaling, but not hyperscaling relations. Therefore these scaling relations provide an important consistency check on the exponent estimates taken from experiments.

The dynamical properties of materials also exhibit critical phenomena[10]. Physically, this can be understood in terms of the long characteristic times which are required for large objects to relax. As a phase transition is approached, ξ becomes large, and a substantial fraction of the moments in the material find themselves as parts of large, correlated domains. As ξ diverges at T_N, the characteristic relaxation time, τ_R, must also diverge, giving rise to a phenomenon known as critical slowing-down. At the critical temperature, domains of all length scales are present. Those which are large are characterized by small values of $q=Q_{ord}-Q$, while small domains correspond to larger values of q. Therefore, $\tau_R(q) \sim q^z$ where z depends only on whether the system is antiferromagnetic or ferromagnetic. This dynamical critical exponent z also obeys a scaling relation of the form:

$$z = a + 1 + \frac{\beta}{\nu}$$ (8)

where a depends simply on whether the order parameter is conserved or not: for a ferromagnet, $a=2$, while for an antiferromagnet, $a = 1$.

2.2. Experimental Considerations

The elastic neutron-scattering cross section is proportional to $M_\perp(Q)^2$. (The $\perp$ subscript refers to the fact that all magnetic neutron scattering, elastic or inelastic, at a particular Q, originates only from those components of the magnetic moment which lie in a plane perpendicular to Q.) The order parameter, $M(Q_{ord})$, can then be measured by performing elastic neutron-scattering measurements and examining the integrated intensity at the ordering wavevector. This scattering is quite strong from most magnetic materials which display long-range order, and so it is easily measured. However the extraction of $M(Q_{ord})$, and the subsequent estimation of the β exponent, can be complicated by an effect known as extinction. Essentially, it arises because the elastic scattering is so strong that it ends up depleting the incident beam of the neutrons which can, in principle, satisfy the Bragg condition. (For a thorough discussion of extinction, see for example reference 11.) Its net effect is to decouple the measured elastic neutron-scattering cross section from $M(Q_{ord})$.

The real difficulty is not the extinction *per se*, but rather the fact that it is temperature-dependent. As $M(Q_{ord})$ (and so the cross section) increases, the depletion of the incident beam becomes more pronounced. If this effect is sufficiently strong, it can render the measurement of the temperature dependence of $M(Q_{ord})$ unreliable, producing anomalously low β estimates if a power law analysis is attempted.

It is difficult to correct for the effects of temperature-dependent extinction on a nominal measurement of $M(Q_{ord}, T)$. However, one can apply consistency checks on a program of measurements to see whether or not it is a serious problem in a particular experiment. This can be done in part by examining the scattering at several equivalent ordering wavevectors. The intensity of the scattering at these equivalent wavevectors may vary due to the magnetic structure factor (in materials with several moments per unit cell) and the magnetic form factor, which causes the scattering to fall off with $|Q|$. Therefore, a consistent estimate of β obtained from the measurements of $M(Q_{ord}, T)$ at several different Q_{ord} over which the strength of the Bragg scattering varies by a large factor (say a factor of 10) is a good indication that extinction effects are not serious. The scaling relations (Eqs. (6-8)) also provide a good consistency check on β, once estimates for other critical exponents have been obtained.

All of the measurements presented in this chapter were performed using a triple-axis neutron spectrometer at a nuclear reactor. The operation of a triple-axis spectrometer is discussed in detail in several reviews[5-8]. This instrument uses a monochromator crystal in the incident beam to select out a particular wavelength of neutron to be directed onto the sample, and a scattered beam analyser crystal to select out a particular wavelength of neutron scattering from the sample. This allows the energy of the incident and scattered neutron beams to be independently controlled, which in essence permits the cross section to be measured over a grid in (Q, E) space. It can also be operated in a mode in which all neutrons coming off the sample in a particular direction are detected, regardless of their energy. This energy-integrated mode of operation is achieved simply by removing the scattered beam analyser.

The elastic neutron-scattering intensity is an interesting quantity to measure even for systems which do not undergo conventional phase transitions. In this case it is most useful to think of the neutron response in terms of spin-correlation functions. The neutron-scattering cross section, for general wavevector, Q, and energy transfer, $h\nu$, is proportional to:

$$S(Q, \nu) \sim \int \delta t \sum_{R_{i,j}} exp[-iQ \cdot (R_i - R_j) + \frac{i\nu t}{2\pi}] < S_i(0)S_j(t) > \qquad (9)$$

where $S_j(t)$ is the moment (or spin) at position R_j and time t. Inelastic scattering

($\nu \neq 0$) gives information on spin correlations between moments at different sites at different times, and hence on the normal modes of the systems, such as spin waves and solitons. Elastic scattering ($\nu=0$) measures infinite-time spin correlations. In practice, the constraints of the energy measurement on the incident and scattered neutron preclude the measurement of truly elastic scattering: the measurement accepts all neutrons with energies within some range of the nominal elastic energy. This range varies depending on the type and configuration of the neutron spectrometer, but is typically between 0.4 THz $> \nu = \frac{E}{h} > 0.02$ THz. A consequence of this is that infinite-time spin correlations, corresponding to a truly static magnetic state, cannot be distinguished from a state which is dynamic on time scales of $\sim 10^{-10}$ seconds or longer. Information on spin dynamics at these relatively slow characteristic times are available from other experimental techniques, such as μSR and NMR. However these techniques do not probe the magnetic response at one particular **Q**, but rather integrate the response over all wavevectors.

Energy-integrated scattering measurements play an important role in understanding phase transitions in that they provide a direct measure of $\chi(\mathbf{Q})$. The integrated cross section can be written as $S(\mathbf{Q}) = \int S(\mathbf{Q}, \nu)\delta\nu \sim \chi(\mathbf{Q})$. A proper measurement of $S(\mathbf{Q})$ must then integrate all of the spectral weight of the inelastic scattering at a particular **Q**. In fact the energy-integrated scattering mode measures $S(\mathbf{Q})$ only approximately: the scattered neutrons lie in a substantially broader range of energy than do the incident neutrons, and so give rise to scattering at slightly different **Q**'s centred around the nominal value. This effect can be minimized by using incident neutrons of relatively high energy, such that the inelasticity of the scattering events are a small perturbation on the energy of the scattered neutrons. The condition that this is satisfied is referred to as the quasi-static approximation. The effect becomes less important as a critical phase transition is approached, since the characteristic time scales for spin relaxation become long, and the energy width of $S(\mathbf{Q}_{\mathrm{ord}}, \nu)$ small.

A quantitative understanding of the energy-integrated neutron scattering measurements requires a detailed treatment of the resolution of the instrument. Measurements taken at a nominal **Q**, in fact accept neutrons scattered over a range of wavevectors in the vicinity of that **Q**. This is due to the finite angular range of the incident and scattered neutrons accepted by the spectrometer, the finite collimation used in each path of the instrument, as well as the quality of the monochromators and sample crystal itself.

Although such effects can in principle be calculated from the detailed parameters which describe each component of the instrument, it is usually preferable to measure the resolution function directly and then convolute it with the assumed form of S(**Q**). This can be done by measuring, what in principle should be, a delta-function response from the sample crystal, such as the order parameter, M($\mathbf{Q}_{ord}$), at low temperatures. Such a measured response is shown in Fig. 2 for the hexagonal insulator CsCoBr$_3$ at 11 K, well below its $T_N \sim 28.4$ K. As will be discussed, this antiferromagnet undergoes three phase transitions to ordered phases which all share the $(\frac{1}{3},\frac{1}{3},1)$ ordering wavevector. Figure 2 shows mesh scans around this position, and it is evident that the single crystal sample is in fact comprised of two closely aligned grains, one of which is roughly five times the volume of the other. It is also clear that the delta-function response is broadened out by the instrumental resolution in a highly anisotropic manner.

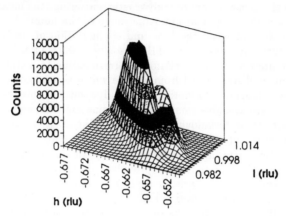

Figure 2: The measured resolution of the triple-axis neutron instrument, operated in an energy-integrating mode, is shown for the (h,h,l) plane near $(\frac{2}{3},\frac{2}{3},1)$ in $CsCoBr_3$. It is obtained by examining the magnetic Bragg scattering well below T_{N1}, where it approximates a delta-function response. The presence of two peaks is due to the mosaic of the $CsCoBr_3$ sample crystal.

Figure 2 shows the measured intensity within the nominal (h,h,l) plane in reciprocal space. Of course the delta-function is also broadened perpendicular to this plane, and the width of the scattering in this direction (usually vertical) must also be accounted for. The measured $S(\mathbf{Q})$ is then quantitatively understood by fitting it to

$$S(\mathbf{Q}) = \sum_{i,j,k} S_A(\mathbf{Q} - \mathbf{Q}_{i,j,k}) R_{i,j,k} \tag{10}$$

where $S_A(\mathbf{Q})$ is the theoretical form of $S(\mathbf{Q})$ and $R_{i,j,k}$ is the measured resolution function. In practice, the full three-dimensional integration is a very computer-intensive operation to perform, and efficient approximate algorithms, in which the vertical resolution is treated analytically while the horizontal resolution is treated numerically, have been devised.

The assumed form of $S(\mathbf{Q}) \sim \chi(\mathbf{Q})$ is most often taken to be the Ornstein-Zernike (OZ) form:

$$\chi(\mathbf{Q}) = \frac{\chi(\mathbf{Q}_{\mathrm{ord}})}{[1 - (\frac{q_{ab}}{\xi_{ab}})^2 - (\frac{q_c}{\xi_c})^2]} \tag{11}$$

where $\mathbf{q} = \mathbf{Q} - \mathbf{Q}_{\mathrm{ord}}$ and the possibility of anisotropic correlation lengths along the c and ab-directions has been allowed for. (This anisotropic form will be needed to describe the anisotropic energy-integrated critical scattering seen in the stacked triangular-lattice antiferromagnets to be described later.) The isotropic OZ form (with $\xi_{ab} = \xi_c$) gives the mean-field solution for $\chi(\mathbf{Q})$ in which the spin correlations fall off as

$$< S_i S_j > \sim \frac{exp[-\frac{r(i-j)}{\xi}]}{r^{\frac{(D-2)}{2}}} \tag{12}$$

where D is the dimensionality of the system and $r(i - j)$ is the distance between the sites of the two spins. Clearly (at least in three dimensions), this form cannot be correct at small $r(i - j)$, giving rise to deficiencies in the description of $\chi(\mathbf{Q})$ at large q. Such deficiencies have been addressed by more sophisticated theoretical treatments for specific model systems[12]. These more sophisticated treatments will not be discussed, as none apply to the model systems of interest here. Also, it can be very difficult to distinguish between the OZ form and these other forms, once resolution effects are considered. However, the choice of the assumed form of $\chi(\mathbf{Q})$ does indeed effect the numerical values of both ξ and $\chi(\mathbf{Q}_{ord})$ extracted from such analysis[6].

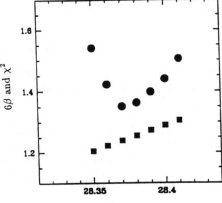

Critical Temperature, T_N

Figure 3. The dependence of β (squares) on T_{N1} is shown along with the corresponding goodness-of-fit parameter (circles). $\chi^2 \sim \Sigma_i [M_{obs}(\mathbf{Q}_{ord}, T_i) - M_{cal}(\mathbf{Q}_{ord}, T_i)]^2$, where $M_{obs}(\mathbf{Q}_{ord}, T_i)$ and $M_{cal}(\mathbf{Q}_{ord}, T_i)$ are the observed order parameter and that calculated using the best fit to the data with a particular T_{N1} value. This plot shows data from CsCoBr$_3$ at the $(\frac{2}{3}, \frac{2}{3}, 1)$ order parameter.

2.3. Exponent Estimates

Estimating the critical exponents from the measured temperature dependence of the relevant observables is relatively straightforward, but requires a cautionary note. Power-law relations are most often derived and demonstrated using log-log plots of the observable as a function of reduced temperature, t. Such plots can be remarkably forgiving, unless the data is of very high quality, spanning several decades in t while still in the asymptotic regime. This quality of data is often technically difficult to obtain. What is more, the data which really pins down the asymptotic critical behaviour, that for which t is smallest, is most sensitive to the uncertainty in T_N.

This uncertainty is the largest contribution to the uncertainties in the exponent estimates, usually substantially larger than the statistical errors associated with the measurements themselves. The dependence on T_N can be explicitly examined by fitting the observable to a power law while holding T_N fixed. One then examines the goodness-of-fit parameter over a reasonable range of T_N's to see where it is minimized and how pronounced the minimum is. This is shown in Fig. 3 for $M(Q_{ord})$ measured in CsCoBr$_3$ near $T_{N1} \sim 28.4$ K. This form of analysis allows a reasonable estimate of the uncertainty in T_N as well as an estimate of the uncertainty in the exponent, in this case β. Ideally such analysis should be done simultaneously for all observables and a consistent value of T_N obtained.

All of the factors mentioned in this section must be carefully considered in order to produce reliable exponent estimates. In addition, one gains considerable confidence in these estimates when the critical properties of a sufficient number of observables (e.g. Eqs. (1-4)) are examined, as one can then use scaling and hyperscaling relations (e.g. Eqs. (6-8)) as consistency checks on the entire program of measurements.

3. Stacked Triangular-Lattice (STL) Antiferromagnets

Two of the magnetic materials to be discussed in this chapter, CsMnBr$_3$ and CsCoBr$_3$, are insulators with a simple hexagonal crystal structure. In both cases the magnetism is associated with the partially-filled 3d shell of the transition metal ion, either Mn^{2+} or Co^{2+}. The crystal structure appropriate to these materials is shown in Fig. 4; it consists of chains of transition metal ions separated by a triad of Br^{1-} ions arranged along the hexagonal c-direction. These chains are then packed together in a triangular array. The Cs^{1+} ions sit out in the unit cell and serve to isolate the chains of magnetic ions from one another. Both of these materials belong to the hexagonal space group $P6_3/mmc$[13,14].

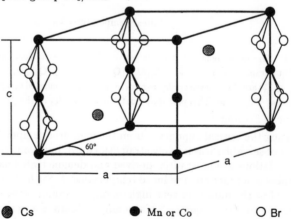

Figure 4. The hexagonal crystal structure of CsMnBr$_3$ and CsCoBr$_3$.

From the point of view of frustration, it is appropriate to picture these materials as comprised of stacked sheets of magnetic moments which reside on a triangular lattice. as it is the combination of simple near-neighbour antiferromagnetic interactions and the triangular lattice which generate the physics of interest.

The microscopic spin Hamiltonians for both materials have been determined from inelastic neutron-scattering measurements, mostly at low temperatures. In both cases it was found that simple nearest-neighbour antiferromagnetic interactions were sufficient to describe the measured spin excitations, both within the triangular basal plane and along the hexagonal c-direction. The triad of Br^{1-} ions between every pair of near-neighbour transition-metal ions provides a very strong superexchange path along c. The interaction path within the triangular plane is longer and more complicated. For this reason, both materials display quasi-one-dimensional magnetic behaviour over an extended temperature range. However, the weak interchain interactions precipitate a phase transition to three-dimensional long-range magnetic order at sufficiently low temperature; $T_N \sim 8.3$ K for $CsMnBr_3$[15] and $T_{N1} \sim 28.4$ K for $CsCoBr_3$[14,16].

Mn^{2+} is a half-filled shell and corresponds, via Hund's rules, to an $S = \frac{5}{2}$ state. As such the anisotropy is primarily dipolar in origin and quite weak. This weak anisotropy confines the quasi-classical moments in $CsMnBr_3$ to the triangular plane, so as to produce an XY-like system below ~ 20 K[17]. The microscopic Hamiltonian for $CsMnBr_3$ is known to be well described[18] by:

$$H = 2J\sum_i \mathbf{S}_i \cdot \mathbf{S}_{i+1} + 2J'\sum_i \sum_{\delta ab} \mathbf{S}_i \cdot \mathbf{S}_{i+\delta ab} + D\sum_i (S_i^z)^2 \tag{13}$$

with $J = 0.21$ THz, $J' = 0.00046$ THz, and $D = 0.0034$ THz. J and J' are the nearest-neighbour, antiferromagnetic exchange constants along c and within the triangular ab plane respectively, and D is the planar anisotropy.

In contrast the moment associated with Co^{2+} in the crystalline environment of $CsCoBr_3$ is a Kramer's doublet, effectively an $S = \frac{1}{2}$ state, in which the moment is strongly constrained to point along c. The symmetry of the magnetic site in $CsCoBr_3$ is thus Ising-like. It is known to be well described[19] by the following microscopic spin Hamiltonian:

$$H = 2J\sum_i [S_i^z S_{i+1}^z + \epsilon(S_i^x S_{i+1}^x + S_i^y S_{i+1}^y)] + 2J'\sum_i \sum_{\delta ab} S_i^z S_{i+\delta ab}^z \tag{14}$$

where $J = 1.62$ THz, $J' = 0.096$ THz, and $\epsilon = 0.137$. Here again J and J' are nearest neighbour, antiferromagnetic exchange constants along c and within the triangular ab plane, respectively, while ϵ is the exchange anisotropy which constrains the spins to order along the c-direction. A weak single-ion-exchange mixing term has been omitted for simplicity.

A classical description of these systems will be used throughout this chapter. However one should bear in mind that the magnetic moments in both of these materials are quantum-mechanical entities. Clearly, this is a more important issue in the description of the $S = \frac{1}{2}$ moments in CsCoBr$_3$ than for the $S = \frac{5}{2}$ moments in CsMnBr$_3$.

4. CsMnBr$_3$: an XY STL Antiferromagnet

The combination of the nearest-neighbour antiferromagnetic exchange interactions and the triangular lattice in the basal plane give rise to interesting behaviour in CsMnBr$_3$ associated with its phase transition at $T_N \sim 8.3$ K. As mentioned, the interactions along c are also antiferromagnetic, however the sign of J in Eq. (13) is actually irrelevant: this interaction is unfrustrated regardless of whether the coupling between the triangular planes is ferromagnetic or antiferromagnetic. It is, however, important that an interaction be present along c, as no conventional phase transition (at least for vector spins) occurs in its absence.

The effect of geometrical frustration on XY spins is different from that on Ising spins and is illustrated in Fig. 5. As the spins are free to rotate anywhere within a plane, they can find an energetic compromise to the conflicting demands of the antiferromagnetic interactions and the symmetry of the triangular lattice. The compromise structure which they take up is the so-called 120° magnetic structure in which neighbouring spins are rotated 120° with respect to each other. However as Fig. 5 demonstrates, there are two degenerate ways in which a local trio of spins can form the 120° structure and minimize their interaction energy. These give rise to two distinct senses-of-rotation of the moments from neighbour to neighbour around the plaquette of nearest-neighbour spins. This sense-of-rotation, or handedness, is most often referred to as chirality. The chirality can take on one of two values, and one local chirality cannot be continuously deformed into the other. As such, it acts as an Ising variable. However the spins can undergo a global rotation within the plane at zero cost in energy due to their XY nature.

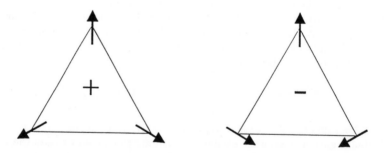

Figure 5.   Geometrical frustration in the XY antiferromagnet on a STL manifests itself by allowing two degenerate local orderings, or chiralities. (labeled here as "+" and "-") of a trio of nearest-neighbour spins.

It was first realized by Kawamura[20] that the order parameter relevant to the magnetically ordered state in such a system, would have to reflect both the XY and

Ising symmetries, and led him to predict the existence of a new universality class, $Z_2 \times S_1$, for such materials. Extensive theoretical work[21,22] relevant to the critical behaviour of this and related systems was carried out, with numerical predictions for the critical exponents coming from finite-size scaling analysis of Monte Carlo simulations. More recently, this view has been challenged by the field-theoretic work of Azaria, Delamotte, and Jolicoeur[23], which suggests that the critical behaviour of such systems is either weakly discontinuous, or characterized by tricritical mean-field behaviour.

4.1. Measurements in Zero Magnetic Field

Motivated by the earliest of this theoretical work, a program of neutron scattering measurements on a single-crystal sample of $CsMnBr_3$ was initiated. The ordered state, below $T_N \sim 8.3$ K, corresponds to a tripling of the chemical unit cell within the triangular plane. The antiferromagnetic ordering does not effect the cell size along c since the chemical unit cell already contains two Mn^{2+} sites in that direction. The ordering wavevectors for this state are therefore of the form $Q_{ord} = (\frac{h}{3}, \frac{h}{3}, l)$ with l odd.

Neutron-scattering scans were carried out within the (h, h, l) scattering plane in order to measure $M(Q_{ord})$ at a number of equivalent ordering wavevectors as well as to measure $\chi(Q)$ near Q_{ord}[24,25]. The temperature dependence of the elastic scattering at the $(\frac{1}{3}, \frac{1}{3}, 1)$ Bragg peak is shown in Fig. 6. This data clearly shows the onset of the order parameter near $T_N \sim 8.3$ K.

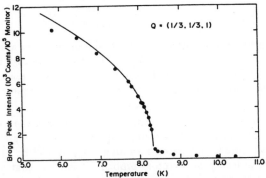

Figure 6. Magnetic Bragg-peak intensity in $CsMnBr_3$ is shown as a function of temperature for $Q=(\frac{1}{3}, \frac{1}{3}, 1)$. The solid line is a fit to the data with $\beta=0.21$[25].

Energy-integrated measurements of the critical fluctuations were performed under similar conditions to those used in the elastic-scattering measurements, but with the analysing crystal removed[25]. Scans were performed at many temperatures above T_N, and representative data is shown in Fig. 7(a). These scans are of the form $(h, h, 1)$ and $(\frac{1}{3}, \frac{1}{3}, l)$ so as to cut through the $(\frac{1}{3}, \frac{1}{3}, 1)$ peak in two symmetry directions and the $(\frac{2}{3}, \frac{2}{3}, 1)$ peak in one symmetry direction. These measurements were complicated by the presence of two closely-aligned grains within our nominally-single crystal sample (labeled by A and B in Fig. 7(b)). This crystalline mosaic was carefully considered within the resolution convolution calculations. The critical scattering lineshape, shown in Fig. 7(a), clearly sharpens as T_N is approached from above along the $(h.h.1)$ direction, indicating the diverging correlation length within the basal plane. This is also true for the lineshape along the $(\frac{1}{3},\frac{1}{3},l)$ direction, however the anisotropy in the correlation lengths, which arises due to the quasi-one-dimensional nature of the exchange interactions, makes the divergence harder to appreciate since ξ_c is large at all temperatures of interest near the phase transition.

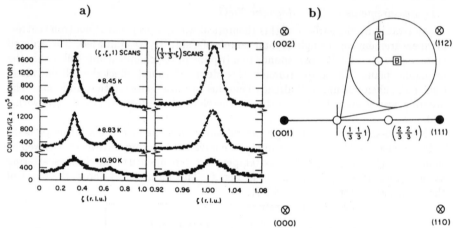

Figure 7. a) Representative energy-integrated critical scattering scans along the $(h, h, 1)$ and $(\frac{1}{3}, \frac{1}{3}, l)$ directions in reciprocal space in CsMnBr₃[25]. The solid lines through the data are fits to the OZ form of the critical scattering, convoluted with the spectrometer resolution. b) A map in reciprocal space of the scans. The inset shows the positions of the two crystallites that comprised the sample (denoted A and B) relative to the scans.

The relevant parameters obtained from fitting this data to the convolution of the asymmetric OZ form, Eq. (11), are shown in Fig. 8. These are log-log plots in which the reduced temperature is calculated assuming a T_N of 8.32 K. Figure 8(a) shows $\chi(Q_{ord})$ at the two ordering wavevectors probed. It is clear that although $\chi(Q_{ord})$ is larger at $(\frac{1}{3},\frac{1}{3},1)$ than at $(\frac{2}{3},\frac{2}{3},1)$, both are diverging with the same power law. The critical behaviour of the inverse correlation lengths is shown in Fig. 8(b). As expected, the correlation length within the basal plane is shorter than that along c, indicating that the correlated domains are cigar-shaped, with their long axes along the hexagonal c-direction, at all temperatures shown. Again, both correlation lengths diverge according to the same power law.

The exponents extracted from fitting these observables to power laws depend on the choice of T_N. The dependence of γ, ν, and β on T_N are shown in Fig. 9 for T_N's in the range ± 0.02 K around the nominal T_N of 8.2 K. Given that this range is the actual stability of the temperature during these measurements, it is reasonable to assign an uncertainty in the estimate of the exponents which is at least double the range covered in Fig. 9. This results in uncertainties in the exponents which range from 6-10 %.

a)

b)

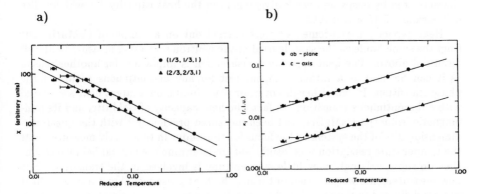

Figure 8. a) A log-log plot of $\chi(\mathbf{Q})$ as a function of reduced temperature for $\mathbf{Q}=(\frac{1}{3}, \frac{1}{3}, 1)$ and $(\frac{2}{3}, \frac{2}{3}, 1)$ in CsMnBr$_3$[25]. The lines are a fit using $\gamma=1.01$. b) A log-log plot of the inverse correlation length $\kappa_1 \sim \frac{1}{\xi}$ as a function of reduced temperature along c and in the ab plane of CsMnBr$_3$. The solid lines here are a fit to the data with $\nu=0.54$.

The neutron-scattering measurements allowed three exponents to be estimated. (These have since been independently confirmed on different crystal samples[26,27].) The scaling and hyperscaling relations (Eqs. (6-8)) allow us to estimate at least two others: the specific heat exponent α, and the dynamical critical exponent z.

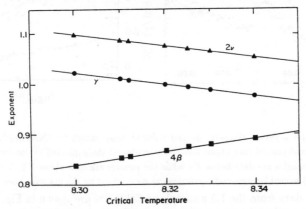

Figure 9. The variation of the fitted critical exponents with the critical temperature, T_N, in CsMnBr$_3$[25]. ν and β have been multiplied by 2 and 4, respectively, for clarity.

Equation (7), a scaling relation, predicts $\alpha=0.38 \pm 0.09$, using the experimental determination of ν. This is consistent with the value estimated by Monte Carlo techniques of 0.35 ± 0.05. These α values are unusually divergent, given that the standard three-dimensional universality classes (Ising, XY, and Heisenberg) are characterized by cusps and weak divergences in the heat capacity, for which α lies in the range -0.15 < α < 0.12.

Heat capacity measurements[28] were carried out on a sample of CsMnBr$_3$ cut from the same boule as the sample used in the neutron scattering measurements[24,25] discussed above. The heat capacity measurement is important for another reason as it can be used as a rather stringent test for any discontinuous nature of the phase transition. It can either discover, or place limits, on a latent heat associated with a discontinuous transition. In fact, the heat capacity of CsMnBr$_3$ and its non-magnetic isomorph CsMgBr$_3$, had been measured previously[15] with the quasi-one-dimensionality of the system as the focus for the study. In these early measurements the temperature resolution was insufficient to determine the critical behaviour, but they do show strikingly how little spin entropy is involved in the transition. The new measurements were carried out using the heat-pulse technique: a precise heat pulse is delivered to the sample and the resulting increase in temperature is inversely proportional to C$_P$. This technique has the advantage over drift and ac heat capacity techniques that it is largely independent of dynamic effects in both the crystal and the apparatus used in the measurement.

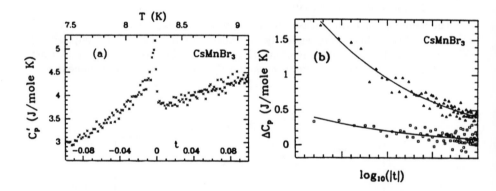

Figure 10. a) C'_P versus temperature and reduced temperature for CsMnBr$_3$[28]. The prime signifies that no background corrections have been made. b) ΔC_P plotted versus the log of |t| in CsMnBr$_3$. The triangles are data below T_N while the squares are data above T_N. This data is the same as that in a) except background corrections as determined by the fits have been made.

Typical C$_P$ data from the 1.5 g sample of CsMnBr$_3$ are shown in Fig. 10(a) over the approximate reduced temperature range -0.1 < t < 0.1. There is no indication of a latent heat associated with the transition, and a very stringent upper limit of 0.002 J/mole can be placed on it. This data was fit to a power-law divergence (Eqs. (4) and (5)) in order to extract α and the critical amplitude ratio, $\frac{C^+}{C^-}$. The data, corrected for background terms, is plotted on a log-log scale against reduced temperature in Fig. 10(b). The solid lines are the results of the fit and clearly describe the data well. The fit is robust in that the extracted critical behaviour is not appreciably sensitive to the precise region in reduced temperature over which the fit is made, or to the inclusion of the correction-to-scaling terms in the power law. Restricting the analysis to data in the range $|t| < 0.05$, produced the estimates $\alpha=0.39 \pm 0.09$ and $\frac{C^+}{C^-}=0.19 \pm 0.10$. Independent work on a different crystal has yielded similar results.[29]

The other exponent which can be calculated using the scaling relations is z. This exponent is measured in inelastic neutron-scattering measurements, with relatively high energy-resolution at the ordering temperature, T_N. Energy scans at fixed $\mathbf{Q}$ were carried out at the $(\frac{1}{3},\frac{1}{3},1)$ ordering wavevector and then at other wavevectors moving away from $\mathbf{Q}_{ord}$ towards $(0,0,1)$[30].

a)   b)

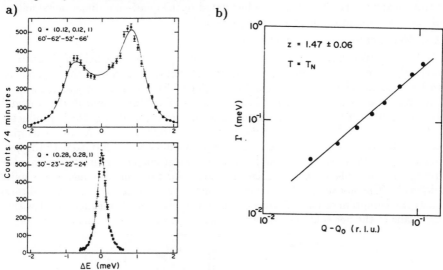

Figure 11. a) Inelastic neutron scattering from CsMnBr$_3$ at T=T$_N$ for two wavevectors[30]. b) A log-log plot of the quasi-elastic Lorentzian line width, Γ, against q at T=T$_N$ near $\mathbf{Q}_{ord} = (\frac{1}{3},\frac{1}{3},1)$. The solid line is a fit to the data with the dynamical critical exponent, z=1.47.

At T_N, correlated domains exist on all length scales: measurements made at $\mathbf{Q}_{ord}$ primarily sense the largest domains, while those made away from $\mathbf{Q}_{ord}$ systematically probe the smaller domains. The energy width of these constant-Q scans is inversely proportional to the characteristic relaxation times for the domains which are probed. The measurements, shown in Fig. 11(a) and (b), demonstrate that the large domains relax slowly, compared with the smaller domains. For q < 0.1 reciprocal lattice units, the quasi-elastic scattering was well described by

$$S(\mathbf{q},\nu) \sim (n(\nu)+1))\frac{\Gamma(\mathbf{q})\nu}{\Gamma(\mathbf{q})^2+\nu^2} \qquad (15)$$

where $(n(\nu) + 1)$ is simply the thermal population factor. The purpose of the fit was to examine the q dependence of $\Gamma(q)$ at T_N in order to estimate the dynamical critical exponent z. A log-log plot of $\Gamma(q)$ vs. q is shown in Fig. 11(b) wherein the exponent $z=1.47 \pm 0.06$ is estimated. As expected from the scaling relation (Eq. (8)), although β and ν are anomalously small for a three-dimensional system, their ratio is not and z takes on the "standard" value for an antiferromagnet, ~ 1.5.

Figure 11(a) shows very different behaviour at large q than at small q. At large q the lineshape is not simply peaked about zero energy transfer, but develops clear inelastic peaks indicative of propagating modes. Surprisingly, for q>0.15 reciprocal lattice units, the energies at which the inelastic peaks occur are equal to, or even slightly larger than the equivalent low-temperature spin-wave energies. One would normally expect a downward renormalization of the spin wave energies as the temperature approaches T_N. This may be related to the quasi-one- dimensional nature of the system, and is not presently understood.

4.2. Measurements in a Magnetic Field

The critical behaviour observed in the absence of a magnetic field in $CsMnBr_3$ is certainly novel. However measurements made in finite field showed the phenomena to be more interesting than originally imagined. Elastic neutron-scattering measurements were carried out on the same sample of $CsMnBr_3$ used in the previous studies, but with the crystal mounted in a magnet cryostat capable of producing a 6.5 T magnetic field perpendicular to the scattering plane[31]. As the scattering plane of interest is the (h, h, l) plane, these experiments were performed with the field aligned along the $(h, 0, 0)$ direction, i.e. a next-nearest-neighbour direction within the hexagonal basal plane. The original motivation for these measurements was quite physical. The so-called 120° magnetic structure shown in Fig. 3 is the natural energetic compromise for XY spins on the triangular net. A magnetic field within the plane of the net could certainly be expected to upset this balance.

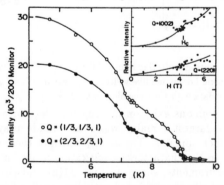

Figure 12. The temperature dependence of the $(\frac{1}{3}, \frac{1}{3}, 1)$ and $(\frac{2}{3}, \frac{2}{3}, 1)$ magnetic Bragg-peak intensities in $CsMnBr_3$ in a magnetic field of 4.2 T applied in the $(1, 0, 0)$ direction[31]. The inset shows the field dependence of the $(2, 2, 0)$ and $(0, 0, 2)$ Bragg peaks at 7 K, which are sensitive to the spin-flopped component of the moment.

Figure 12 shows the elastic neutron-scattering intensity at the $(\frac{1}{3}, \frac{1}{3}, 1)$ and $(\frac{2}{3}, \frac{2}{3}, 1)$ ordering wavevectors as a function of temperature in the presence of a 4.2 T applied magnetic field. This figure should be compared with Fig. 4, the order parameter in zero applied field, which rises smoothly and continuously from zero at $T_N(H = 0) \sim$ 8.3 K. Clearly, the presence of the field has stabilized a new magnetically ordered phase at intermediate temperatures and fields. Two phase transitions are now seen; one at $T_{N1} = 7.15 \pm 0.10$ K and the other at $T_{N2} = 9.00 \pm 0.10$ K.

Similar measurements were repeated at many field values less than 6.5 T. The resulting phase diagram, Fig. 13, shows that the zero-field transition at $T_N \sim 8.3$ K is in fact a tetracritical point. This type of novel multicritical point was predicted to occur for XY antiferromagnets on tetragonal lattices in zero magnetic field[32], but a neutron-scattering search in an appropriate candidate material, Fe_2As, failed to show any anomalous critical behaviour[33]. Tetracritical behaviour was however experimentally observed in stressed $LaAlO_3$ and $GdAlO_3$.[34]

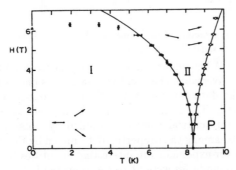

Figure 13. The phase diagram of $CsMnBr_3$ in a magnetic field applied along $(1, 0, 0)$ is shown[31]. P denotes the paramagnetic phase, II the spin-flop phase, and I the triangular phase. The lines are the results of a fit to the phase boundaries near the tetracritical point, for the purpose of extracting the cross-over exponents Ψ_{P-II} and Ψ_{II-I}.

The low-temperature phase below T_{N1} corresponds to a minor modification of the 120°, six sublattice, magnetic structure due to the presence of the field. The new intermediate phase shares a periodicity with the low-temperature phase, as both of these phases possess the $(\frac{1}{3}, \frac{1}{3}, 1)$ and $(\frac{2}{3}, \frac{2}{3}, 1)$ ordering wavevectors. Further measurements made at the $(\frac{4}{3}, \frac{4}{3}, 1)$, $(\frac{1}{3}, \frac{1}{3}, 3)$, $(0, 0, 2)$, and $(2, 2, 0)$ wavevectors identify the intermediate phase as having spin-flop character. The six sublattices collapse to four, with two of the three sublattices within a given basal plane layer becoming equivalent. The spins on these sublattices flop almost perpendicular to the field direction and cant towards the applied field. These structures are shown schematically in Fig. 13. In addition, a Monte Carlo simulation[35], to be discussed below, has produced spin configurations appropriate to these two phases using the microscopic Hamiltonian of $CsMnBr_3$ (Eq.(13)). The spin configurations in the 120° structure and the spin-flop structure are shown in Fig. 14. A transition between two such ordered states was in fact predicted by a zero-temperature classical calculation, taking into account the quantum renormalization of the spin $\frac{5}{2}$ moments. This transition is predicted to occur at a critical field of H_C =6.1 T[36]. It is marked on Fig. 14 as a triangle and can be seen to agree very well with the extrapolated zero-temperature behaviour of CsMnBr$_3$.

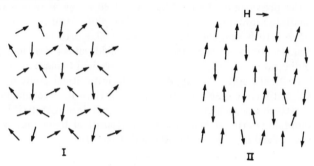

Figure 14. The magnetic structure of the triangular phase (I) and the spin-flop phase (II) as determined from Monte Carlo simulations of CsMnBr$_3$[35].

This new multicritical behaviour has now been incorporated into the description of the chiral XY universality class and is regarded as one of its properties by Kawamura[37]. All of the data suggest that both phase boundaries are continuous over the range measured. The form these phase boundaries take as they asymptotically approach the tetracritical point is controlled by cross-over exponents, Ψ_i, which are expected to be among the properties determined by the universality class. Therefore these cross-over exponents are interesting quantities to study. With this in mind the measured boundaries were fit to the expected form:

$$\frac{T_i(H) - T_N(H = 0)}{T_N(H = 0)} \sim (H^2)^{\frac{1}{\Psi_i}} \qquad (16)$$

for the purpose of estimating Ψ_i. However these estimates are extremely sensitive to the range of data over which the fit is attempted and the choice of $T_N(H = 0)$. The results of one such fit are shown as the solid lines in Fig. 13, and this fit produced the estimates Ψ_{P-II} =1.21 ± 0.07 and Ψ_{I-II} =0.75 ± 0.05. If $T_N(H = 0)$ is allowed to vary by as little as a few hundredths of a degree, the two Ψ estimates can change considerably, but always in such a way that $\Psi_{P-II} + \Psi_{I-II} \sim 2$. A subsequent study[38], based on susceptibility measurements attempted a more precise determination of the cross-over exponents, and in fact claimed Ψ_{P-II}=1.02 ± 0.05 and Ψ_{I-II} =1.07 ± 05. These estimates are in excellent agreement with their expected values within Kawamura's picture. However it is very difficult to distinguish between the two lines of phase transitions for fields less than ~ 1 T. The author has digitized the data from which these cross-over exponents were produced and has reanalysed it, allowing a variation of ~ 0.05 K in $T_N(H = 0)$. This is shown in Fig. 15. While the cross-over exponent estimates claimed by this study are correct for T_N=8.32 K, a variation in T_N of only 0.05 K results in relative changes in the exponent estimates of roughly 25%. Consequently, the only firm result on the cross-over exponents at present is the constraint $\Psi_{P-II} + \Psi_{I-II} \sim 2$.

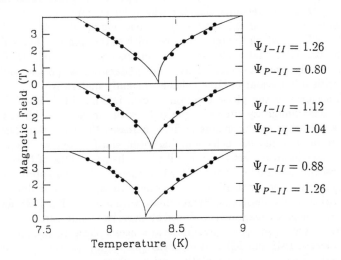

$\Psi_{I-II} = 1.26$

$\Psi_{P-II} = 0.80$

$\Psi_{I-II} = 1.12$

$\Psi_{P-II} = 1.04$

$\Psi_{I-II} = 0.88$

$\Psi_{P-II} = 1.26$

Figure 15. Fits to the phase boundaries of CsMnBr$_3$ determined from susceptibility measurements are shown[38]. As can be seen, variation in T_N(H=0) by $\pm$ 0.05 K results in $\sim$ 25% variation of the cross-over exponents.

An attempt has also been made to perform a critical-behaviour analysis of the uniform susceptibility data itself (as opposed to on the phase diagram derived from such measurements, as described above). Unfortunately, this work[39] incorrectly interpreted the predicted cross-over exponents as producing a zero in the uniform susceptibility at T_N. In fact the uniform susceptibility displays a break in its slope at T_N.

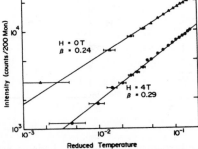

Figure 16. Log-log plots of the $(\frac{1}{3}, \frac{1}{3}, 1)$ magnetic Bragg-peak intensity versus reduced temperature for zero applied field and in a field of 4 T applied in the $(1, 0, 0)$ direction in CsMnBr$_3$[31]. The lines are results of power-law fits to the data with the critical exponent β as indicated.

The application of the magnetic field completely changes the character of the highest-temperature phase transition. The transition is now from the paramagnetic to the spin-flop state. Theoretically, it is expected that this transition should belong to the 3D XY universality class for which, for example, $\beta \sim 0.35$. Figure 16 shows measurements[31] of $M(\mathbf{Q}_{ord})$ at the $(\frac{1}{3}, \frac{1}{3}, 1)$ position in zero field and also in an applied field of 4T, for which the temperature width of the spin-flop phase is approximately 1.3 K. It is clear that while the β exponent increases on application of the magnetic field, it only reaches ~ 0.29. This is not understood at present and a full study of the critical behaviour in large fields is currently underway.

The form of the phase diagram, shown in Fig. 13, is itself remarkable. The P-II phase boundary curves to the right with increasing field; that is to say that the application of the magnetic field increases the transition temperature. The nature of this unusual behaviour has been clarified by two Monte Carlo studies which sought to understand the full phase diagram. One of these used the actual microscopic parameters which enter into the Hamiltonian appropriate to CsMnBr$_3$, Eq.(13), and therefore simulated a very anisotropic, quasi-one-dimensional system[35]. The other used an isotropic Hamiltonian, with $J = J'$ in Eq. (13), for computational ease[40]. The phase diagrams produced by these two studies are shown in Fig. 17. The simulation with isotropic interactions produces a phase diagram in which both phase boundaries, P-II and I-II, have negative slopes for all values of temperature (Fig. 17 (a)). However, the simulation using the quasi-one-dimensional Hamiltonian, produced a phase diagram in which the P-II line of transitions has positive slope, at least at small values of the applied magnetic field (Fig. 17 (b)). Therefore, the positive slope of the P-II line of transitions on the phase diagram is identified as being a consequence of the quasi-one-dimensional nature and fluctuations of CsMnBr$_3$.

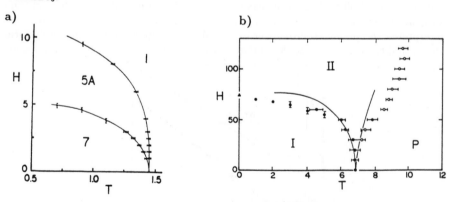

Figure 17. Magnetic phase diagrams produced by Monte Carlo simulations using a) the isotropic Hamiltonian, Eq. (13) with $J = J'$[40], and b) the quasi-one-dimensional Hamiltonian, Eq (13), appropriate to CsMnBr$_3$[35]. The lines in b) show the experimentally determined phase boundaries[31] scaled to agree at H=0.

4.3. Exponent Estimates: Kawamura vs. Azaria et al.

A consistent picture has emerged experimentally with regard to the character of the transition at $T_N(H = 0)$ for CsMnBr$_3$. There is no indication of a discontinuous phase transition. Further, several experimental groups have independently measured critical exponents which are both novel, in that they are clearly distinct from the exponents which characterize the well known three-dimensional universality classes (Ising, XY, and Heisenberg), and show none of the signs of being "effective" exponents, i.e. outside of the asymptotic regime.

The measured exponents are listed in Table I, and are compared with the predictions from the three theoretical perspectives; that of Kawamura[21,22], that of Azaria, Delamotte, and Jolicoeur (assuming the tricritical mean-field theory scenario)[23], and those predicted for the 3D XY model[41]. As can be seen, the predicted exponent values of Kawamura and those of Azaria, Delamotte, and Jolicoeur are remarkably close to each other. Given that the experimental uncertainties on the exponents are typically $\sim 10\%$ of the value itself, the only real possibility of distinguishing between the two scenarios seems to lie in the critical heat capacity exponent, α. Here the experimental evidence tends to favour the Kawamura scenario.

Critical Exponents for CsMnBr$_3$

	Experiment	$Z_2 \times S_1$ [21,22]	Tricritical[23]	3DXY[41]
β	0.22±.02[24], 0.21±.02[25], 0.24±.02[31], 0.25±.01[26],	0.25±.015	0.25	0.346±.009
γ	1.01±.08[25] 1.10±.05[27]	1.14±.05	1.0	1.316±.009
ν	0.54±.03[25] 0.57±.03[27]	0.55±.02	0.5	0.669±.007
α	0.39±.09[28] 0.40±.05[29]	0.35±.05	0.5	-0.007±.009
$\dfrac{C^+}{C^-}$	0.19±0.10[28] 0.32±0.20[29]	0.38±0.20	–	–
z	1.46±.06[30]	–	–	–

Table 1. Measured and predicted critical exponents for CsMnBr$_3$ are shown. The exponents are defined in Eqs. (1-4) and (8).

As the zero-field exponents themselves do not clearly settle the issue, the resolution of this theoretical conflict may well reside in other critical phenomena; the universal amplitude ratios, the cross-over exponents, or other features associated with the full magnetic phase diagram. These aspects of the critical phenomena are not as well investigated at present, either theoretically or experimentally, as the zero-field exponent values. These seem to be the most promising areas of future research on this interesting material.

5. The Helical Rare-Earth Metals: Ho, Dy, and Tb

The symmetry of the order parameter in $CsMnBr_3$ has both an XY and an Ising nature due to the chiral local ordering of the XY moments on a triad of near neighbours. This also occurs for XY magnetic systems which exhibit a helical or spiral magnetic structure. Such structures are known to occur in the rare-earth metals, Ho, Dy, and Tb[42]. The spiral structure, and hence the chirality, does not arise from geometrical effects, as in $CsMnBr_3$, but rather due to the combination of long-range, oscillatory, RKKY interactions between the well-localized 4f magnetic moments and the hexagonal close-packed (HCP) lattice on which the moments reside. In fact, this combination leads to very rich phase diagrams in all of the magnetic rare-earth elements.

At high temperatures Ho, Dy, and Tb all exhibit a helical or spiral state: the moments are confined to the basal plane in ferromagnetic sheets, but the ordering direction rotates by some small angle from one basal plane to the next, so as to form a long wavelength, incommensurate helix along c. This structure is shown schematically in Fig. 18. (More recently it has been found that there are many subtleties associated with these structures. See for example references 42 and 43.) Since the magnetic structure shown in Fig. 18 possesses a chirality, in that it can wind into either a clockwise or counter-clockwise helix along c, the symmetry of the order parameter in this state has both XY and Ising character as well.

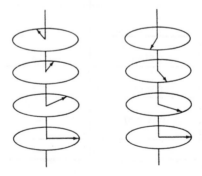

Figure 18. A schematic representation of the two chiral domain states exhibited by the spiral rare-earth metals. Ferromagnetically-coupled moments lie within the HCP basal plane. and their ordering direction rotates by some small angle from one basal plane to the next.

The critical properties of the helical rare-earth metals have been of considerable theoertical interest. These elements were originally argued to show novel critical properties due to the presence of four independent components to the order param-eter, described by the $O(n=4)$ model[44]. (This view was challenged by an analysis which showed that it lead to a discontinuous phase transition[45].) Kawamura[20-22] includes them in the $Z_2 \times S_1$ universality class, whereas Azaria et al.[23] argue that these systems could undergo a weakly discontinuous transition, display tricritical mean-field theory behaviour, or $O(n=4)$ behaviour.

The helical state in Tb only exists over ~ 10 K, below which it undergoes a discontinuous transition to a ferromagnetic state. The structure is stable in both Ho and Dy over a wide temperature range, and of these, the transition to the spi-ral state in Ho ($T_N \sim 132$ K) is better studied. (Naturally-occurring Dy possesses a large neutron-absorption cross-section which makes neutron measurements diffi-cult.) However, even in Ho, the experimental results appear to conflict with each other.

The order of the spiral phase transition, that is whether it is a continuous or discontinuous transition, is not fully established. One dilatometry study on Ho shows evidence for a discontinuous transition in a-axis behaviour, but not in c-axis behaviour on the same crystal[46]. Other studies report no evidence of a discontinuity[28,47-49]: of particular importance is a careful heat capacity measure-ment which failed to observe latent heat and set an upper limit of 0.03 J/mole on any possible latent heat associated with the transition[28]. It seems clear that if the transition is discontinuous, the discontinuity occurs at very small values of the order parameter.

Measurements of the sublattice magnetization exponent β are quite difficult in the rare-earth magnets due to the large moments (~ 8 μ_B) which give rise to potentially large extinction effects. All neutron measurements on the helical rare-earth metals probe ordering wavevectors of the form $(h, k, l \pm \tau)$, where $\tau = \frac{\theta}{\pi}$ and θ is the turn angle between the moment directions on adjacent basal planes, within the helical structure. Elastic neutron-scattering measurements on both Dy and Ho, have consistently produced the estimate $\beta \sim 0.39$[47,50,51]. Energy-integrated neutron measurements of the critical scattering above T_N have been used to estimate γ in both Ho and a sample of Dy which had been isotopically enriched so as to reduce its neutron-absorption cross-section[49]. Measurements are typically made by cutting through the ordering wavevector, in this example $(0, 0, 2 \pm \tau)$, along l and also perpendicular to this direction. A consistent picture of the critical scattering in Ho and Dy has arisen from this study, and it is consistent with an additional independent study on a different sample of Ho to be discussed below[52]. The results for Ho are $\gamma = 1.05 \pm 0.07$, and $\nu = 0.57 \pm 0.04$[49], while those for Dy are $\gamma = 1.14 \pm 0.10$ and $\nu = 0.57 \pm 0.05$[49]. These values are in good agreement with the estimates for γ and ν in CsMnBr$_3$. The problem is that these values are *not* consistent with the estimates for β in Ho using the scaling relations.

Heat capacity measurements on Ho and Dy have not produced a consistent picture[28,48,53-54]. While relatively early estimates of α in both elements yielded an exponent of $\alpha \sim 0.23$, a more recent study (on the same single crystal of Ho used in the previously mentioned neutron study) has shown that the α estimate is not robust to correction-to-scaling terms[28]. This is perhaps not so surprising, as the RKKY interaction between the magnetic moments in the rare earths are long range, falling off with distance only as r^{-3}, in contrast to the short range, nearest-neighbour-only interactions in $CsMnBr_3$. Therefore a rather narrow critical regime may be expected in the rare-earth metals. Measured and predicted exponents for Ho and Dy are listed for easy comparison in Table 2.

Critical Exponents for Dy and Ho

	Experiment	$Z_2 \times S_1$[21,22]	O(n=4)[44]	Tricritical[23]
Ho:β	0.39±.04[47] 0.37±.10[52]	0.25±.015	0.39	0.25
Dy:β	0.39±.04[51] 0.34±.01[50]	0.25±.015	0.39	0.25
Ho:γ	1.14±.10[49]	1.14±.05	1.39	1.0
Dy:γ	1.05±.07[49]	1.14±.05	1.39	1.0
Ho:ν	0.57±.04[49]	0.55±.02	0.70	0.5
Dy:ν	0.57±.05[49]	0.55±.02	0.70	0.5
Ho:α	0.27±.02[49] 0.10±.02[28] 0.22±.02[28]	0.35±.05	−0.17	0.5
Dy:α	0.24±.02[53] 0.18±.08[54]	0.35±.05	−0.17	0.5

Table 2. Measured and predicted critical exponents for Ho and Dy are shown. The exponents are as defined in Eqns. (1-4). Two different estimates of α were obtained in the same study[28], due to the sensitivity of the data to correction-to-scaling terms.

A very recent study[52] of Ho, using both resonant magnetic X-ray as well as neutron scattering techniques, may well have cast some light on why a consistent picture of the critical properties of the rare-earth metals has not emerged. These new measurements made use of the inherent high angular resolution afforded by synchrotron radiation and the high transverse (perpendicular to Q) resolution possible with neutron scattering at relatively small scattering angles, to discover the presence of a "second length scale" near T_N. In addition to the previously measured correlation length, which is characteristic of the average ordered domain size, another much longer characteristic length scale appears just above T_N. This length scale is sufficiently large, and therefore the scattering due to it sufficiently narrow in Q-space, that high resolution is required to observe it. Figure 19. shows high-resolution, energy-integrated neutron scattering around $(0, 0, \tau)$ at $T=T_N+0.28$ K. The data from this transverse scan has been fit to a two-component lineshape representing the presence of a relatively short correlation length (the broad component) and a relatively long correlation length (the narrow component). This transverse neutron-scattering measurement is sensitive to both the broad and the narrow components of the critical scattering, whereas the resonant magnetic X-ray scattering experiment, with its intrinsically high resolution and small resolution volume, is sensitive to to the narrow component only. The transverse widths of the two components of the critical scattering from the X-ray and neutron-scattering studies are shown in Fig. 20.

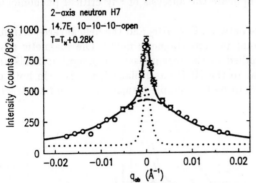

Figure 19. Transverse neutron-scattering scan through the $(0, 0, \tau)$ position in the paramagnetic state of Ho[52]. This scan emphasizes the two-component nature of the scattering present above T_N. The broad component is fit to a Lorentzian (dashed line), while the narrow component is fit to a Lorentzian squared (dotted line).

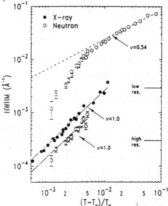

Figure 20. The half-width at half-maximum (HWHM) of the broad and narrow components of the critical scattering in Ho above T_N are shown[52]. The solid lines are fits of the data to $1/\xi_{ab} \sim \kappa_{ab} \sim ((T\text{-}T_N)/T_N)^\nu$. The two sets of neutron measurements correspond to those taken at high and low resolution.

The second, longer length scale is thought to arise from random strain fields localized at or near the sample surface. The relatively short correlation length appears to be diverging at T_N with a power-law exponent consistent with previous neutron scattering measurements, $\nu = 0.54 \pm 0.04$. The long correlation length also appears to diverge at T_N with roughly double the power-law exponent. It is possible that the relative importance of the long length scale compared to the short length scale is sample-dependent, especially if it arises from strain field effects. Irrespective of its physical origin, the presence of a second, much longer, length scale near T_N will certainly complicate the analysis of any critical phenomena which is unaware of its presence.

One difference between CsMnBr₃ and the helical rare-earth metals is the nature of the critical point in zero magnetic field. The magnetic phase diagram of Ho has been well studied in the recent past, for magnetic fields applied both within and perpendicular to the HCP basal plane (Fig. 21). In both cases the zero field transition appears to be to a simple transition, i.e. *not* a tetracritical point[55].

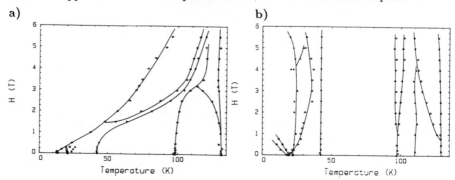

Figure 21. The magnetic phase diagram for Ho with a magnetic field applied along the a) b-axis and b) the c-axis. This phase diagram was determined from magnetization measurements[55].

6. CsCoBr₃ and CsCoCl₃: Ising-like STL Antiferromagnets

As discussed in section 3, CsCoBr₃ is well described as an Ising-like, effective spin-$\frac{1}{2}$ antiferromagnet on a stacked triangular lattice[19]. The nature of the frustration within the triangular plane is quite different than in the case of the XY-like moments in CsMnBr₃. The Ising system has no simple energetic compromise like the 120° structure (Fig. 5), so that chirality is irrelevant to the three-dimensionally ordered states which it displays.

6.1. Theoretical Models

The classical Ising model with AF interactions on a STL has been theoretically investigated and is known to display two magnetically ordered states, both of which include partial disorder[56-58]. As the temperature is lowered, an ordered state is formed in which two out of three spins within a triad of near neighbours form an up-down pair, while the third remains paramagnetic. This state is labelled $(M, -M, 0)$. At a lower temperature it enters another partially paramagnetic state labelled as $(\frac{M}{2}.\frac{M}{2}.-M)$. in which all three sites within a triad of near neighbours possess an ordered moment, but the moments at two of the sites are half that of the third and are aligned antiparallel to the third.

The higher temperature transition from the paramagnetic state to the $(M.-M.0)$ state has recently been the subject of much theoretical attention. Originally, an argument related to the symmetry of the Ginzburg-Landau-Wilson Hamiltonian was made which put this phase transition in the same universality class as the 3D XY model[56]. This view was challenged by a conventional Monte Carlo study which indicated behaviour close to that of tricritical mean-field theory[59]. Two new histogram Monte Carlo studies, with full finite-size scaling analysis, have shown that there is in fact no evidence for tricritical behaviour[60,61]. The exponent estimates given by these two new studies are quite close to each other and to those expected from the 3D XY model[41]. However both studies show deviations of the exponent estimates in the same direction away from the predicted exponents. One of these studies suggests the possibility of novel critical phenomena associated with this frustrated model[60].

6.2. Experimental Studies of CsCoBr₃ and CsCoCl₃

Both CsCoBr₃ and CsCoCl₃ appear to display the magnetically-ordered phases discussed in section 6.1 $((M.-M.0)$ and $(\frac{M}{2},\frac{M}{2},-M))$[14,62]. Both of these ordered structures give rise to a tripling of the unit cell within the basal plane. As discussed in section 3, the ordering along c is also antiferromagnetic so that the ordering wavevectors are of the form $(\frac{h}{3},\frac{h}{3}.l)$, where l is odd. The intensity of the elastic neutron-scattering at the $(\frac{2}{3}.\frac{2}{3}.1)$ Bragg position in CsCoBr₃ is shown as a function of temperature in Fig. 22[63]. In fact *three* ordered phases are seen. The material undergoes its high temperature transition at $T_{N1} \sim 28.4$ K, and its lowest temperature transition near $T_{N3} \sim 13$ K. However, the Bragg intensity changes anomalously near 16 K, and then flattens out until T_{N3} is reached. This is fairly strong evidence for the presence of an intermediate phase occuring between ~ 13 K and $T_{N2} \sim 16$ K.

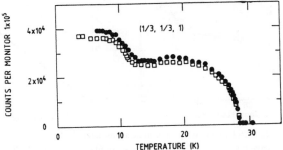

Figure 22. Elastic neutron-scattering intensity as a function of temperature at the $(\frac{1}{3},\frac{1}{3}.1)$ ordering wavevector in CsCoBr₃[63]. Measurements are shown in both zero applied magnetic field (solid circles) as well as a field of 2.2 T applied along the moment direction.

The critical behaviour associated with the transition from the paramagnetic to the ordered state at T_{N1} in both $CsCoBr_3$ and $CsCoCl_3$ has been the subject of many experimental investigations. Unfortunately a consistent picture has yet to emerge. Three determinations of the exponent β have been made. Mekata and Adachi[62] obtained the estimate 0.35 ± 0.02 for $CsCoCl_3$, a value which is consistent with the predictions of the 3D XY model. Yelon et al.[14] obtained 0.31 ± 0.02 for $CsCoBr_3$, a value which is consistent with both the predictions of the 3D Ising model and with the new Monte Carlo study which suggests the possibility of novel critical phenomena. Finally, Farkas et al.[63] obtained 0.22 ± 0.02 for $CsCoBr_3$, which is close to that expected from tricritical mean-field theory. The reason for this wide range of experimental estimates is not presently clear. It may be related to the presence of low-energy soliton excitations[64–67], which have a pronounced temperature dependence near T_{N1}, and could lead to difficulties in separating the elastic and inelastic neutron-scattering contributions. Another possible reason is that the spin correlations may not follow the simple OZ form, as was shown to be the case in Ho[52].

Recently a careful heat capacity measurement[68] has been made on the same single crystal sample of $CsCoBr_3$ studied by Farkas et al.[63] Data from this study are shown in Fig. 23. These measurements were made using very similar heat-pulse techniques as those used on $CsMnBr_3$ and Ho described earlier[28]. A clear anomaly is seen only near $T_{N1} \sim 28.4$ K (Fig. 23(a)), with no evidence (Fig. 23 (b)) of the lower temperature transitions seen with neutron scattering from the same crystal. In fact this is consistent with an earlier heat capacity measurement[14]. These measurements show that the relative entropy associated with the lower transitions are exceedingly small.

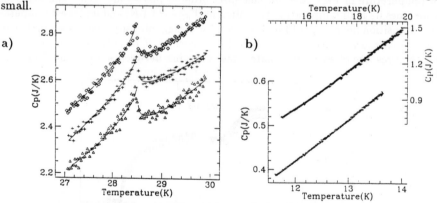

Figure 23. a) Three independent specific heat scans near $T_{N1} \sim 28.4$ K in $CsCoBr_3$[68]. For clarity the data sets have been offset. The solid curves are fits to the data. b) The specific heat near the transitions identified in neutron-scattering experiments at $T_{N2} \sim 16$ K and $T_{N3} \sim 12$ K.

The heat capacity associated with the transition at T_{N1} was analyzed along similar lines as $CsMnBr_3$[28]. The fits were found to be robust to the inclusion of correction-to-scaling terms and the best fits along with the data are shown in Fig.

24. This analysis yields an estimate of $\alpha = -0.025 \pm 0.004$ and $\frac{C^+}{C^-} = 1.07 \pm 0.02$. The α estimate is close to that expected from the 3D XY model[41], but shows a discrepancy in the same direction as the recent Monte Carlo study[60]. Perhaps the most remarkable comparison with these new measurements is not to theory, but rather to heat capacity measurements on the superfluid transition in liquid ^{4}He, which give $\alpha = -.026 \pm 0.004$ and $\frac{C^+}{C^-} = 1.112 \pm 0.022$[69]. The superfluid transition in liquid ^{4}He is also expected to belong to the 3D XY universality class. Table 3 shows the comparison of these new heat capacity measurements on CsCoBr$_3$ to those on liquid ^{4}He as well as the relevant theoretical estimates.

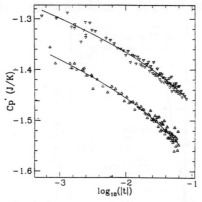

Figure 24. The specific heat versus logarithm of $|t|$ for data from CsCoBr$_3$ as well as the best fit[68]. Upward triangles are for $t < 0$, while downward triangles are for $t > 0$. The negative values reflect the negative power law amplitudes in Eq. (4).

	Experiment[68]	Liquid ^{4}He[69]	3DXY[41]	Monte Carlo[60]
α	−0.025±.004	−0.026±.004	−.007±.009	−.05±.03
$\dfrac{C^+}{C^-}$	1.07±.02	1.12±.022	—	—

Table 3. Critical heat capacity exponent and amplitude ratio for CsCoBr$_3$ compared to theoretical predictions and experimental values obtained for liquid ^{4}He.

6.3. The Role of Solitons in the Formation of the Ordered State in CsCoBr$_3$

The nature of the high-temperature ordered state in CsCoBr$_3$ and CsCoCl$_3$ is particularly interesting. Perhaps the most striking manifestation of geometrical frustration in these systems is the fact that the transition at T_{N1} is from a paramagnetic state to a long-range ordered structure in which the magnetic moment at one out of three sites is disordered. Both experiment and simulation have sought to examine how such an ordered state can form out of the paramagnetic state.

The magnetic excitation spectrum in both of these spin $\frac{1}{2}$, quasi-one-dimensional magnetic insulators has been well studied and is known to consist of two sets of excitations as shown in Fig. 25 for CsCoBr$_3$[19,64]. These are the spin-wave excitations which exist in a "bow tie" continuum of states, all at relatively high energy, separated from the ground state by a large energy gap at all wavevectors. Such excitations can be thought of as corresponding to single spin flips. The energy gap relevant to these excitations is $\sim$ 155 K in CsCoBr$_3$. Clearly, these excitations cannot play a significant role in the phase transition which occurs at only 28.4 K.

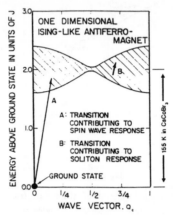

Figure 25. The magnetic excitation spectrum of a quasi-one-dimensional Ising-like antiferromagnet, for which Eq. (14) is the appropriate Hamiltonian. The separation of the centre-of-gravity of the spin-wave continuum from the ground state is $\sim$ 155 K in CsCoBr$_3$.

In addition it is well established that at finite temperature, transitions within the band of spin-wave states are possible[64-67]. These excited state transitions are known to give rise to a high density of low-energy excitations in both CsCoBr$_3$ and CsCoCl$_3$. These excitations are referred to as solitons and can be thought of physically as anti-phase domain walls propagating along the quasi-one-dimensional c-axis. The role of these solitons within the high-temperature ordered state has been investigated in several studies. Neutron-scattering work by Boucher et al.[67] on CsCoCl$_3$ demonstrated the existence of solitons well below T_{N1}. Kikuchi and Ajiro[70] used NMR to study the effects of the soliton response on the ^{132}Cs resonance, and their measurements were consistent with a disordering of $\frac{1}{3}$ of the magnetic sites as the high-temperature ordered phase was approached from below.

Recent neutron-scattering measurements on CsCoBr$_3$ have specifically examined how the soliton scattering changes as T_{N1} is approached from above, in order to determine if solitons have a substantial role in mediating the transition[66]. Inelastic neutron scattering at fixed energies of ($\nu = \frac{E}{\hbar}$) 0.2 THz and 0.05 THz are shown in Fig. 26 for wavevectors of the form $(h, h, 1)$. This low-energy inelastic scattering is taken at the one-dimensional zone centre ($h.k.l = odd$), and the scan passes through both an ordering wavevector, $Q_{ord} = (\frac{4}{3}.\frac{4}{3}.1)$, as well as the special point $Q_{1D} = (1.19.1.19.1)$. The significance of Q_{1D} is that at this wavevector the Fourier transform of the basal plane interactions, J', is identically zero, so that (excluding resolution effects) one should obtain a measure of the response of the one-dimensional system at Q_{1D}.

Scattering of the form displayed in Fig. 26 was taken for relatively low energy transfers, from 0 to 0.5 THz. The top part of Fig. 26 shows scattering at $\nu = 0.2$ THz, while the basal-plane component of wavevector, h, is scanned. This h-dependence is typical for scattering at energy transfers, $\nu \geq 0.2$ THz, and is as expected for a one-dimensional soliton-bearing system. The scattering remains flat across this $l=1.0$ ridge in wavevector space and decreases in intensity as the temperature is lowered. This contrasts sharply with the behaviour seen at lower energy transfers, such as the $\nu=0.05$ THz data shown in the lower part of Fig. 26, in which at temperatures as high as 43 K (~ 1.5 T_{N1}) some dispersion is seen within the basal plane, and the intensity reaches a maximum at Q_{ord}.

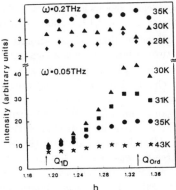

Figure 26. Constant energy scans of the form ($h.h.1$) are shown in CsCoBr$_3$[66]. The top part of the panel shows scans at the relatively large energy transfer. $\nu =0.2$ THz, while the lower part of the panel shows $\nu=0.05$ THz. Q_{ord} is at ($\frac{4}{3}.\frac{4}{3}.1$), while Q_{1D} is at ($1.19.1.19.1$).

The soliton contribution to the phase transition was examined by studying the sum over the soliton part of the relaxation function, $F(Q,\nu)$, both close to and away from Q_{ord}. This quantity is related to the dynamic structure function, $S(Q,\nu)$, by:

$$S(Q,\nu) = \frac{h\nu}{k_B T}(1 - exp(-\frac{h\nu}{k_B T}))^{-1}F(Q,\nu) \qquad (17)$$

(In this form the thermal population factor is separated from the relaxation function.) The sum of $F(Q,\nu)$ over the low-energy soliton response is shown in Fig. 27, both at Q_{ord} and at Q_{1D}, and is clearly very different for these two wavevectors. At Q_{ord} the integrated soliton response is larger than that at Q_{1D} even at 43 K. Most strikingly , it appears to be heading towards a divergence at T_{N1}. These data show the soliton scattering to condense into the Bragg peaks of the three-dimensional ordered state below T_{N1}.

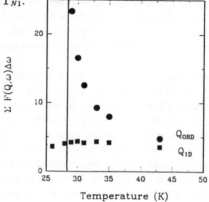

Figure 27. The temperature dependence of the soliton part of the relaxation function, $F(Q, \nu)$ at Q_{ord} and Q_{1D}[66]. The soliton component appears headed towards a divergence at Q_{ord} and T_{N1}, while that at Q_{1D} is largely unaffected. $T_{N1} \sim 28.4$ K is indicated by the vertical line.

The physical picture suggested by these measurements is as follows. The development of the high-temperature ordered state is a dynamic process which involves locking pairs of AF chains into anti-phase domain singlets. This process leaves the third nearest-neighbour AF chain with no mean exchange field acting on it and hence it maintains its paramagnetic nature below T_{N1}. This dynamic locking-in process involves a *local* reorganization of the soliton density: instead of being uniformly distributed over all three near-neighbour chains, the solitons are present on only one of the three chains. All indications are that the high-temperature ordered state below T_{N1} is truly three-dimensional in nature. The data shows that the three-dimensional correlated regions form well above T_{N1} and grow as the temperature is lowered. This can be achieved by a phase separation of the AF chains into soliton-rich and soliton-depleted regions; i.e. on the scale of one correlation length along c, a given chain changes over from a paramagnetic state (associated with a locally-high density of solitons) to one of the two possible ordered domain states. Consequently, these measurements suggest that regions of high soliton density are localized along the chains and are correlated with like-regions along neighbouring chains. This picture has recently received support from a histogram Monte Carlo study of the Ising STL antiferromagnet, in which the real-space spin and soliton configurations were examined[60].

7. Antiferromagnetism on a Network of Corner-Sharing Tetrahedra

The geometrical frustration in the STL antiferromagnets is two-dimensional in nature, since the antiferromagnetic interactions along the third dimension, the stacking direction, are unfrustrated. These materials undergo conventional phase transitions (in that they are characterized by sharp Bragg peaks, a divergent $\chi(Q)$, etc.) with novel properties. Naively one expects that if the interactions were geometrically frustrated in all dimensions, the manifestations of the frustration would be more pathological. This seems to be the case, at least for one class of magnetic materials which exhibit such frustration, the pyrochlores.

7.1. The Pyrochlores

As already discussed in section 1, the three-dimensional analogue of a triangle is a tetrahedron. The frustration inherent in the combination of antiferromagnetic interactions and the tetrahedral geometry can be appreciated by considering four antiferromagnetically coupled near neighbour spins on a single tetrahedron. (For simplicity the Ising case will be discussed, although there would be interesting consequences of geometrical frustration for vector spins as well.) As shown in Fig. 28 (a) arranging any two spins into an up-down pair does not allow the other spins to align antiferromagnetically without frustrating two of the six interactions.

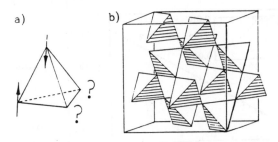

Figure 28. a) The frustration inherent between four antiferromagnetically-coupled spins on a single tetrahedron is illustrated. b) The three-dimensional network of corner-sharing tetrahedra formed by one of the metal-ion sublattices in the pyrochlores is shown.

The pyrochlores can be thought of as formed by a network of corner-sharing tetrahedra. Many pyrochlores[71] have the chemical formula $A_2B_2O_7$ and most of these belong to the cubic space group Fd3m. Both the A and B metal sublattices form an infinite three-dimensional network of corner sharing tetrahedra, as shown in Fig. 28(b). This type of network forms in other materials as well. However, as will be elaborated on below, the rare-earth based pyrochlores can be made very pure in powder form, while this is not the case for other related magnets.

Susceptibility measurements on a number of $A_2B_2O_7$ pyrochlores show interesting effects at low temperature. Figure 29 shows the zero-field-cooled and field-cooled susceptibilities for $Y_2Mo_2O_7$[72] and $Tb_2Mo_2O_7$[73] respectively. Above a particular "freezing" temperature, T_f, (~ 20 K for $Y_2Mo_2O_7$ and ~ 25 K for $Tb_2Mo_2O_7$) the zero-field-cooled and field-cooled susceptibilities are identical. However, below T_f they show a marked history dependence. History dependence is usually associated with the freezing of spin configurations into short-range ordered droplets whose conformation and morphology depend on the prior environment of the material. It is a common feature of spin glasses and other disordered materials at low temperatures[74]. However, in spin glasses it is a *combination* of frustrated interactions *and* structural randomness which gives rise this behaviour. The rare-earth pyrochlores discussed here are of high chemical quality. Neutron and X-ray powder diffraction on samples of both $Y_2Mo_2O_7$ and $Tb_2Mo_2O_7$ show no evidence of structural disorder. Cation disorder can be ruled out on the basis of a bond-length argument: the rare-earth ion simply does not fit in the transition metal site[75]. Non-stoichiometric oxygen content seems to be the most likely form of structural disorder, but diffraction profile refinements indicate that any such non-stoichiometry must be below the 1% level. Small single pyrochlore crystals of the form $A_2B_2O_7$ have been grown, but their lattice parameters are quite different from those of the same material in powder form. As the powders are known to be relatively free of structural disorder, this indicates that the single crystals produced to date possess substantial disorder.

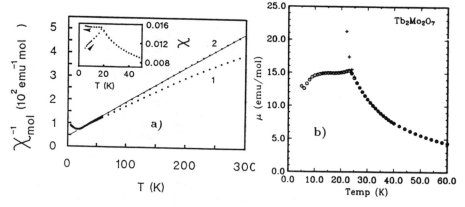

Figure 29. The field-cooled (FC) and zero-field cooled (ZFC) susceptibilities measured in (a) $Y_2Mo_2O_7$[72] (inset of (a)) and (b) $Tb_2Mo_2O_7$[73]. The two sets of points in (a), labeled 1 and 2, differ in that data set 2 has been corrected for diamagnetism. In both cases, history dependence appears below a characteristic T_f, in that the FC susceptibilty is larger than the ZFC susceptibility.

7.2. Neutron-Scattering Studies of $Tb_2Mo_2O_7$

Neutron-scattering studies have been carried out on powder samples of several pyrochlores. The most extensive measurements have been made on $Tb_2Mo_2O_7$, for the simple technical reason that the large magnetic moment ($\sim 9\ \mu_B$ at the Tb^{3+} site) gives rise to an intense neutron-scattering signal[73,76]. The Mo^{4+} site also possesses a magnetic moment, but it is substantially smaller than that at the Tb^{3+} site. Since a neutron-scattering experiment measures dynamical spin-correlation functions of the form $< S_i(0)S_j(t) >$, the contribution to the neutron-scattering signal from Tb^{3+}-Tb^{3+} correlations is much larger than that from Mo^{4+}-Mo^{4+} correlations, or Tb^{3+}-$Mo4+$ correlations. The experiment is then primarily sensitive to Tb^{3+}-Tb^{3+} correlations and one can reasonably think of the magnetism associated with the Mo^{4+} site as providing an additional interaction path for the Tb^{3+} moments.

Neutron-scattering measurements on a powder sample of $Tb_2Mo_2O_7$ are shown in Fig. 30[76]. Relatively low-energy resolution elastic-scattering measurements were made at an incident neutron energy of 3.52 THz. Representative scans are shown as a function of temperature in Fig. 30(a). Strong diffuse scattering, reminiscent of a liquid structure factor, is seen to develop continuously below $\sim$ 50 K, with the growth of this scattering being most pronounced below $\sim$ 30 K. The measured intensity is seen to increase with decreasing temperature at all wavevectors studied, and at low temperatures strong diffuse peaks are seen near Q=1.0 and 2.0 $\AA^{-1}$ with a minimum between them near $\sim$ 1.5 $\AA^{-1}$. Interestingly, the width in Q space of the diffuse peaks is approximately temperature-independent over the range shown, even though the strength of the scattering is increasing. This indicates that while the static magnetic moment (on the scale of the inverse of the frequency resolution of the measurement) within a correlated volume is growing below $\sim$ 50 K, the spatial extent of a correlated region is not. The half width of the diffuse peaks can be taken as a measure of the correlation length, which is estimated at 5 ± 1 $\AA$. A correlated droplet of Tb^{3+} spins therefore includes spins on nearest-neighbour tetrahedra only.

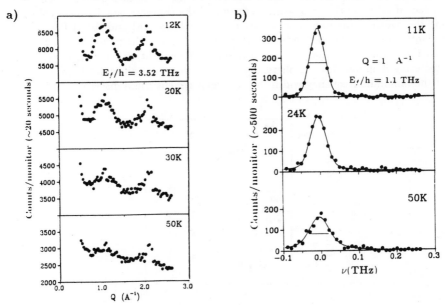

Figure 30. a) Low-resolution elastic neutron-scattering scans of $Tb_2Mo_2O_7$ are shown for four temperatures[76]. Strong diffuse scattering, with an approximately temperature-independent wavevector width develops continuously below $\sim$ 50 K. b) High-resolution inelastic, constant-|Q| scans are shown for three temperatures in $Tb_2Mo_2O_7$ at $|Q| = 1.0\AA^{-1}$.

Constant $|Q|$, high-resolution energy scans were performed near the maximum of the strongest diffuse peak at $Q=1.0$ Å^{-1}, and near the minimum of the profile at $Q=1.5$ Å^{-1}. Representative frequency scans at $Q=1.0$ Å^{-1} (corrected for background) are shown in Fig. 30 (b). The distributions are dominated by a quasi-elastic peak which increases with intensity as the temperature is lowered. In the vicinity of T_f (~ 25 K), there is evidence that the inelastic scattering in the wings of the distribution ($\nu \geq 0.10$ THz) is time-dependent on the very long time scale of ~ 15 h following a relatively small temperature change. This suggests that for temperatures around T_f, extremely slow spin dynamics are relevent and true equilibrium behaviour may not be reached over the course of a measurement.

The measured lineshape is well described by a thermal population factor times a Lorentzian in energy:

$$S(Q,\nu) = \frac{h\nu}{k_B T}[1 - exp(-\frac{h\nu}{k_B T})]^{-1}\frac{a}{\nu^2 + \nu_C^2} \tag{18}$$

Equation (18) is similar to Eq. (17) with a Lorentzian form of the relaxation function, $F(Q, \nu)$. This expression, convoluted with the resolution of the spectrometer, was fit to the data and these fits are shown as the solid lines in Fig. 30(b). The extracted values of the $\nu=0$ peak intensity and the characteristic frequency describing the width of the lineshape, ν_C are shown in panels (a) and (b) of Fig. 31. At the minimum of the Q-dependent scattering, $Q=1.5$ Å^{-1}, the frequency width of the distribution is only weakly temperature-dependent, if at all. This is expected for scattering dominated by incoherent processes: these ν_C values are then indicative of the minimum frequency width that this experiment could discern.

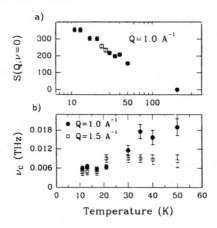

Figure 31. a) Fit values for S($Q=1.0$ Å^{-1}, $\nu=0$) in Tb$_2$Mo$_2$O$_7$ shown as a function of temperature (on a logarithmic scale)[76]. b) The fit values of ν_C, the characteristic energy width, are shown as a function of temperature at the maximum and minimum of the magnetic scattering in Tb$_2$Mo$_2$O$_7$[76].

Examination of Fig. 31 shows that the diffuse-peak intensity approximately doubles between 50 K and 11 K, evolving smoothly as a function of temperature. Over this entire range its behaviour is well described by $S(Q, \nu = 0) \sim -ln(\frac{T}{T_0})$, where $T_0 \sim 200$ K. There is no anomaly visible in $S(Q, \nu = 0)$ near T_f, although data sets near T_f, at 24 K and 27 K, show some indication of non-equilibrium behaviour and are labeled separately from the others. Even so, both are consistent with the logarithmic dependence mentioned above.

Below ~ 25 K, the magnetic scattering at Q=1.0 $\overset{\circ}{A}{}^{-1}$ is characterized by roughly the same values of ν_C as that at Q=1.5 $\overset{\circ}{A}{}^{-1}$, indicating that the frequency width of the magnetic scattering has dropped below the resolution limit. Above ~ 25 K, ν_C at Q=1.0 $\overset{\circ}{A}{}^{-1}$ rises well above that at Q=1.5 $\overset{\circ}{A}{}^{-1}$, demonstrating that the moments at these temperatures are fluctuating on a time scale comparable to that of the inverse frequency (or energy) resolution of the measurement.

The picture that emerges from these measurements is that the spin-glass like anomaly observed in the dc susceptibility at ~ 25 K is a dynamical phase transition. The Tb^{3+} spins on the corner-sharing tetrahedral lattice order locally into short-range ordered droplets at relatively high temperatures (compared to T_f). The characteristic size of the short-range ordered regions does not increase with decreasing temperature. However the characteristic time scale for the fluctuating moments slows down as the temperature is lowered, with a dramatic spin freezing occuring below $T_f \sim 25$ K.

7.3. Experimental Studies of Other Pyrochlores

There have been several recent experimental studies which have produced very interesting results, which are likely of relevence to the above discussion of the rare-earth-based pyrochlores.

The first of these is a single-crystal neutron-scattering study of $CsNiCrF_6{}^{77}$. This material contains two magnetic species, Ni^{2+} and Cr^{3+}, randomly distributed over a network of corner-sharing tetrahedra. Consequently this material possesses considerable chemical disorder. The results of this study are very similar to those described above for $Tb_2Mo_2O_7{}^{76}$, however the single-crystal nature of the sample allowed the directional dependence of the scattering to be fully investigated (as opposed to the averaging over all directions which occurs in scattering studies on powders). Energy-integrated scans along the cubic high-symmetry directions, [001], [111], and [110], are shown in Fig. 32, along with a fit to the data based on the scattering from four antiferromagnetically-coupled spins on a *single*, isolated tetrahedron. In this model, the four neighbouring spins have picked out an *approximately* collinear, antiferromagnetic arrangement. Although the description of the data by this model is not perfect, this study also leads to a low-temperature description of the magnetic structure that bears many similarities to that of a liquid.

A study of the magnetic susceptibility and heat capacity of two rare-earth-based pyrochlores $Gd_2Mo_2O_7$ and $Sm_2Mo_2O_7$ has also been carried out[72]. The high-temperature susceptibility indicates that ferromagnetic interactions predominate in both materials, yet both display history-dependent susceptibilities not unlike

$Y_2Mo_2O_7$[72] and $Tb_2Mo_2O_7$[73], although at substantially higher temperatures. What is interesting, however, is that an analysis of the heat capacity in both materials suggests that the history dependence may arise from a local freezing of the Mo^{4+} pyrochlore sublattice.

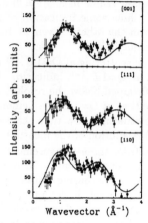

Figure 32. Energy-integrated neutron-scattering scans of $CsNiCrF_6$ shown along three cubic-symmetry directions[77]. The fit lines are based on a model of an approximately collinear antiferromagnetic arrangement of spins on a single tetrahedron.

Finally, it should be pointed out that not all magnets with the pyrochlore structure display spin-glass type behaviour. Two which have been well studied in powder samples are $Nd_2Mo_2O_7$[73] and FeF_3[78]. $Nd_2Mo_2O_7$ undergoes a transition to ferrimagnetic long-range order with $T_C \sim 93$ K, while a second transition occurs at lower temperatures to an antiferromagnetic structure. The behaviour associated with these transitions is rather conventional, likely due to the presence of ferromagnetic interactions (which are unfrustrated on any lattice) and to the relatively long-range nature of the interactions.

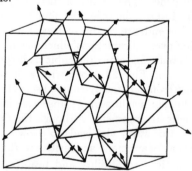

Figure 33. The proposed magnetic structure of the Fe moments in the long-range ordered state of the pyrochlore FeF_3[78].

Measurements on FeF_3 show that it undergoes a phase transition to a long-range ordered state near $T_N \sim 18$ K. An interesting feature of this ordered state is that the scattering data is consistent with a local spin structure which is somewhat analogous to the 120° spin structure seen in $CsMnBr_3$, discussed in section 3. In this spin structure, shown in Fig. 33, the spins point along cubic [111] directions and thus make an angle of 109° with each other. Analysis of the Bragg peak intensities indicate an anomalously small value of $\beta \sim 0.18 \pm 0.02$. The authors of this study also carried out a histogram Monte Carlo calculation of classical Heisenberg spins on a network of corner-sharing tetrahedra, coupled with antiferromagnetic nearest-neighbour interactions and ferromagnetic third-nearest-neighbour interactions[78]. These interactions account for both the ordered structure, as well as the anomalously low value of β. Recently, however, a new histogram Monte Carlo work suggests that the transition may be discontinuous[79].

7.4. Relevant Theory

Anderson was the first to consider the problem of antiferromagnetically coupled spins on a lattice of corner-sharing tetrahedra[4]. He argued that no long-range ordered state should exist for classical Ising spins at low temperatures. Recent mean-field theory and Monte Carlo simulations support this conclusion. A similar picture arose from Villain's consideration of Heisenberg spins on this lattice[80]. He coined the term "cooperative paramagnet" to describe the low-temperature state of such a system. Monte Carlo simulations have also given support to these ideas[81]. Finally, initial work on quantum Heisenberg spins on a lattice of corner-sharing tetrahedra suggests an unconventional dimer ordering to occur[82]. The measurements described above on $Tb_2Mo_2O_7$[73,76], $Y_2Mo_2O_7$[72], and $CsNiCrF_6$[77], are all qualitatively consistent with the picture put forward by these theories.

8. Concluding Remarks

It is hoped that the reader has been left with a sense of the richness of the phenomena which this class of magnetic materials displays. This chapter has reviewed some of the recent progress which has been made experimentally in characterizing, and also in delineating some of the experimental limitations to understanding, this phenomena. It is clear that there is much for the experimentalists in this field to pursue.

Experimental studies of cooperative phenomena depend strongly on the availability of high-quality single-crystal samples. These single-crystal samples of the STL antiferromagnets have been available for some time, whereas such samples do not currently exist for the pyrochlore antiferromagnets. The availability of these samples would allow a more detailed experimental investigation. Although they were not discussed in this chapter, this is also true for studies of the kagomé antiferromagnets. Furthermore, the role of impurites in geometrically-frustrated magnetic systems has yet to be investigated in depth. Magnets such as $CsMnBr_3$ and $CsCoBr_3$ are quasi-one-dimensional, and non-magnetic impurities break the system up into finite, weakly-coupled chains. These materials seem to undergo conventional phase transitions in the presence of small amounts of disorder[83],[84]. However, T_N is observed to be very sensitive to small impurity concentrations and the liklihood of impurity-concentration gradients in the sample make a quantitative understanding of these phase transitions very difficult. Impurity effects in the pyrochlore antiferromagnets will undoubtedly be the focus of much attention in the near future, due to the spin-glass nature of the apparent magnetic ground state in materials such as $Y_2Mo_2O_7$ and $Tb_2Mo_2O_7$. Although these materials are characterized as being disorder-free at the level which can be discerned in powder diffraction profile refinements, it is possible that the nature of the three-dimensional frustration which occurs in the pyrochlores is such that the system is sensitive to impurities at concentrations well below those which effect magnetism in non-frustrated geometries.

9. Acknowledgements

It is a pleasure to thank my collaborators with whom much of the work described in this chapter was performed: D. P. Belanger, A. Bunker, H. R. Child, M. F. Collins, A. Farkas, D. Gibbs, M. J. P. Gingras, J. E. Greedan, G. Helgesen, M. Hagen, A. Harrison, J. P. Hill, C. Kallin, R. Kiefl, J. Z. Larese, T. E. Mason, J. N. Reimers, R. B. Rogge, G. Shirane, C. V. Stager, Z. Tun, T. Thurston, J. Wang, and Y. S. Yang. It is also a pleasure to thank A. J. Berlinsky, A. Caillé, H. Kawamura, G. Luke, M. Mekata, M. L. Plumer, A. P. Ramirez and Y. J. Uemeura for many helpful discussions. Finally I am indebted to A. A. M. Gaulin and K. M. Hughes for a critical reading of the manuscript.

10. References

1. G. Toulouse, *Commun. Phys.* **2** (1977) 115.
2. G. H. Wannier, *Phys. Rev.* **79** (1950) 357.
3. I. Syozi, in *Phase Transitions and Critical Phenomena*, eds. C. Domb and M. S. Green (Academic, New York, 1972), p. 269.
4. P. W. Anderson, *Phys. Rev.* **102** (1956) 1008.
5. J. Als-Nielsen, in *Phase Transitions and Critical Phenomena* , eds. C. Domb and M. S. Green (Academic Press, London, 1976), p. 87.
6. R. A. Cowley, in *Methods of Experimental Physics, Vol 23, Part C*, eds. K. Skold and D. L. Price (Academic Press, Orlando, 1987), p. 1.
7. For a general review of neutron scattering see articles in *Methods of Experimental Physics, Vol 23, Parts A, B and C*, eds. K. Skold and D. L. Price (Academic Press, Orlando, 1987).
8. M. F. Collins, *Magnetic Critical Scattering* (Oxford University Press, New York, 1989)
9. A. Aharony and G. Ahlers, *Phys. Rev. Lett.* **44** (1980) 782.
10. B. I. Halperin and P. C. Hohenberg, *Phys. Rev.* **177** (1969) 952.
11. V. F. Sears, *Neutron Optics* (Oxford University Press, New York, 1989).

12. M. E. Fisher and R. J. Burford, *Phys. Rev.* **156** (1967) 583.
13. J. Goodyear and D. J. Kennedy, *Acta Crystallogr.* **B28** (1974) 1640.
14. W. B. Yelon, D. E. Cox and M. Eibschutz, *Phys. Rev. B* **12** (1975) 5007.
15. M. Eibshutz, R. C. Sherwood, F. S. L. Hsu and D. E. Cox, in *Magnetism and Magnetic Materials (Denver, 1972) Proc. 18th Annual Conf. on Magnetism and Magnetic Materials, AIP Conf. Proc. No. 10*, eds. C. D. Graham Jr. and J. J. Rhyne (AIP, New York, 1973), p. 684.
16. M. Melamud, H. Pinto, J. Makovsky and H. Shaked, *Phys. Stat. Sol.* **63** (1974) 699.
17. B. D. Gaulin and M. F. Collins, *Can. J. Phys.* **62** (1984) 1132; *J. Phys. C* **19** (1986) 5483.
18. B. D. Gaulin, M. F. Collins and W. J. L. Buyers, *J. Appl. Phys.* **61** (1987) 3409.
19. S. E. Nagler, W. J. L. Buyers, R. L. Armstrong and B. Briat, *Phys. Rev. B* **27** (1983) 1784.
20. H. Kawamura, *J. Phys. Soc. Jpn.* **56** (1987) 474.
21. H. Kawamura, *J. Appl. Phys.* **63** (1988) 3086.
22. H, Kawamura, *Phys. Rev. B* **38** (1988) 4916 and private communication
23. P. Azaria, P. Delamotte and T. Jolicoeur, *Phys. Rev. Lett.* **64** (1990) 3175.
24. T. E. Mason, M. F. Collins and B. D. Gaulin, *J. Phys. C* **20** (1987) L945.
25. T. E. Mason, B. D. Gaulin and M. F. Collins, *Phys. Rev. B* **39** (1989) 586.
26. Y. Ajiro, T. Nakashima, Y. Unno, J. Kadowaki, M. Mekata and N. Achiwa, *J. Phys. Soc. Jpn.* **57** (1988) 2648.
27. H. Kadowaki, S. M. Shapiro, T. Inami and Y. Ajiro, *J. Phys. Soc. Jpn.* **57** (1988) 2640.
28. J. Wang, D. P. Belanger and B. D. Gaulin, *Phys. Rev. Lett.* **66** (1991) 3195.
29. R. Deutschmann, H. von Lohneysen, J. Wosnitza, R. K. Kremer, and D. Visser, *Euro. Phys. Lett.* **17** (1992) 637.
30. T. E. Mason, Y. S. Yang, M. F. Collins, B. D Gaulin, K. N. Clausen and A. Harrison, *J. Magn. Magn. Mater* **104-107** (1992) 197.
31. B. D. Gaulin, T. E. Mason, M. F. Collins and J. F. Larese, *Phys. Rev. Lett.* **62** (1989) 1380.
32. M. Kerszberg and D. Mukamel, *Phys. Rev. B* **18** (1978) 6283.
33. L. M. Corliss, J. M. Hastings, W. Kunnmann, R. J. Begum, M. F. Collins, E. Gurewitz and D. Mukamel, *Phys. Rev. B* **25** (1982) 245.
34. K. A. Muller, W. Berlinger, J. E. Drumheller and J. G. Bednorz, in *Multicritical Phenomena*, eds. R. Pynn and A. Skjeltorp (Plenum, New York, 1983), p. 143.
35. T. E. Mason, M. F. Collins and B. D. Gaulin, *J. Appl. Phys.* **67(9)** (1990) 5421.
36. A. V. Chubukov, *J. Phys. C* **21** (1988) L441.
37. H. Kawamura, A. Caillé and M. Plumer, *Phys. Rev. B* **41** (1990) 4416.
38. T. Goto, T. Inami and Y. Ajiro, *J. Phys. Soc. Jap.* **59** (1990) 2328.
39. T. E. Mason, C. V. Stager, B. D. Gaulin and M. F. Collins, *Phys. Rev. B* **42** (1990) 2715.

40. M. L. Plumer and A. Caillé, *Phys. Rev. B* **42** (1990) 10388.
41. G. A. Baker, B G. Nickel and D. I. Meiron, *Phys. Rev. B* **17** (1978) 1365; J. C. LeGuillou and J. Zinn-Justin, *Phys. Rev. Lett.* **39** (1977) 95.
42. J. Jensen and A. R. Mackintosh *Rare Earth Magnetism* (Oxford University Press, New York, 1991).
43. D. Gibbs, J. Bohr, J. D. Axe, D. E. Moncton and K. L. D'Amico, *Phys. Rev. B* **34** (1986) 8182.
44. P. Bak and D. Mukamel, *Phys. Rev. B* **13** (1976) 5086.
45. Z. Barak and M. B. Walker, *Phys. Rev. B* **25** (1982) 1969.
46. D. A. Tindall, M. O. Steinitz and M. L. Plumer, *J. Phys. F* **7** (1977) L263.
47. J. Eckert and G. Shirane, *Solid State Commun.* **19** (1976) 911.
48. K. D. Jayasuriya, S. J. Campbell and A. M. Stewart, *J. Phys. F* **15** (1985) 225.
49. B. D. Gaulin, M. Hagen and H. R. Child, *J. Phys. (Paris)* **Colloq.** **49** (1988) C8-327.
50. E. Loh, C. L. Chien and J. C. Walker, *Phys. Lett. A* **49** (1974) 357.
51. P. de V. du Plessis, C. F. van Doorn and D. C. van Delden, *J. Magn. Magn. Mat.* **40** (1983) 91.
52. T. R. Thurston, G. Helgesen, D. Gibbs, J. P. Hill, B. D. Gaulin and G. Shirane, *Phys. Rev. Lett.* **70** (1993) 3151; T. R. Thurston, G. Helgesen, D. Gibbs, J. P. Hill, B. D. Gaulin and P. J. Simpson, submitted to *Phys. Rev. B*.
53. K. D. Jayasaria, S. J. Campbell and A. M. Stewart, *Phys. Rev. B* **31** (1985) 6032.
54. F. L. Lederman and M. B. Salaman, *Sol. St. Comm.* **15** (1974) 1373.
55. F. Willis, N. Ali, M. O. Steinitz, M Kahrizi and D. A. Tindall, *J. Appl. Phys.* **67** (1990) 5277.
56. D. Blankschtein, M. Ma, A. N. Berker, G. S. Grest and C. M. Soukoulis, *Phys. Rev. B* **29** (1984) 5250.
57. R. R. Netz and A. H. Berker, *Phys. Rev. Lett.* **66** (1991) 377.
58. S. N. Coppersmith, *Phys. Rev. B* **32** (1985) 1584.
59. O. Heinonen and R. B. Petschek, *Phys. Rev. B* **40** (1989) 9052.
60. A. Bunker, B. D. Gaulin and C. Kallin, *Phys. Rev. B* **48** (1993) 15861.
61. M. L. Plumer, A. Mailhot, R. Ducharme, A. Caillé and H. T. Diep, *Phys. Rev. B* **47** (1993) 14312.
62. M. Mekata and K. Adachi, *J. Phys. Soc. Jpn.* **44** (1978) 806.
63. A. Farkas, B. D. Gaulin, Z. Tun and B. Briat, *J. Appl. Phys.* **69** (1991) 6167.
64. S. E. Nagler, W. J. L. Buyers, R. L. Armstrong and B. Briat, *Phys. Rev. B* **27** (1983) 1784; *Phys. Rev. B* **28** (1983) 3873; W. J. L. Buyers, M. J. Hogan, R. L. Armstrong and B. Briat, *Phys. Rev. B* **33** (1986) 1727.
65. S. K. Satija, G. Shirane and H. Yoshizawa, *Phys. Rev. Lett.* **44** (1980) 1548.
66. Z. Tun, B. D. Gaulin, R. B. Rogge and B. Briat, *J. Magn. Magn. Mater.* **104-107** (1992) 1045.
67. J. P. Boucher, L. P. Regnault, J. Rossat-Mignod, Y. Henry, J. Bouillot and W. G. Stirling, *Phys. Rev. B* **31** (1985) 3015.
68. J. Wang, D. P. Belanger and B. D. Gaulin, *to appear in Phys. Rev. B*.

69. G. Ahlers, *Rev. Mod. Phys.* **52** (1980) 489.
70. H. Kikuchi and Y. Ajiro, in *Cooperative Dynamics in Complex Physical Systems*, ed. H. Takayama (Springer, Berlin, 1989), p. 40.
71. J. E. Greedan, in *Landolt-Börnstein* **27 g**, (Springer-Verlag, Berlin, Heidelberg, 1992).
72. N. P. Raju, E. Gmelin and R. K. Kremer, *Phys. Rev. B* **46** (1992) 5405.
73. J. E. Greedan, J. N. Reimers, C. V. Stager and S. L. Penny, *Phys. Rev. B* **43** (1991) 5682.
74. K. Binder and A. P. Young, *Rev. Mod. Phys.* **58** (1986) 801.
75. J. N. Reimers and J. E. Greedan, *J. Solid State Chem.* **72** (1988) 390.
76. B. D. Gaulin, J. N. Reimers, T. E. Mason, J. E. Greedan and Z. Tun, *Phys. Rev. Lett.* **69** (1992) 3244.
77. M. J. Harris, M. P. Zimkin, Z. Tun, B. M. Wanklyn, and I. P. Swainson, (1994) *preprint.*
78. J. N. Reimers, J. E. Greedan and M. Bjorgvinsson, *Phys. Rev. B* **45** (1992) 7295.
79. A. Mailhot and M. L. Plumer, *Phys. Rev. B* **48** (1993) 9881.
80. J. Villain, *Z. Phys. B* **33** (1978) 31.
81. J. N. Reimers, *Phys. Rev. B* **45** (1992) 7287.
82. A. B. Harris, A. J. Berlinsky and C. Bruder, *J. Appl. Phys.* **69** (1991) 5200.
83. D. Visser, A. Harrison and G. J. McIntyre, *J. Phys. (Paris)* **Colloq. 49** (1988) C8-1255.
84. R. B. Rogge, Y. S. Yang, Z. Tun, B. D. Gaulin, J. A. Fernandez-Baca, R. M. Nicklow and A. Harrison, *J. Appl. Phys.* **73** (1993) 6451; M. Mekata, S. Okamoto, S. Onoe, S. Mitsuda and H. Yoshizawa, *J. Magn. Magn. Mat.* **90&91** (1990) 267.

INDEX

Almeida-Thouless line 244
anisotropy 16,23,24
anyonic excitation 145
axial anisotropy 4,35
background field method 79,106,112
Bethe Ansatz 123
bimodal J_{ij} distribution 243,251
BL structure 211,212
body-centered cubic lattice 199
centered honeycomb lattice 162,171,174,187-190
centered square lattice 162,178-181
charge density wave 250
checkerboard lattice 168,169
chiral order 147,149
chirality 6,8,129,147
chiral universality class 7,8
continuous Ising spin 216
correlation length 94
corner-sharing tetrahedra 318
critical condition 170
critical exponents 7,9,30,44,288,307,310,315
critical phenomena 287
critical temperature 167,174,183,205,206,211
critical value of J_2 134,136
crossover exponent 24
$CsCoBr_3$ 291,292,294,295,312
$CsCoCl_3$ 312
$CsMnBr_3$ 25,26,30,36,41,43
$CsMnCl_3$ 1,3,38,43
$CsMnI_3$ 37,38
C_{3V} symmetry 59,66,104
current term 63,74,92